W9-CSF-171

Springer Series on
Atoms+Plasmas

22

Editor: I. I. Sobel'man

Springer
Berlin
Heidelberg
New York
Barcelona
Budapest
Hong Kong
London
Milan
Paris
Singapore
Tokyo

Springer Series on

Atoms+Plasmas

Editors: G. Ecker   P. Lambropoulos   I. I. Sobel'man   H. Walther
Managing Editor: H. K. V. Lotsch

V. S. Lebedev I. L. Beigman

Physics of Highly Excited Atoms and Ions

With 59 Figures

 Springer

Dr. Vladimir S. Lebedev
Dr. Israel L. Beigman
Optical Division, P. N. Lebedev Physical Inst.
Russian Academy of Sciences
Leninsky Prospect 53
117924 Moscow / Russia

Series Editors:

Professor Dr. Günter Ecker
Ruhr-Universität Bochum, Lehrstuhl Theoretische Physik I, Universitätsstraše 150,
D-44801 Bochum, Germany

Professor Peter Lambropoulos, Ph. D.
Max-Planck-Institut für Quantenoptik, D-85748 Garching, germany, and
Foundation for Research and Technology - Hellas (F.O.R.T.H.),
Institute of Electronic Structure & Laser (IESL) and
University of Crete, PO Box 1527, heraklion, Crete 71110, Greece

Professor Igor I. Sobel'man
Lebedev Physical Institute, Russian Academy of Sciences,
Leninsky Prospekt 53, 117924 Moscow, Russia

Professor Dr. Herbert Walther
Sektion Physik der Universität München, Am Coulombwall 1,
D-85748 Garching/München, Germany

Managing Editor:

Dr.-Ing. Helmut K. V. Lotsch
Springer-Verlag, Tiergartenstraše 17, D-69121 Heidelberg, Germany

Library of Congress Cataloging–in–Publication Data
Lebedev, V. S. (Vladimir S.) Physics of highly excited atoms and ions / V. S. Lebedev, I. L. Beigman.
p. cm. – (Springer series on atoms + plasmas; 22) Includes bibliographical references and index.
ISBN 3-540-64234-X (alk. paper)
1. Atomic spectroscopy. 2. Rydenberg states. 3. Ions–Spectra.
I. Beigman, I. L. (Israel L.), 1940– . II. Title. III. Series.
QC454.A8L4   1998   539.7–dc21   98-35427

ISSN 0177-6495
ISBN 3-540-64234-X Springer-Verlag Berlin Heidelberg New York

This work is subject to copyright. All rights are reserved, whetwer tho whole or part of the material is concerned, specifically the rights of translation, reprinting, reuse of illustrations, recitation, broadcasting, reproduction on microfilm or in other ways, and storage in data banks. Duplication of this publication or parts thereof is permitted only under the provisions of the German Copyright Law of September 9, 1965, in its current version, and permission for use must always be obtained from Springer-Verlag. Violations are liable for prosecution act under German Copyright Law.

© Springer-Verlag Berlin Heidelberg 1998 . Printed in Germany

The use of general descriptive names, registered names, trademarks, etc. in this publication does not imply, even in the absence of a specific statement, that such names are exempt from the relevant protective laws and regulations and therefore free for general use.

Data conversion: PTP - Protago • TeX • Production, Berlin
Cover: design & production, Heidelberg
SPIN: 10663787   54/3020 - 5 4 3 2 1 0 - Printed on acid-free paper

QC454
A8 L4
1998
PHYS

Preface

This monograph is devoted to the basic aspects of the physics of highly excited (Rydberg) states of atom's. After almost twenty years, this remains a hot topic of modern atomic physics. Such studies are important for many areas of physics and its applications including spectroscopy, astrophysics and radio astronomy, physics of electronic and atomic collisions, kinetics and diagnostics of gases, and low- and high-temperature plasmas. Physical phenomena in radiative, collisional, and spectral-line broadening processes involving Rydberg atoms and ions are primarily determined by the peculiar properties and exotic features of highly excited states.

The growth of interest and research activity in the physics of Rydberg atoms over the last two decades was stimulated by an extremely rapid development of high-resolution laser spectroscopy, methods of selective excitation and detection of highly excited states, atomic-beam techniques as well as radio astronomy. This has facilitated significant progress in the different directions of the physics of highly excited atoms being of fundamental and practical importance. In particular, evident advances were achieved in studies of the structure and spectra of highly excited atoms, their behavior in static electric and magnetic fields, interactions with electromagnetic radiation, spectral-line broadening and the shift of Rydberg series, collisions with electrons, ions, atoms, and molecules, etc. The principle objective of the present book is to reflect the most important physical approaches and efficient theoretical techniques in the modern physics of highly excited atoms and ions.

These approaches provide the basis for understanding the structure and spectra of Rydberg atoms (ions) as well as their interactions with neutral and charged particles and electromagnetic fields. It is important to stress that the theoretical description of collisional and radiative processes involving Rydberg atoms and ions requires new physical approaches as compared to the case of atoms and ions in the ground or low-excited states. We present both standard theoretical methods in the physics of highly excited states and a number of quantum, semiclassical, and pure classical methods, which have found wide application only in recent years. Moreover, we consider both analytical and numerical methods, which combine the simplicity of a qualitative description of Rydberg states and relevant collisional and broadening processes with reliable quantitative results.

On the basis of the identified physical approaches we present an up-to-date summary of the knowledge in the field of collisional and spectral-line broadening processes involving Rydberg atoms and ions. Major emphasis is put on recent achievements in this field and original results that are not covered in available reviews or monographs.

Moscow, May 1998

Vladimir S. Lebedev
Israel L. Beigman

Contents

1. Introduction

Before going into the formal developments of the physical approaches and methods in the physics of highly excited states, we shall give a brief account of the main features and the most important physical properties of Rydberg atoms and ions. The objective of this chapter is to introduce the basic parameters of highly excited atoms and ions, such as their orbital radius and geometrical area, the ionization potential, transition frequencies and wavelengths, period and the mean velocity of electron motion, quantum defect, etc. We shall analyze their dependencies on the principal quantum number n and the ionic charge Z and shall present the characteristic scales of the identified physical parameters for typical values of the principal quantum numbers observed in laboratory and astrophysical conditions. At the end of this introductory chapter we shall also discuss briefly the subject and specific goals of the present book.

1.1 Physical Properties and Features of Rydberg Atoms and Ions

The Rydberg atom is the atom in which one of its electrons is in the highly excited state with large magnitude of the principal quantum number n. As a result, the Rydberg atom has a number of peculiar physical properties (see [1.1–3]) as compared to the atoms in the ground or low excited states. In particular, the Rydberg atom is characterized by a very small ionization potential $I_n \propto 1/n^2$ and small average velocity of a highly excited electron $v_n \propto 1/n$, very large dimension $r_n \propto n^2$, and a period of orbital electron motion around the parent ion core $T_n \propto n^3$ on the atomic scale. One more specific feature of the Rydberg atom is a great number of closely spaced energy levels. The radiation lifetime of the Rydberg atom increases drastically with increasing principal quantum number. It is proportional to n^3 and n^5 for the highly excited states with small ($l \ll n$) and large ($l \sim n$) values of the orbital angular momentum l, respectively. At the same time, the highly excited atom turns out to be very sensitive to the influence of various perturbations and external fields. The cross sections for collisions of Rydberg atoms with charged and neutral targets are much greater than those for collisions involving atoms in the ground or low excited states. Thus, the Rydberg atom is a rather unusual object having simultaneously both pure quantal and classical properties and a number of exotic features. On the whole, however, the structure and basic features of any atom in the highly excited state are similar to the Hydrogen atom since the interaction between the Rydberg electron and parent ionic core is primarily determined by their Coulomb interaction at large distances.

Characteristic magnitudes of the principal quantum number n observed in laboratory and cosmic space vary within a large range. Usually, in the rarefied gas cells and in the beams of the Rydberg atoms, n is of the order of 10–100. At the same time, in the recent years the Rydberg atoms with very high magnitudes of the principal quantum number $n \sim 500$–1000 are also becoming available in laboratories. In the interstellar medium the principal quantum numbers of atoms in the Rydberg states are observed up to ~ 1000, whereas the typical magnitudes are about 100–300. The very large range of n achieved in modern laboratory experiments and obtained from astrophysical observations leads to extremely large variations in the values characterizing the physical properties of highly excited states.

Let us consider the basic physical properties of the Rydberg atoms and discuss their dependencies on the principal quantum number n. All simple expressions presented below will also contain explicitly the dependencies on the charge Z of the ionic parent core ($Z = 1$ for a neutral atom). They clarify the main features of ions in the highly excited states and are important, in particular, for understanding the peculiar properties of multicharged ions ($Z \gg 1$).

a) **Characteristic Radius r_n and Geometrical Area S_n.** The radius and geometrical area of the Rydberg atom or ion increase drastically with n growing in accord with the well-known relations

$$r_n \sim n^2 a_0 / Z, \qquad S_n \sim \pi a_0^2 n^4 / Z^2, \tag{1.1}$$

Here $a_0 = \hbar^2/me^2$ is the Bohr radius ($a_0 = 0.529 \cdot 10^{-8}$ cm), $\hbar$ is the Plank constant, e and m are the charge and mass of electron, respectively. It is important to stress that the characteristic dimensions of Rydberg atoms correspond to macroscopic sizes ($> 1\mu$) for $n > 100$. For the principal quantum number $n \sim 1000$ now observed in both laboratory and cosmic conditions, the radius r_n of the Rydberg atom turns out to be of the order of 0.1 mm. This value exceeds the characteristic radius of the ground-state atom by more than 10^6 times. As is evident from these expressions, the radius and geometrical area of the Rydberg ion are inversely proportional to its ionic charge Z in the first and second powers, respectively. One can see also that the Z-dependencies of the r_n and S_n values are weaker than the corresponding n-dependencies.

b) **Period of Electron Motion T_n and Mean Orbital Velocity v_n.** In accord with the correspondence principle many properties of an atom (ion) with large value of the principal quantum number $n \gg 1$ may be reasonably described in terms of classical mechanics. In particular, an important characteristic of the Rydberg atom (ion) is the period T_n of classical orbital motion of the highly excited electron around the parent ionic core and the mean orbital velocity v_n. They are given by the following expressions:

$$T_n \sim 2\pi(a_0/v_0)n^3/Z^2, \qquad v_n \sim Zv_0/n, \tag{1.2}$$

where $v_0 = e^2/\hbar$ is the atomic unit of velocity ($v_0 = 2.188 \cdot 10^8$ cm·s^{-1}). Thus, the mean orbital velocity of a highly excited electron v_n is small compared with the characteristic velocities of atomic electrons in the ground or low excited states, while the period T_n of its orbital motion increases rapidly with n growing, and decreases with an increase of the ionic charge Z. At the same time, the T_n and v_n values depend weakly on the orbital quantum number l that determine the form of the classical electron orbit.

c) **Energy Spectrum and Radiative Properties of the Hydrogen Atom and Hydrogen-like Ions.** In spite of important specific features in the line spectra and structure of one or another atom or ion in the highly excited state, their basic energy and spectral characteristics (such as the ionization potential I_n, frequency $\omega_{n,n\pm1}$ and wavelength $\lambda_{n,n\pm1}$ for transitions between neighboring states and some others) may be estimated with the use of the well-known expressions for the hydrogen-like states. The nonrelativistic Schrödinger equation with the Coulomb interaction $U(r) = -Ze^2/r$ leads to the simple expression for the electronic energy of the Hydrogen atom ($Z = 1$) or hydrogen-like ion ($Z > 1$)

$$E_n = -Z^2 Ry/n^2, \tag{1.3}$$

where $Ry = me^4/2\hbar^2$ is the Rydberg constant ($2Ry = 27.212$ eV is the atomic unit of energy). The energy levels with the given magnitude of the principal quantum number n are degenerated over the orbital angular momentum l ($l = 0, 1, 2, ..., n-1$), its z-projection m, and the z-projection σ of the electron spin ($s = 1/2$) provided only electrostatic interaction between the electron e and nuclear with charge Ze is taken into account. The total statistical weight g_{tot} of the given n-level and the density of states per unit energy interval $\rho(E_n) = g_s g_n \, |dE_n/dn|^{-1}$ are given by

$$g_{tot} = g_n g_s = 2 \sum_{l=0}^{n-1} g_l = 2n^2, \tag{1.4}$$

$$\rho(E_n) = \frac{n^5}{Z^2 Ry} = \frac{2m^{3/2} Z^3 e^6}{\hbar^3 (2|E_n|)^{5/2}}, \tag{1.5}$$

where $g_l = 2l + 1$ and $g_s = 2$ are the statistical weights of the l- and s-sublevels, respectively. Thus, the specific feature of the Rydberg atom with high principal quantum number n is a great number of closely spaced energy levels.

The line spectra corresponding to transitions from the highly excited state to the ground (or low excited) state of a neutral atom lie in the visible or ultra-violet diapason. They are shifted to the shorter wavelength region for ions and correspond to the X-ray diapason for multicharged ions ($Z \gg 1$). At the same time, the transition frequencies between two closely spaced Rydberg states are small. For example, the transition frequency $\omega_{n,n\pm1} = |E_n - E_{n\pm1}|/\hbar$ between two levels with neighboring magnitudes of the principal quantum numbers n and $n \pm 1$ can be written as

$$\omega_{n,n\pm1} = \frac{Z^2 Ry}{\hbar}\left[\frac{1}{n^2} - \frac{1}{(n\pm1)^2}\right] \approx \frac{2Z^2 Ry}{\hbar n^3}. \tag{1.6}$$

This is the Kepler frequency of the Rydberg electron, which corresponds to the classical period $T_n = 2\pi/\omega_{n,n\pm1}$ of its orbital motion (1.2). The corresponding wavelength $\lambda_{n,n\pm1} = 2\pi c/\omega_{n,n\pm1}$ increases drastically with n-growing. It is embedded in the micron-region for $n \sim 10$ and in the centimeter region for $n \sim 100$. Radioastronomy observations of states with $n \sim 300$ and more are usually carried out in the meter diapason of wavelength.

An account of the spin-orbit interaction of a highly excited electron leads to the appearance of the relativistic correction $E_{nJ}^{(1)}$ to the energy of Rydberg levels

$$E_{nJ} = -\frac{Z^2 Ry}{n^2} + E_{nJ}^{(1)}, \qquad E_{nJ}^{(1)} = -\frac{\alpha^2 Z^4 Ry}{n^3}\left(\frac{1}{J+1/2} - \frac{3}{4n}\right)(1.7)$$

and to the fine-structure splitting of states with different magnitudes $J = l+1/2$ and $J' = |l-1/2|$ of the total angular momentum of electron ($\mathbf{J} = \mathbf{l} + \mathbf{s}$) but with the same n and l values

$$\omega_{J'J} = \Delta E_{J'J}/\hbar = \frac{\alpha^2 Z^4 Ry}{\hbar n^3 l(l+1)}. \tag{1.8}$$

Here $\alpha = e^2/\hbar c = 1/137$ is the fine structure constant. One can see that the relativistic correction decreases rapidly with n increasing and turns out to be particularly important for multicharged ions with large Z. The second term in (1.7) for $E_{nJ}^{(1)}$ is negligible for Rydberg states with $n \gg 1$. It should be noted also that energies of states with equal magnitudes of the total angular momentum J, but different orbital momenta l are the same. This is a specific feature of Hydrogen and hydrogen-like ions. Difference of these energies is connected with radiation corrections and is usually negligible for Rydberg states.

The radiation lifetime of an atom (ion) in the highly excited n-state ($n \gg 1$) averaged over the lm-sublevels may be successfully described by a simple expression

$$\tau_n \approx \frac{n^5}{3A_0 Z^4 \ln(n/1.1)}, \qquad A_0 = \frac{8\alpha^3 (v_0/a_0)}{3\pi\sqrt{3}}, \tag{1.9}$$

Table 1.1. Characteristic properties of the Rydberg states. Numerical values are given for neutral atom ($Z = 1$); numerical values of the fine structure correction correspond to $J = 1/2$

Physical value	Expression	Unit	$n = 10$	$n = 100$	$n = 1000$
Ionization potential I_n	$Z^2 Ry/n^2$	eV	$1.36 \cdot 10^{-1}$	$1.36 \cdot 10^{-3}$	$1.36 \cdot 10^{-5}$
Characteristic radius r_n	$a_0 n^2/Z$	cm	$5.3 \cdot 10^{-7}$	$5.3 \cdot 10^{-5}$	$5.3 \cdot 10^{-3}$
Geometrical area S_n	$\pi a_0^2 n^4/Z^2$	cm^2	$8.8 \cdot 10^{-13}$	$8.8 \cdot 10^{-9}$	$8.8 \cdot 10^{-5}$
Characteristic velocity v_n	$Z v_0/n$	cm/s	$2.18 \cdot 10^7$	$2.18 \cdot 10^6$	$2.18 \cdot 10^5$
Transition frequency $\omega_{n,n\pm1}$	$2\frac{Z^2 Ry}{\hbar n^3}$	s^{-1}	$4.1 \cdot 10^{13}$	$4.1 \cdot 10^{10}$	$4.1 \cdot 10^7$
Wavelength $\lambda_{n,n\pm1}$	$2\pi c/\omega_{n,n\pm1}$	cm	$4.6 \cdot 10^{-3}$	4.6	$4.6 \cdot 10^3$
Period of classical motion T_n	$2\pi/\omega_{n,n\pm1}$	s	$1.5 \cdot 10^{-13}$	$1.5 \cdot 10^{-10}$	$1.5 \cdot 10^{-7}$
Fine structure correction $E_{nJ}^{(1)}$	$-\frac{\alpha^2 Z^4 Ry}{n^3(J+1/2)}$	eV	$7.3 \cdot 10^{-7}$	$7.3 \cdot 10^{-10}$	$7.3 \cdot 10^{-13}$
Averaged radiation lifetime τ_n	$\frac{n^5}{3 A_0 Z^4 \ln(n/1.1)}$	s	$8.4 \cdot 10^{-5}$	17	$3.7 \cdot 10^3$

which takes into account the total contribution of all dipole $nl \rightarrow n', l \pm 1$ transitions. Here $A_0 = 7.9 \cdot 10^9$ s^{-1} is the typical value of the Einstein coefficient for low-excited states). The main contribution to the total probability $A_n = 1/\tau_n$ of spontaneous radiation is determined by transitions to the final levels with small n'-magnitudes ($n' = 1-2$) and to the neighboring level with $n' = n-1$. Radiation transitions between intermediate n'-levels and the given n-level turns out to be less effective. Thus, the radiation lifetime increases drastically with n-increasing and becomes greater than 10s for states with $n > 100$.

In order to obtain a detailed picture about the dependencies of basic physical properties of highly excited atoms on the principal quantum number we present in Table 1.1 some of the previously mentioned characteristics for three typical values of $n = 10$, 100, and 1000.

d) Structure of Rydberg Levels for Nonhydrogenic Atoms. Structure of energy levels and spectra of nonhydrogenic Rydberg atoms have a number of important features due to a presence of the short-range part of the interaction between the highly excited electron and parent ion core, as well as to the polarization interaction (and other correction terms to pure Coulomb interaction) at large distances. Let us briefly discuss this problem for the simplest case of an atom (ion) having one valence electron in the open shell (for example, alkali-metal atoms, for which there are particularly extensive experimental and theoretical material). It is well known that the energy of such an atom (ion) in the highly excited state with the given magnitudes of the principal n and orbital l quantum numbers is described by the simple expression

$$E_{nl} = -Z^2 Ry/n_*^2 \,, \qquad n_* = n - \delta_l \,, \tag{1.10}$$

proposed first by Rydberg. Here n_* is the effective principal quantum number and δ_l is the quantum defect of the nl-state. This expression differs from the case of Hydrogen atom only by the presence of quantum defect δ_l which characterizes the non-Coulomb part of the electron-core interaction for non-hydrogenic Rydberg atoms. The values of the quantum defect are determined by the specific type of Rydberg atom, and their dependence on the principal quantum number n is weak enough and can be neglected in the zero approximation. At the same time, the δ_l values fall strongly with an increase of the orbital angular momentum l. It is due to a strong decrease of the probability for the Rydberg electron placing in the short-range distances near the parent ion core as well as to a decrease of the long-range polarization parts of the electron-core interaction with l growing; a few Rydberg nS-, nP-, and nD-states (as well as nF-states for the heavy atoms) are usually significantly separated from the practically hydrogen-like quasi-degenerate manifold of the nl-states with $2-3 \leq l \leq n-1$, for which $\delta_l \sim 0$. Whereas the transition frequency between two Rydberg nonhydrogenic nl- and $n'l'$-states with the given magnitudes of both the principal and orbital quantum numbers $\omega_{n'l',nl} \approx 2Z^2 Ry|\delta_l - \delta_{l'} + n' - n|/\hbar n^3$ may be considerably smaller than between two neighboring hydrogen-like levels due to mutual compensation of the quantum defect values.

The quantum defects of Rydberg alkali-metal atoms are presented in Table 1.2.

As is apparent from this table, the quantum defect values grow rapidly with an increase of the atomic number of neutral atoms. This behavior remains the same for all other elements. However, the structure of highly excited levels and spectra of Rydberg atoms with two (or many) electrons in the open shell have a number of peculiarities such as the presence of autoionizing states and correlation electron-electron effects, etc. For example, an important fea-

Table 1.2. The quantum defects of alkali-metal atoms

Atom	δ_s	δ_p	δ_d	δ_f
Li	0.399	0.053	0.002	
Na	1.347	0.854	0.0145	0.0016
K	2.178	1.712	0.267	0.010
Rb	3.135	2.65	1.34	0.0164
Cs	4.057	3.58	2.47	0.033

ture of Rydberg alkali-earth atoms (Be, Mg, Ca, Sr, and Ba) is the presence of a few level series with different ionization limits and strong interseries interactions in the region, where the moduli of their quantum defects become close to each other. Structure and spectra of Rydberg alkali-earth atoms were the subject of intensive spectroscopical studies and theoretical calculations on the basis of the multichannel quantum defect theory [1.4] during recent years. Some results in this field were reflected in the book by *Gallagher* [1.2].

It should be noted that, on the whole, the quantum defects of Rydberg ions with $Z \gg 1$ are less important as compared to the Rydberg neutral atom ($Z = 1$) since in this case the relative role of the non-Coulomb part of the electron-core interaction turns out to be less pronounced.

1.2 Scope of the Book

The objective of the book is to familiarize the reader with the recent achievements, general approaches, and theoretical methods in the physics of highly excited atoms and ions. The main emphasis is put on the basic approaches and modern techniques for describing the structure and properties of highly excited atoms (ions), radiative transitions between Rydberg states, photoionization and photorecombination processes. We also present an up-to-date summary of knowledge on collisional processes involving highly excited atoms and other particles (electrons, ions, atoms, and molecules) as well as on the spectral-line broadening and shift of Rydberg atomic series in gases, low- and high-temperature plasmas. The predominant part of the results presented in the book were published only in the original papers.

We paid particular attention to the comparison of theoretical results with available experimental data. However, the experimental results are included mainly to illustrate the applicability of theoretical methods and we do not discuss specially available experimental methods and techniques in the physics of Rydberg states. Moreover, it is evident that it would be impossible to reflect all the aspects of this intensively developed field of atomic physics within the framework of one book. Therefore, we do not consider the physical principles for production and detection of Rydberg atoms, methods of their photoexcitation in electric fields and pulsed field ionization, microwave excitations and ionization, effects of interaction of Rydberg atoms with the blackbody radiation and their behavior in electric and magnetic fields. These methods, as well as spectroscopy of Rydberg atoms having one- and two- electrons in the open shell, the problems of autoionizing Rydberg states and interseries interaction in bound states were the subject of a recent comprehensive book by *Gallagher* [1.2]. We also would recommend to the reader a number of review articles devoted to different aspects of the physics of Rydberg atoms. Those are cited in the list of references to this introductory chapter. For example, a detailed description of basic methods in the one- and multichannel quantum defect theory was given in [1.4]. Recent progress in multichannel

Rydberg spectroscopy of complex atoms was reflected in an extensive review article [1.5]. There are several reviews devoted to radiative properties of Rydberg states in resonant cavities [1.6], to high-resolution Rydberg spectroscopy [1.7] and to interaction between the highly excited atoms and electromagnetic fields including some close problems of classical and quantum chaos [1.8], [1.9]. In addition to the previously mentioned books [1.1], [1.2], the new topics in the Stark and Zeeman effects in a Hydrogen atom were reflected in [1.10]. The dynamics of the wave packets from highly excited atomic states was discussed in [1.11]. There is also an extensive material on astrophysical studies of Rydberg atoms and ions, which were reflected in [1.12]. Finally, the last achievements in the physics of electronic and atomic collisions involving Rydberg atoms has been reflected in a recent extensive review [1.3].

Let us now discuss the structure and the content of our book, which includes 9 chapters. Chapter 2 is devoted to an account of general methods for describing an isolated Rydberg atom (ion). We consider a pure classical description of Rydberg electron motion, standard quantum approach (based on the Schrödinger equation), semiclassical approach in the action variables, the JWKB-method as well as the asymptotic approach based on the analysis of the exact quantum expressions in the limiting case of large quantum numbers. In particular, we present here an extensive material for the wave functions of the Rydberg atom in the coordinate and momentum representations, efficient method of distribution functions, Wigner and Green's functions. In Chap. 3, our attention is generally focused on the last achievements in calculations of the form factors and dipole matrix elements for the different types of bound-bound and bound-free transitions and their applications to the radiative transitions involving Rydberg states. In Chap. 4, we concentrate on the formulation of basic theoretical methods and physical approaches to collisions involving Rydberg atoms, which take into account the peculiar properties of highly excited states. We present various versions of perturbation theory, pure classical and semiclassical methods, close coupling method, binary-encounter theory and quantum impulse approximation. The material of this chapter is a basis for all collisional and broadening problems considered in the next chapters.

Chapters 5–8 contain a systematic description of major directions and modern techniques in the collision theory of Rydberg atoms and ions with atoms, molecules, electrons, and ions. We consider the main physical mechanisms of such collisions and the most important elementary processes involving highly excited states. A particular attention is paid to analysis of the probabilities, cross sections and rate coefficients behavior of identified processes in dependence on the principal quantum number, energy defects of transition, relative velocity of colliding particles, gas or plasma temperature, etc. The final results are given in the relevant form convenient for use by both specialists in the collision theory, and experimentalists. We present here a great number of simple analytical expressions and results of numerical calculations for a wide class of elementary processes and we make a detailed comparison between theory and experiment. The general theoretical methods described in these chapters will be a basis for further analysis in Chap. 9 of spectral-line broadening and shift of Rydberg atomic series induced by collisions with neutral and charged particles. We discuss here the results obtained within the framework of the impact-parameter method with the Fermi pseudopotential, in the impulse approximation and in the adiabatic quasimolecular approach. The contributions of different physical mechanisms to the width and shift of highly excited levels induced by collisions with the rare gas and alkali-metal atoms are considered in detail. Particular attention is paid to the effects of the potential and resonance scattering of a quasi free electron by a perturbing atom and to the perturber-core effects in the broadening and shift of the Rydberg atomic series. We also consider the impact broadening of Rydberg levels by the plasma-free electrons.

The book is oriented primarily towards the theoreticians and experimentalists in atomic and molecular physics, in plasma physics, in spectroscopy and astrophysics as well as students of the corresponding specialities. We hope that it would be of some interest to physicists in other fields. Thus, we found it more convenient to use usual units throughout the book, and we write in explicit form the Plank constant $\hbar$, charge e and mass m of electron and some combination of them (for example, Bohr radius and velocity $a_0 = \hbar^2/me^2$ and $v_0 = e^2/\hbar$ and the Rydberg constant $Ry = me^4/2\hbar^2$).

2. Classical and Quantum Description of Rydberg Atom

This Chapter is devoted to an isolated highly excited atom which is not affected by any external interactions or fields. We start with a purely classical consideration of highly excited electron motion, and then give a review of various methods for describing the Rydberg atom wave functions in the coordinate and momentum representations. In addition to quantum expressions, we describe the semiclassical methods including the standard JWKB approximation for the radial and angular wave functions as well as the action variables approach. Special attention is paid to the derivation of explicit expressions for the Coulomb Green's function, the density matrix, classical and quantum distribution functions of the Rydberg electron momentum and coordinate. The methods presented in this chapter are widely used in modern theory of Rydberg states. They will be the basis for the majority of the problems considered in this book.

2.1 Classical Motion in a Coulomb Field

We present two classical approaches for the description of the electron motion around a fixed center of charge Ze with the potential energy

$$U(|\mathbf{r}|) = -Ze^2/r, \tag{2.1}$$

where $\mathbf{r}$ is the radius-vector of electron. The first one deals with the usual motion trajectory. We introduce below the main physical parameters of the elliptic orbits and give a simple parametric solution of the Kepler problem for dependencies of the electron coordinates and velocities on time. Another approach is based on the action function describing the trajectory manifolds. This approach is aimed at the further use of perturbation theory and it is natural to get a correspondence with quantum mechanics.

2.1.1 Orbital Electron Motion

The electron motion in the central field is confined to a plane perpendicular to a total angular momentum vector $\mathcal{L}$. Therefore, it is convenient to present a general consideration of the Kepler problem for the Coulomb potential (2.1) in the polar coordinate frame (r, ϕ, ζ) with the origin 0 placed to the charge center Ze and the ζ-axis directed along the $\mathcal{L}$-vector. The general solution of this problem may be directly derived from the conservation laws of energy and angular momentum:

$$\mathcal{L}_\zeta \equiv \mathcal{L} = mr^2 \dot{\phi} = const, \tag{2.2}$$

$$E = \frac{m}{2} \left[\dot{r}^2 + r^2 \dot{\phi}^2 \right] + U(r) = const. \tag{2.3}$$

Substituting (2.2) into (2.3), separating the variables and performing integration, we have

$$t(r) = \int \frac{dr}{\sqrt{(2/m)\left[E - U(r)\right] - \mathcal{L}^2/m^2 r^2}} + const, \tag{2.4}$$

$$\phi(r) = \int \frac{\left(\mathcal{L}^2/r^2\right) dr}{\sqrt{2m\left[E - U(r)\right] - \mathcal{L}^2/r^2}} + const. \tag{2.5}$$

Here E is the total energy, and $\mathcal{L}$ is the magnitude of the angular momentum. The first formula determines the electron separation r from the charge center Ze as an implicit function on time t. The second one gives the relation between r and ϕ, i.e. it determines the trajectory equation.

As it directly follows from (2.5), the form of trajectory for the Coulomb field (2.1) is given by

$$\phi(r) = \arccos \left(\frac{(\mathcal{L}/r) - (Ze^2 m/\mathcal{L})}{\sqrt{2mE + (Ze^2 m/\mathcal{L})^2}} \right) + const. \tag{2.6}$$

Formula (2.5) can be rewritten as

$$\frac{\mathfrak{p}}{r} = 1 + \epsilon \cos \phi, \quad \mathfrak{p} = \frac{\mathcal{L}^2}{Ze^2 m}, \quad \epsilon = \sqrt{1 - \frac{2|E|}{m} \left(\frac{\mathcal{L}}{Ze^2} \right)^2} \tag{2.7}$$

that directly demonstrates the elliptic form of the electron orbit for finite motion with energy $E < 0$. The quantities $\mathfrak{p}$ and ϵ ($\epsilon < 1$ at $E < 0$) are the parameter and eccentricity of the orbit, respectively. Here the initial condition for the ϕ angle was chosen so that the integration constant $const = 0$. This means that the point with $\phi = 0$ corresponds to the nearest separation $r_1 = \mathfrak{p}/(1 + \epsilon)$ (perihelion) of the electron from the Coulomb center (i.e., focus of the orbit), while the point with $\phi = \pi$ corresponds to the aphelion $r_2 = \mathfrak{p}/(1 - \epsilon)$, i.e. to the maximal possible separation from the centre.

The parameters of the electron orbit can also be determined by the values of the semimajor axis a and the semiminor axis b of the ellipse

$$a = \frac{\mathfrak{p}}{1 - \epsilon^2} = \frac{Ze^2}{2|E|}, \quad b = \frac{\mathfrak{p}}{\sqrt{1 - \epsilon^2}} = \frac{\mathcal{L}}{\sqrt{2m|E|}}. \tag{2.8}$$

Since the semimajor axis is dependent only on the electron binding energy $|E|$ it is convenient to express the minimal and maximal electron separations from the Coulomb centre in the form

$$r_{\min} \equiv r_1 = a\left(1 - \epsilon\right), \qquad r_{\max} \equiv r_2 = a\left(1 + \epsilon\right). \tag{2.9}$$

As is apparent from (2.9), for a fixed value of the binding energy $|E|$, the form of the electron orbit is close to a circle if the value of the angular momentum $\mathcal{L}$ is large, i.e., when $\epsilon \ll 1$ and, hence, $r_1 \approx r_2 \approx a$. In the other limiting case of small angular momenta, for which the eccentricity of the orbit is close to unity $\epsilon \to 1$, we have $r_1 \to 0$ and $r_2 \to 2a$, so that the elliptic orbit becomes strongly drawn out.

The period T of the classical electron motion and corresponding angular frequency $\omega = 2\pi/T$ are independent of the angular momentum value $\mathcal{L}$ and are expressed through the binding energy $|E|$ (or through the value a of the semimajor axis of the ellipse) by the relation

$$\omega = \frac{2\pi}{T} = a^{-3/2}\sqrt{\frac{Ze^2}{m}} = \frac{(2\,|\,E\,|)^{3/2}}{Ze^2\sqrt{m}}. \tag{2.10}$$

In accordance with (2.8,10), the averaged velocity v_a in its orbital electron motion is given by

$$v_a = 2\pi a/T = \left(2\,|\,E\,|\,/m\right)^{1/2}. \tag{2.11}$$

For a number of problems involving classical electron motion in the Coulomb field it is useful to a have simple parametric expression describing the dependence of its coordinate r on time t. In accordance with (2.1,8), for the elliptic trajectory general expression, (2.4) takes the form

$$t = \left(\frac{ma}{Ze^2}\right)^{1/2} \int \frac{r\,dr}{\sqrt{a^2\epsilon^2 - (r-a)^2}}. \tag{2.12}$$

Using natural substitution of variables $r - a = -a\epsilon\cos u$, this formula can be rewritten as

$$t = \left(\frac{ma^3}{Ze^2}\right)^{1/2} \int \left(1 - \epsilon\cos u\right)du + const \tag{2.13}$$

Calculating this integral and using expression (2.10) we see that the parametric representation of the dependence of r on t is given by

$$r = a(1 - \epsilon\cos u), \qquad \omega t + \varsigma = \omega\tau = \Theta = u - \epsilon\sin u. \tag{2.14}$$

In the second relation of (2.14), t is the time measured from a fixed origin, while τ is the time measured from that instant when the electron was at the perihelion; ς is the phase of the classical motion, and Θ is the mean anomaly. Formulas (2.14) allows us also to express the Cartesian coordinates $\xi = r\cos\phi$ and $\eta = r\sin\phi$ of electron and the projections $v_\xi = \dot{\xi}$ and $v_\eta = \dot{\eta}$

of the electron velocity as elementary functions of the identified parameter u :

$$\xi = a(\cos u - \epsilon), \qquad \eta = a(1 - \epsilon^2)^{1/2} \sin u,$$

$$v_\xi = -v_a \frac{\sin u}{1 - \epsilon \cos u}, \qquad v_\eta = v_a \frac{(1 - \epsilon^2)^{1/2} \cos u}{1 - \epsilon \cos u}. \tag{2.15}$$

Thus, to obtain the position and the velocity of the electron at a given time t, one has to solve the second equation of (2.14) for the value of the u parameter, with t given (for fixed magnitudes of ω, ς, and ϵ) and to substitute it to the first equation of (2.14) and to (2.15).

As usual, the orientation of the Cartesian coordinate frame $0\xi\eta\zeta$ with respect to a fixed standard frame $0xyz$ is determined by the Euler angles $\alpha\beta\gamma$. The intercept of the $0xy$ and $0\xi\eta$ planes is named the line of nodes. The angle α is the angle formed by the angular momentum vector $\mathcal{L}$ and the z-axis, and $\pi/2 - \gamma$ is the angle formed by perihelion and the line of nodes. The six quantities

$$E, \ \mathcal{L}, \ \alpha, \ \beta, \ \gamma, \ \tau \tag{2.16}$$

define the electron position and its velocity uniquely; all identified values except of the last one (time τ) are conserved.

2.1.2 Action Variables

A charged particle in the Coulomb field is a dynamical system for which the Hamilton-Jacobi equation and the Schrödinger equation are separable. Action variables are convenient for the consideration of perturbations of such a dynamical system. The system is particularly simple since classical canonical transformations may be carried out independently in each of the separated variables and their conjugate momenta. For a system of N degrees of freedom the coordinates u_k, $k = 1, \ldots, N$, are called angle variables and their conjugate momenta I_k are called action variables. We use the notation following *Landau* and *Lifshitz* [2.1], containing the angle variables which change from 0 to 2π; actual angles often prove more convenient than fractions of angles.

If q_k and p_k, are generalized coordinates and momenta for which the system is separable and if $H_0(q_k, p_k)$ is the unperturbed Hamiltonian, the Hamilton-Jacobi equation for the time-independent action function S is

$$H_0\left(q_k, \frac{\partial S}{\partial q_k}\right) = E \tag{2.17}$$

and it has the solution

$$S(q_1, \ldots, q_N) = \sum_{k=1}^{N} S_k(q_k; \alpha_1, \ldots, \alpha_N). \tag{2.18}$$

Each S_k is a function of the N integration constants α_j, and of q_k only. The conjugate momenta are given by

$$p_k = \frac{\partial S}{\partial q_k} = \frac{\partial}{\partial q_k} S_k, \qquad k = 1, \ldots, N. \tag{2.19}$$

We define a second set of canonical momenta and the action variables, by the relation

$$I_k = \frac{1}{2\pi} \oint p_k dq_k, \tag{2.20}$$

where the integrals are taken around complete cycles of the variables q_k. Equations (2.20) give us N relations between the N integration constants and the action variables from which we may obtain the integration constants as functions of action variables. Thus, (2.18) becomes

$$S(q_1, \ldots, q_N) = \sum_{k=1}^{N} S_k(q_k; I_1, \ldots, I_N), \tag{2.21}$$

and the angle variables conjugate to the action variables are defined by

$$u_k = \frac{\partial S}{\partial I_k} = \sum_{j=1}^{N} \frac{\partial}{\partial I_k} S_j. \tag{2.22}$$

It can be shown that with q_j going through a complete cycle, the change in u_k is $2\pi\delta_{kj}$, where δ_{kj} is the Kronecker delta, so that the system is periodic in the angle variables, with period being 2π. Relations (2.19, 22) define a canonical transformation, the new variables being u_k and I_k only, so that the equations of motion become

$$\dot{u}_k = \frac{\partial H}{\partial I_k} = const, \qquad u_k = \omega_k t + \zeta_k,$$

$$\dot{I}_k = -\frac{\partial H}{\partial u_k} = 0, \qquad I_k = const \tag{2.23}$$

for $k = 1, \ldots, N$. As is apparent from (2.23), the I_k variables are the integrals of motion. The fundamental frequencies ω_k of the system are defined by the relation

$$\omega_k = \frac{\partial H}{\partial I_k}. \tag{2.24}$$

In configuration space the motion is multiply periodic in the u_k but in general it is not periodic in time and completely fills an N-dimensional region. Because of this an arbitrary function of the generalized coordinates may be expressed as a multiple Fourier series in the u_k, with Fourier components depending only upon the action variables I_k. This forms a basis of the Heisenberg's form of the correspondence principle.

In many cases of physical interest, for example, symmetric potentials, relations exist between the fundamental frequencies of the form

$$\sum_{k=1}^{N} n_{ik}\omega_k = 0, \qquad i = 1, \ldots, P < N, \tag{2.25}$$

where the n_{ik} are integers. The system then degenerates and the motion in configuration space now fills an $(N - P)$-dimensional region. If this is the case, another set of action-angle variables may be found, such that

$$
\begin{aligned}
u_k &= const, & k &= 1, \ldots, P, \\
u_k &= \omega_k t + \zeta_k, & k &= P+1, \ldots, N, \\
I_k &= const, & k &= 1, \ldots, N.
\end{aligned}
\tag{2.26}
$$

Consider now a special case of a particle of mass m and coordinate vector $\mathbf{r} = (r, \theta, \varphi)$ moving in a spherical potential $U(r)$, so that in spherical coordinates the Hamilton function is

$$
H(r, p) = \frac{1}{2m}\left(p_r^2 + \frac{1}{r^2}p_\theta^2 + \frac{1}{r^2 \sin^2 \theta}p_\varphi^2 \right) + U(r),
\tag{2.27}
$$

where the generalized momenta $(p_r, p_\theta, p_\varphi)$, conjugate to (r, θ, φ) are given by

$$
p_r = m\,\dot{r}, \quad p_\theta = mr^2\dot{\theta}, \quad p_\varphi = mr^2\,\dot{\varphi}\sin\theta.
\tag{2.28}
$$

The Hamilton-Jacobi equation for the action function $S(r, \theta, \varphi)$ can be written as

$$
\left[\left(\frac{\partial S}{\partial r}\right)^2 + \frac{1}{r^2}\left(\frac{\partial S}{\partial \theta}\right)^2 + \frac{1}{r^2 \sin^2 \theta}\left(\frac{\partial S}{\partial \varphi}\right)^2\right] - 2m\left[E - U(r)\right] = 0.
$$

On writing S in the form (2.18) and separating the variables, we obtain

$$
\begin{aligned}
&\frac{d}{d\varphi}S_\varphi = a_\varphi, \\
&\left(\frac{d}{d\theta}S_\theta\right)^2 + a_\varphi^2 / \sin^2 \theta = a_\theta^2, \\
&\left(\frac{d}{d\theta}S_r\right)^2 + a_\theta^2/r^2 - 2m\left(E - U(r)\right) = 0,
\end{aligned}
\tag{2.29}
$$

where a_φ, a_θ and the energy E are the integration constants. Remembering (2.19), and substituting expressions (2.28) into (2.29), we find that a_φ is the angular momentum about the z-axis and that

$$
\begin{aligned}
a_\theta &= \mathcal{L} = mr\left[(r\dot{\theta})^2 + (r\,\dot{\varphi}\sin\theta)^2\right]^{1/2}, \\
a_\varphi &= \mathcal{L}\cos\beta,
\end{aligned}
\tag{2.30}
$$

where $\mathcal{L}$ is the magnitude of the angular momentum. The action variables may be obtained from the definitions (2.20) on integrating around one cycle:

$$
\begin{aligned}
I_\varphi &= a_\varphi, \\
I_\theta &= a_\theta - a_\varphi, \\
I_r &= \frac{1}{2\pi}\oint \left[2m(E - U(r)) - (a_\theta/r)^2\right]^{1/2} dr.
\end{aligned}
\tag{2.31}
$$

The last integral requires careful definition at the extremes where the argument of the square root is zero.

From (2.31) it is evident that the energy is only a function of I_r and $a_\theta = I_\theta + I_\varphi$, so that $\omega_\varphi = 0$. This is because the motion is coplanar. By taking linear combinations we define a new set of action variables:

$$I_1 = I_r + I_\theta + I_\varphi, \quad I_2 = I_\theta + I_\varphi, \quad I_3 = I_\varphi, \tag{2.32}$$

and the corresponding angle variables ($u_k = \partial S/\partial I_k$):

$$u_1 = \omega m \int [2m(E - U(r)) - (I_2/r)^2]^{-1/2} dr, \quad \omega = \partial E/\partial I_1,$$

$$u_2 = I_2 \left[\int (I_2^2 - I_3^2/\sin^2\theta)^{-1/2} d\theta - \int [2m(E - U(r)) - (I_2/r)^2]^{-1/2} r^{-2} dr \right],$$

$$u_3 = \varphi - I_3 \int (I_2^2 - I_3^2/\sin^2\theta)^{-1/2} \sin^{-2}\theta d\theta. \tag{2.33}$$

These new action variables are the integrals of motion and the action function of particle takes the form

$$S = \sum_i I_i u_i. \tag{2.34}$$

For the particular case of the attractive Coulomb potential (2.1) the integral (2.31) gives

$$E = -m(Ze^2)^2/2I_1^2. \tag{2.35}$$

Thus, the energy is independent of I_2 and I_3, so that for the Coulomb potential $\omega_l = 0$ and $\omega_m = 0$; this shows that the orbit form a closed curve.

For the attractive Coulomb potential, the action-angle variables are related to the six variables of (2.16) by the following expressions:

$$\mathcal{L} = I_2, \quad \alpha = u_3, \quad \cos\beta = I_3/I_2, \quad \gamma = u_2, \quad \omega\tau = u_1. \tag{2.36}$$

Thus, the action variables I_k and u_k ($k = 1, 2, 3$) define the plane and axes of the particle orbit in terms of the Euler angles $\alpha\beta\gamma$ as well as the very position of this particle on the orbit.

We end this section by mentioning that the action variables introduced here are the classical analogs of the quantum numbers n, l, m. It is clear that for a quantal system $I_1 = n\hbar$, where n is the principal quantum number. In the Bohr-Sommerfeld model, the quantum atom is represented by elliptic orbits of specific I_2 and I_3 through similar relations $I_2 = l\hbar$ and $I_3 = m\hbar$.

2.2 Wave Functions: Coordinate Representation

2.2.1 Quantum Wave Function of Hydrogen-like States

The Schrödinger equation for hydrogenlike ions with nuclei having charge Z and infinite mass, is

$$-\frac{\hbar^2}{2m}\nabla^2\psi - \frac{Ze^2}{r}\psi = E\psi, \tag{2.37}$$

where m is the electron mass, and ∇^2 is the Laplace operator, which is in the spherical coordinate system

$$\nabla^2 = \frac{1}{r^2}\frac{\partial}{\partial r}\left(r^2\frac{\partial}{\partial r}\right) + \frac{1}{(r\sin\theta)^2}\left[\sin\theta\frac{\partial}{\partial\theta}\left(\sin\theta\frac{\partial}{\partial\theta}\right) + \frac{\partial^2}{\partial\varphi^2}\right]. \tag{2.38}$$

a) **Discrete Spectrum.** The solution of (2.37) may be written in the form:

$$\psi = \mathcal{R}_{nl}(r)Y_{lm}(\theta,\varphi), \quad Y_{lm}(\theta,\varphi) = \Theta_{lm}(\cos\theta)\frac{e^{im\varphi}}{\sqrt{2\pi}}, \tag{2.39}$$

where $Y_{lm}(\theta,\varphi)$ is the spherical function. The $\mathcal{R}_{nl}$ and Y_{lm} functions are orthonormal systems, therefore

$$\int_0^\infty \mathcal{R}_{n'l'}\mathcal{R}_{ln}r^2dr = \delta_{nn'}, \quad \int Y^*_{l'm'}Y_{lm}d\Omega = \delta_{ll'}\delta_{mm'}. \tag{2.40}$$

Note that the radial $\mathcal{R}_{nl}$ functions of discrete spectrum are to be assumed real throughout the book. The expressions of these functions in terms of the Legendre polynomial P_l^m and hypergeometric function $F(a,b;c;x)$ are given by (see, for example, [2.2, 3])

$$\Theta_{lm}(\theta) = (-1)^m\left[\frac{(2l+1)(l-m)!}{2(l+m)!}\right]^{1/2}P_l^m(\cos\theta),$$

$$P_l^m(z) = (-2)^{-m}\frac{(l+m)!}{m!(l-m)!}(1-z^2)^{m/2}$$
$$\times F\left[l+m+1, m-l; m+1; (1-z)/2\right]. \tag{2.41}$$

The radial function $\mathcal{R}_{nl}(r)$ can be written as

$$\mathcal{R}_{nl} = \frac{1}{(2l+1)!}\left(\frac{(n+l)!}{(n-l-1)!2n}\right)^{1/2}\left(\frac{2Z}{na_0}\right)^{3/2}$$
$$\times e^{-Zr/na_0}\left(\frac{2Zr}{na_0}\right)^l F\left[-(n-l-1); 2l+2; \frac{2Zr}{na_0}\right], \tag{2.42}$$

where $a_0 = \hbar^2/me^2$ is the Bohr radius. Its expression in terms of the Laguerre $L_{n+l}^{2l+1}(z)$ polynomial has a form

$$\mathcal{R}_{nl} = -\left[\frac{(n-l-1)!}{[(n+l)!]^3 2n}\right]^{1/2}\left(\frac{2Z}{na_0}\right)^{3/2}e^{-Zr/na_0}\left(\frac{2Zr}{na_0}\right)^l L_{n+l}^{2l+1}\left(\frac{2Zr}{na_0}\right), \tag{2.43}$$

$$L_n^m(z) = (-1)^m\frac{(n!)^2}{m!(n-m)!}F(-(n-m); m+1; z).$$

Thus, the radial function is the polynomial consisting of the $(n - l)$-terms. It has only one term at $l = n - 1$:

$$R_{n,n-1} = [(2n)!]^{-1/2} \left(\frac{2Z}{na_0}\right)^{3/2} e^{-Zr/na_0} \left(\frac{2Zr}{na_0}\right)^{n-1}. \qquad (2.44)$$

The general expression (2.43) takes the following form as $r \to \infty$:

$$R_{nl} \to (-1)^{n-l-1}[(2n)(n + l)!(n - l - 1)!]^{-1/2} \left(\frac{2Z}{na_0}\right)^{3/2} e^{-Zr/na} \left(\frac{2Zr}{na_0}\right)^{n-1}, \qquad (2.45)$$

and at $r \to 0$

$$R_{nl} \to \frac{1}{(2l + 1)!} \left[\frac{(n + l)!}{(n - l - 1)!2n}\right]^{1/2} \left(\frac{2Z}{na_0}\right)^{3/2} \left(\frac{2Zr}{na_0}\right)^{l}. \qquad (2.46)$$

The last result is valid at $Zr/a_0 \ll 1$. The Tricomi expansion (see, for example [2.4]) of the hypergeometric function over the Bessel functions allows wider application. The first term of the expansion is

$$F(a; b + 1; x) \to e^{x/2} \Gamma(b + 1)(nx)^{-b/2} J_b(2x), \quad n = b/2 - a + 1/2, \qquad (2.47)$$

where $J_b(z)$ is the Bessel function. In fact, the expansion parameter is $(x/n^{2/3})^{1/2}$. As a result, the Tricomi expansion for radial wave function yields:

$$R_{nl} \to 4(Z/na_0)^{3/2} \frac{J_{2l+1}\left(\sqrt{8Zr/a_0}\right)}{\sqrt{8Zr/a_0}}. \qquad (2.48)$$

This expression is valid at $Zr/a_0 \ll n^{5/3}$.

b) Continuous Spectrum. The radial function of the continuum for the Coulomb potential with charge $\pm Ze$ may be written in the form :

$$R_{kl} = \frac{C(k)}{(2l + 1)!}(2kr)^l e^{-ikr} F(\pm Z\frac{i}{ka_0} + l + 1; 2l + 2; 2ikr), \qquad (2.49)$$

where k is the wave number of electron, and $C(k)$ is the normalizing factor. Here the upper sign corresponds to the attractive potential, the lower one to the repulsive potential. If the "scale k" is used

$$\int_0^\infty R_{k'l}^* R_{kl} r^2 dr = \delta(k - k'), \qquad (2.50)$$

then $C(k)$ is given by

$$C(k) = (2/\pi)^{1/2} k \exp\left(\pm \pi Z/2\chi\right) \mid \Gamma(l + 1 \mp Z\frac{i}{\chi}) \mid =$$

$$\qquad (2.51)$$

$$= 2 \left|\frac{Z\chi}{1 - \exp\left(\pm \pi Z/2\chi\right)}\right|^{1/2} (k/Z)^l \prod_{s=0}^{l} (s^2 + (Z/\chi)^2)^{1/2}, \quad \chi = ka_0.$$

The radial function $\mathcal{R}_{El}$ in the "energy scale" $\delta(E - E')$ corresponds to the factor

$$C(E) = C(k)(m/\hbar^2 k)^{1/2}, \quad \mathcal{R}_{El} = (m/\hbar^2 k)^{1/2}\mathcal{R}_{kl}, \tag{2.52}$$

where $E = \hbar^2 k^2/2m$. Asymptotical behavior of this function at $r \to \infty$

$$\mathcal{R}_{kl} \to (2/\pi)^{1/2}\frac{1}{r}\sin\left(kr \pm Z\frac{\ln(2kr)}{ka_0} - \frac{\pi}{2}l + \zeta_l\right),$$

$$\zeta_l = \arg\left(\Gamma(l + 1 \mp Z\frac{i}{\kappa})\right), \tag{2.53}$$

is expressed in terms of the scattering phase shifts ζ_l. Its behavior at small r is given by

$$\mathcal{R}_{kl} \underset{r \to 0}{\to} \frac{C(k)}{(2l + 1)!}(2kr)^l.$$

The analysis of the wave function behavior at small energies (or momenta) turns out to be more convenient in the "scale of energy". In the case of the attractive potential, the Tricomi expansion yields:

$$\mathcal{R}_{El} \to 4\sqrt{Z}(me/\hbar^2)\frac{J_{2l+1}\left(\sqrt{8Zr/a_0}\right)}{\sqrt{8Zr/a_0}}. \tag{2.54}$$

For the repulsive Coulomb potential, the result is expressed through the Bessel function $I_b(z)$:

$$\mathcal{R}_{El} \to 4\sqrt{Z}(me/\hbar^2)e^{-Z\pi/ka_0}\frac{I_{2l+1}\left(\sqrt{8Zr/a_0}\right)}{\sqrt{8Zr/a_0}}. \tag{2.55}$$

These expressions are valid at $kr \ll (Z/ka_0)^{2/3}$.

The wave functions $\mathcal{R}_{kl}$ introduced above are the eigenfunctions with the given values of the orbital angular momentum l. The wave function corresponding to a given direction of the $\mathbf{k}$ wave vector is more adequate for the scattering problem. It can be built in terms of the radial function and the scattering phase shifts ζ_l

$$\psi_{\mathbf{k}}^+(\mathbf{r}) = (2\pi)^{-3/2}\frac{1}{2k}\sum_{l=0}^{\infty} i^l(2l + 1)e^{i\zeta_l}\mathcal{R}_{kl}(r)P_l\left(\widehat{\mathbf{kr}}\right). \tag{2.56}$$

Asymptotically this function reduces to a superposition of a plane wave and an outgoing spherical wave. It is normalized on the momentum space delta function $\delta(\mathbf{k} - \mathbf{k}')$. Additional symmetry of the Coulomb field allows us to obtain its explicit expression. In parabolic coordinates this function can be written as

$$\psi_{\mathbf{k}}^+ = (2\pi)^{-3/2}e^{-(\pi Z/2\chi)}\Gamma(1 + iZ/\chi)e^{i(\chi/2Z)(\xi-\eta)}F(-iZ/\chi; 1; i\chi Z\eta), \quad \chi = ka_0, \tag{2.57}$$

where the coordinates ξ and η are:

$$\xi = (r + z)/Za_0, \quad \eta = (r - z)/Za_0, \quad \varphi = \arctan(y/x). \quad (2.58)$$

The z-axis is directed along the $\mathbf{k}$ vector. Asymptotically, at $r \to \infty$ we have for the wave function $\psi_{\mathbf{k}}^+$:

$$\psi_{\mathbf{k}}^+ \to (2\pi)^{-3/2} \left\{ \left[1 - \frac{(a_0/Z)^2}{ikr^3(1 - \cos\theta)} \right] \exp\left\{ ikz + \frac{iZ}{\chi}\ln\left[kr(1 - \cos\theta)\right] \right\} + \right.$$
$$\left. + \frac{f(\theta)}{r}\exp\left[ikr + \frac{iZ}{\chi}\ln(2kr) \right] \right\}, \quad (2.59)$$

and for the scattering amplitude $f(\theta)$:

$$f(\theta) = -\frac{Z}{2k^2 a_0 \sin^2(\theta/2)} \exp\left[-\frac{iZ}{\chi}\ln(\sin\theta/2) \right] \frac{\Gamma(1 + iZ/\chi)}{\Gamma(1 - iZ/\chi)}. \quad (2.60)$$

The first term in (2.59) is the incident wave which is distorted by the logarithmic phase due to the long range of the Coulomb field. The second term is the divergent wave, and the amplitude $f(\theta)$ gives the well-known Rutherford formula for the differential cross section of scattering into the solid angle $d\Omega$

$$d\sigma = (Za_0)^2 \frac{d\Omega}{4(ka_0)^4 \sin^4(\theta/2)}. \quad (2.61)$$

2.2.2 JWKB Approximation

There are two approaches for investigating the behavior of the wave functions at large quantum numbers. The first one is based on the semiclassical description of atomic electron. The second approach uses the asymptotic expressions of hypergeometric functions at large indices. The example of the second approach is formula (2.48) based on the Tricomi expansion (2.47). Below we would concentrate on the mostly used semiclassical approach first of all.

Let the Schrödinger equation be written in the form:

$$\psi'' + k^2(x)\psi = 0, \quad k^2(x) = (2m/\hbar^2)[E - U(x)], \quad (2.62)$$

and let the potential $U(x)$ be slowly changed on scales of the order of the particle wavelength (k^{-1}). Then ψ may be written in the form

$$\psi = ak^{-1/2}\sin\left(\int_{x_1}^x k\,dx + \varphi\right), \quad k^2 > 0,$$

$$\psi = |k|^{-1/2}\left[A_1 \exp\left(\int |k|\,dx\right) + A_2 \exp\left(-\int |k|\,dx\right)\right], \quad k^2 < 0.$$
$$(2.63)$$

The classically allowed region corresponds to the case of $k^2 > 0$. The x_1 point is the "turning" point, in which $k^2(x_1) = 0$. The phase φ is defined by the behavior of k^2 at point x_1. It is equal to $\varphi = \pi/4$, if k^2 is a linear function in the vicinity of the x_1 point. The constant of a is defined by the normalization condition. If there are only two "turning" points (x_1 and x_2), and the classical motion is periodical, then

$$a^2 = \left(\int_{x_1}^{x_2} \frac{dx}{2k(x)} \right)^{-1} = \frac{2m\omega}{\pi\hbar} = \frac{4m}{\hbar T}, \tag{2.64}$$

where m is the particle mass, ω and $T = 2\pi/\omega$ are the frequency and period of the classical motion. The classically forbidden region corresponds to the case of $k^2 < 0$. Usually this region makes small contribution to the matrix elements. However, there are some cases when the classically forbidden region appears relevant (see, e.g., [2.5]). Below we shall consider the allowed region alone.

a) Radial Wave Function. In the JWKB approximation the radial wave function is given by

$$\mathcal{R}_{nl} \approx \left(\frac{2Z}{\pi n^3 a_0^2} \right)^{1/2} \frac{\cos \Phi_r}{r k_r^{1/2}}, \quad \Phi_r = \int_{r_1}^{r} k_r dr - \pi/4, \tag{2.65}$$

where the phase of the semiclassical wave function Φ_r can be expressed in analytical form for the Coulomb field

$$\Phi_r = k_r r - n \arcsin \left[(1 - Zr/n^2 a_0)/\epsilon \right] -$$

$$-(l+1/2) \arcsin \left[\left(1 - (l+1/2)^2 a_0/Zr \right)/\epsilon \right] + [n - (l+1/2)] \frac{\pi}{2} - \frac{\pi}{4},$$

$$k_r^2 = \frac{2m}{\hbar^2} \left[E - U_l(r) \right], \quad U_l(r) = -\frac{Ze^2}{r} + \frac{\hbar^2}{2m} \frac{(l+1/2)^2}{r^2}, \tag{2.66}$$

and the left and right turning points are given by

$$r_{1,2} = n^2 (a_0/Z)(1 \pm \epsilon), \quad \epsilon = \left[1 - (l+1/2)^2 / n^2 \right]^{1/2}. \tag{2.67}$$

In (2.66) the quantum factor $l(l+1)$ in the centrifugal potential is replaced by $(l+1/2)^2$, that corresponds to the Langer correction in the JWKB approximation (see [2.2, 6] for more details).

In the case of $\Delta n \ll n$ and $\Delta l \ll l, l' \ll n$ the first term of expansion of the phase difference $\Delta\Phi_r = \Phi_r(n + \Delta n, l + \Delta l) - \Phi_r(n, l)$ over $\Delta n = n' - n$ and $\Delta l = l' - l$ takes the form

$$\Delta\Phi_r = \frac{E_{n'} - E_n}{2Ry} \int\limits_{r_1}^{r} \frac{dr}{a_0^2 k_r} - (l + 1/2)\, \Delta l \int\limits_{r_1}^{r} \frac{dr}{r^2 k_r}$$

$$= -\Delta n \left\{ \arcsin\left[(1 - Zr/n^2 a_0)/\epsilon \right] + k_r r - \pi/2 \right\} - \tag{2.68}$$

$$- \Delta l \left\{ \arcsin\left[\left(1 - (l + 1/2)^2 a_0/Zr\right)/\epsilon \right] + \pi/2 \right\}.$$

This expression is fruitful for calculation of transition matrix elements between closely spaced Rydberg states.

b) Restricted Interference Approximation. The normalizing factor (2.64) can be obtained from the semiclassical expression (2.65) neglecting fast oscillations of the wave function square corresponding to the doubled phase and restricting integration only to the classically allowed region. These approximations, often used for calculations of transition matrix elements, improve the results (see, for example, [2.6]). They can be formulated as a restricted interference approximation [2.7]. Let us introduce functions:

$$\psi_{nl}^{(\pm)} = \frac{c_{nl}}{\sqrt{k(r)}} e^{\pm i\Phi(r)}. \tag{2.69}$$

Then the semiclassical function from (2.65) can be written in the form:

$$r\mathcal{R}_{nl} = \frac{1}{2}\, \mathbf{Re}\left\{ \psi_{nl}^{(+)}(r) + \psi_{nl}^{(-)}(r) \right\}, \tag{2.70}$$

where the normalization constant is fixed by equation:

$$\frac{1}{2}\, \mathbf{Re}\left\{ \int\limits_{r_1}^{r_2} \psi_{nl}^{(+)}(r)\, \psi_{nl}^{(-)}(r)\, dr \right\} = 1. \tag{2.71}$$

Illustrations presented in [2.7] show that the semiclassical functions (2.65, 70) satisfactorily agree with the hydrogen quantum wave functions (even for $n = 1,\ 2$) except for the regions near the turning points. The principle of restricted interference consists of the calculation of the matrix elements of any physical value $\widehat{Q}$ (including normalization constant, which corresponds to $\widehat{Q} = 1$) according to formula:

$$\left\langle nl \left| \widehat{Q} \right| n'l' \right\rangle = \frac{1}{2}\, \mathbf{Re}\left\{ \int\limits_{r_1}^{r_2} \psi_{nl}^{(+)}(r)\, Q(r)\, \psi_{n'l'}^{(-)}(r)\, dr \right\}. \tag{2.72}$$

This prescription means neglecting the fast oscillating terms of the kind $\psi_{nl}^{(+)} \psi_{n'l'}^{(+)}$. It should also be noted that prescription (2.72) is equivalent to the semiclassical matrix element in the action variable representation (2.79).

c) **Angular Wave Function.** Let us consider now the angular wave function Θ_{lm} [see (2.39, 41)]. Introducing the function $\chi(x)$ by the relation:

$$\Theta_{lm}(x) = \left(1 - x^2\right)^{-1/2} \chi(x), \qquad x = \cos\theta, \tag{2.73}$$

and using the semiclassical approach for $\chi(x)$, the wave function (2.73) can be written as

$$\Theta_{lm}(x) = \left(\frac{2l+1}{\pi\left(1-x^2\right)k}\right)^{1/2} \cos\left(\int_{x_1}^{x} k\,dx - \pi/4\right), \tag{2.74}$$

Here

$$k^2(x) = \frac{(l+1/2)^2 - m^2/\left(1-x^2\right)}{1-x^2},$$

and the "turning" x_1 and x_2 points are defined by equations: $k(x_1) = k(x_2) = 0$, so that

$$x_{1,2} = \mp[1 - m^2/(l+1/2)^2]^{1/2}.$$

These expressions may be simplified in the case of $l \gg m$:

$$\Theta_{lm}(\cos\theta) \approx \sqrt{\frac{2}{\pi}} \frac{\cos\left[(l+1/2)\theta + (2m-1)\pi/4\right]}{(\sin\theta)^{1/2}}. \tag{2.75}$$

Formula (2.75) is not valid at small θ. In this case, when $\theta \to 0$, $l \to \infty$, and $l\theta$ is fixed, we have (see, for example, [2.8])

$$\Theta_{lm}(\cos\theta) \approx \sqrt{(l+1/2)}J_m\left[(l+1/2)\theta\right],$$
$$P_l^{-m}(\cos\theta) \approx (l+1/2)^{-m}J_m\left[(l+1/2)\theta\right]. \tag{2.76}$$

2.2.3 Semiclassical Approach in Action Variables

The variables of action introduced above (see Sect. 2.1.2) are especially convenient for using classical mechanics in quantum-mechanical calculations. Quantization conditions for variable I_i are

$$I_1 = n\hbar, \quad I_2 = (l+1/2)\hbar, \quad I_3 = m\hbar, \tag{2.77}$$

and the semiclassical atomic wave function in terms of the action variables is

$$\psi = e^{iS/\hbar}, \qquad S = \sum_i I_i u_i. \tag{2.78}$$

It is important to stress that the use of the semiclassical expression (2.78) for the wave function in quantum matrix element leads directly to the correspondence principle. According to quantum mechanics the matrix element of the $Q(\mathbf{r})$ operator is equal to

$$\langle n'l'm'| Q\,(\mathbf{r})\,|nlm\rangle = \frac{1}{2\pi} \int\limits_{0}^{2\pi} d\mathbf{u}\exp\left\{\frac{i}{\hbar}\,(S'-S)\right\}Q(\mathbf{u})$$

$$\hspace{6cm}(2.79)$$

$$= \frac{1}{2\pi} \int\limits_{0}^{2\pi} d\mathbf{u}\exp\left\{\frac{i}{\hbar}\,(\Delta n u_1 + \Delta l u_2 + \Delta m u_3)\right\}Q(\mathbf{u}).$$

We see, that the Fourier transform of the Q value in the action variables does correspond to the transition matrix element.

a) Quantum Defect. The concept of quantum defect is natural in the action variables. Let the highly excited electron potential $U(r)$ differs from the Coulomb one $U_c = -Ze^2/r$ by small value $\Delta U(r)$ $(\Delta U(r) \ll U(r))$. The quantization condition for the I_1 variable may be written as:

$$I_1 = n\hbar = \frac{1}{2\pi}\oint [2m(E - U_c - \Delta U - (l+1/2)^2\hbar^2/r^2)]^{1/2}dr \approx$$

$$\hspace{6cm}(2.80)$$

$$\approx n_*\hbar - \frac{1}{2\pi}\oint \frac{m\Delta U\,(r)}{p_c\,(r)}dr = (n_* + \delta)\hbar,$$

where

$$p_c(r) = [2m(E - U_c - (l+1/2)^2\hbar^2/r^2)]^{1/2},$$

and δ is the quantum defect. It is important to stress that the integral in (2.80) is taken over full circle of the classical variable r. Hence it is equal to the doubled usual semiclassical integral between two turning points.

The use of the relation

$$du_1 = \frac{dr}{p_c\,(r)}m\frac{\partial E}{\partial I_1},$$

which directly follows from (2.33), yields simple expression for the quantum defect:

$$\delta = -\frac{1}{2\pi\hbar}\oint \frac{m\Delta U\,(r)}{p_c\,(r)}dr = -\frac{1}{2\pi}\int\limits_{0}^{2\pi}\frac{\Delta U\,(r)}{\hbar\omega_n}du_1.\hspace{2cm}(2.81)$$

Here the frequency of the classical motion ω_n corresponds to the transition frequency between neighboring levels. The expression (2.81) may be rewritten in the form which clarifies the relation with the usual quantum theory of perturbation:

$$\hbar\omega_n\delta = \frac{2Z^2Ry}{n^3}\delta = -\int\limits_{0}^{2\pi}\frac{\Delta U\,(r)}{2\pi}du_1.\hspace{2cm}(2.82)$$

The left-hand side of this equation is the correction ΔE to the level energy, and the right-hand side is the diagonal element of the perturbation ΔU with the semiclassical wave functions. Therefore, at small δ, the formula (2.82) corresponds to the relation $\Delta E = \langle n | \Delta U | n \rangle$ in the usual quantum perturbation theory. However, for highly excited states the validity region of (2.82) is significantly wider than the quantum one. The classical validity condition $\Delta S \ll S$ corresponds to $\delta \ll n$, in contrast to the quantum one $\Delta E \ll \hbar \omega$. It should be noted that usually the quantum defect is positive, since the difference between real potential of the Rydberg electron and the Coulomb potential $\Delta U < 0$.

2.3 Wave Functions: Momentum Representation

2.3.1 Hydrogenlike Wave Functions

The momentum representation is often used in various Rydberg state theories. It is convenient to use the wave vector $\mathbf{k} = \mathbf{p}/\hbar$ for further applications. The wave function is defined by the relation

$$G(\mathbf{k}) = (2\pi)^{-3/2} \int \exp(-i\mathbf{k} \cdot \mathbf{r}) \psi(\mathbf{r}) d\mathbf{r}. \tag{2.83}$$

a) Discrete Spectrum. The form (2.39) for $\psi(\mathbf{r})$ gives a similar form for $G(\mathbf{k})$:

$$G_{nlm}(\mathbf{k}) = g_{nl}(k) Y_{lm}(\theta, \varphi), \quad g_{nl}(k) = (-i)^l \sqrt{\frac{2}{\pi}} \int\limits_0^\infty \mathcal{R}_{nl}(r) j_l(kr) r^2 dr, \tag{2.84}$$

where $j_l(z) = (\pi/2z)^{1/2} J_{l+1/2}(z)$ is the spherical Bessel function, and the radial part g_{nl} satisfies the normalization condition:

$$\int\limits_0^\infty g_{n'l}^*(k) g_{nl}(k)(k) k^2 dk = \delta_{nn'}.$$

For the Coulomb field g_{nl} is

$$g_{nl}(k) = \left(\frac{a_0}{Z}\right)^{3/2} \left[\frac{2}{\pi} \frac{(n-l-1)!}{(n+l)!}\right]^{1/2} n^2 2^{2(l+1)} l! \frac{(-ix)^l}{(x^2+1)^{l+2}} C_{n-l-1}^{l+1}\left(\frac{x^2-1}{x^2+1}\right), \tag{2.85}$$

where $x = nka_0/Z$, and C_{n-l-1}^{l+1} is the Gegenbauer polynomial, which may be expressed in terms of the hypergeometric function:

$$C_{n-l-1}^{l+1}\left(\frac{x^2-1}{x^2+1}\right) = \frac{(2l+1+n)!}{n!(2l+1)!} F\left(-n, n+2l+2; l+3/2; \frac{1}{x^2+1}\right). \tag{2.86}$$

Of course, the wave function in the momentum representation may be chosen real. The phase factor in (2.85) provides the validity of relation (2.84). In the special case of $l = 0$, we have

$$g_{n0}(k) = \sqrt{\frac{2}{\pi}} \left(\frac{a_0 n}{Z}\right)^{3/2} \frac{4}{(x^2+1)^2} \frac{\sin(n\beta)}{\sin(\beta)}, \qquad \beta = \arccos\left(\frac{x^2-1}{x^2+1}\right). \quad (2.87)$$

The functions $G(\mathbf{k})$ are the solutions of the integral equation, which may be obtained by the Fourier transform of the Schrödinger equation:

$$\left(\frac{\hbar^2 k^2}{m} - 2E\right) G(\mathbf{k}) = Z \left(\frac{e}{\pi a_0}\right)^2 \int d\mathbf{k} \frac{G(\mathbf{k})}{|\mathbf{k} - \mathbf{k}'|}. \quad (2.88)$$

Separating the momentum angle yields the equation for the radial functions $g_{nl}(k)$:

$$\left(\frac{\hbar^2 k^2}{m} - 2E\right) g_{nl}(k) = 2 \left(\frac{Z}{\pi k}\right) \int_0^\infty g_{nl}(k') Q_l \left(\frac{k^2 + (k')^2}{2kk'}\right) k' dk', \quad (2.89)$$

where Q_l is the Legendre function of the second kind, related to the usual Legendre polynomial:

$$Q_l(z) = \mathcal{P} \int_{-1}^1 dt \frac{P_l(t)}{(z-t)}. \quad (2.90)$$

The symbol "$\mathcal{P}$" means the principal value of the integral, i.e.,

$$\mathcal{P} \int_a^b \frac{f(x)\,dx}{z-x} = \lim_{\eta \to 0} \left(\int_a^{z-\eta} \frac{f(x)\,dx}{z-x} + \int_{z+\eta}^b \frac{f(x)\,dx}{z-x}\right). \quad (2.91)$$

b) **Continuous Spectrum.** The momentum space wave function corresponding to the Coulomb wave function of the scattering problem $\psi_{\mathbf{k}}^+(\mathbf{r})$ may be obtained by the Fourier transformation from (2.57):

$$G(\mathbf{k}') = (2\pi)^{-3/2} \int \exp(-i\mathbf{k}' \cdot \mathbf{r}) \psi_{\mathbf{k}}^+(\mathbf{r}) d\mathbf{r} =$$

$$= -\frac{1}{2\pi^2} \lim_{\eta \to 0} \frac{d}{d\eta} \left(\frac{\left[(k')^2 - (k+i\eta)^2\right]^{-iZ/\chi}}{\left[(\mathbf{k}-\mathbf{k}')^2 + \eta^2\right]^{1-iZ/\chi}}\right). \quad (2.92)$$

In the case of $Z/ka_0 = Z/\chi \ll 1$ this expression is reduced to the momentum wave function in the Born approximation:

$$G^B(\mathbf{k}') = \delta(\mathbf{k}' - \mathbf{k}) - \left(\frac{Z}{\pi^2}\right) \frac{1}{(k - k' + i0)(k + k')(\mathbf{k} - \mathbf{k}')^2}. \quad (2.93)$$

Here the first term corresponds to the plane wave $(2\pi)^{-3/2} \exp(i\mathbf{k}\mathbf{r})$ and the second one describes the outgoing wave. Formula (2.93) contains the pole, and the standard designation "i0" in the denominator shows the path of integration according to the known relation

$$\frac{1}{x - x_0 + i0} = \mathcal{P}\frac{1}{x - x_0} - i\pi\delta\left(x - x_0\right).$$
(2.94)

2.3.2 Momentum Wave Functions with Quantum Defect

As it was already mentioned the concept of quantum defect is very useful for describing the energy spectra. The Coulomb wave functions taking into account quantum defect were used for atomic calculations in [2.9]. The main difficulty is the divergence of the regular Coulomb function at $r \to \infty$ with a noninteger principal quantum number at $r \to 0$. The recommendation [2.9] is to cut off the divergent terms. The detailed analysis of the quantum defect method and various applications are given in [1.4]. We present here the momentum space wave function with the quantum defect following the result of [2.10]:

$$g_{nl}(k) = -\left(\frac{a_0}{Z}\right)^{3/2}\left(\frac{2}{\pi}\frac{\Gamma\left(n_* - 1\right)}{\Gamma\left(n_* + l + 1\right)}\right)^{1/2} n_* 2^{2(l+1)}(l+1)!$$

$$\times \frac{(-ix)^l}{(x^2 + 1)^{l+2}} J(n_*, l + 1, z),$$
(2.95)

where $x = n_* k a_0 / Z$, and the J-function is determined by the recurrent relationship

$$J(n_*, l + 1, z) = -\frac{1}{2(2l + 2)}\frac{\partial}{\partial z}J(n_*, l, z)$$

with the initial value at $l = 0$

$$J(n_*, 0, z) = -\frac{n_* \sin[n_*(\beta - \pi)]}{\sin(\beta - \pi)} - \frac{\sin(n_*\pi)}{\pi}\int\limits_0^1 \frac{(1 - s^2)s^{n_*}}{1 - 2zs + s^2}ds,$$

$$z = (x^2 - 1)/(x^2 + 1), \quad n_* = n - \delta.$$

2.4 Density Matrix and Distribution Function

2.4.1 Classical Distribution Functions

The classical distribution functions turn out to be useful in many Rydberg problems. They are an effective tool in the theory of the Rydberg atom, in both the classical and semiclassical approaches. The distribution function yields the probability density W of some coordinates and momenta set being equal to the fixed values for the kind of motions given by magnitudes of the other variables or additional conditions. The present approach is based on the canonical distribution function, which is a constant in the phase space.

At first, we consider a single particle ensemble in the action variables (I_i, u_i) (homogeneous ensemble). In these variables the distribution function can be written as

$$W\, d\mathbf{u}\, d\mathbf{I} = d\mathbf{u}\, d\mathbf{I}/(2\pi\hbar)^3. \tag{2.96}$$

We choose a natural normalization: there is one particle in the phase volume $(2\pi\hbar)^3$. The coordinate distribution function with fixed I_1, I_2, I_3 values in the spherical coordinates r, θ, φ can be obtained from (2.96) with the help of the corresponding delta functions and further integration:

$$W_{I_1, I_2, I_3}(r, \theta, \varphi) r^2 \sin\theta\, dr d\theta d\varphi = d\mathbf{u} \int \delta(\mathbf{I}/\hbar - \mathbf{I}'/\hbar) \frac{d\mathbf{I}'}{(2\pi\hbar)^3}. \tag{2.97}$$

Using the expressions for variables u_1, u_2, u_3 in terms of r, θ, φ [see (2.29–33)], the coordinate distribution function can be written as

$$W_{I_1, I_2, I_3}(r, \theta, \varphi) r^2 \sin\theta\, dr d\theta d\varphi$$
$$= \frac{r^2 \sin\theta\, dr d\theta d\varphi}{\{\pi a^2 r[\epsilon^2 - (1 - r/a)^2]^{1/2}\}\{2\pi^2[\sin^2\theta - (I_3/I_2)^2]^{1/2}\}}, \tag{2.98}$$
$$a = Ze^2/2|E| = I_1^2/mZe^2, \qquad \epsilon^2 = 1 - I_2^2/I_1^2,$$

where a and ϵ are the semimajor axis and eccentricity, respectively. The W_{I_1, I_2, I_3} function is defined in the range between the reversal points. But in contrast to (2.33), in which double integration (back and forth) is suggested, each point in formula (2.98) is taken into account only once. The I_i values satisfy the following conditions:

$$-I_2 \leq I_3 \leq I_2, \qquad 0 \leq I_2 \leq I_1. \tag{2.99}$$

Bearing in mind applying distribution functions to atomic physics, it is convenient to measure the action variables in values of $\hbar$, namely $n = I_1/\hbar$, $l = I_2/\hbar$, and $m = I_3/\hbar$. Taking into account that the function (2.98) is independent on φ, one can rewrite expression (2.98) in terms of the identified variables in accord with [2.11]

$$W_{nlm}(r, \theta) r^2 \sin\theta\, dr d\theta$$
$$= \frac{r^2 \sin\theta\, dr d\theta}{\{\pi a^2 r[\epsilon^2 - (1 - r/a)^2]^{1/2}\}\{\pi[\sin^2\theta - (m/l)^2]^{1/2}\}} \tag{2.100}$$
$$a = n^2\hbar^2/mZe^2, \qquad \epsilon^2 = 1 - (l/n)^2.$$

Integration of (2.100) over the electron coordinates (r, θ, φ) is equal to 1 [one particle in the phase space $(2\pi\hbar)^3$]. Additional integration over possible m values gives the phase space volume [in units of $(2\pi\hbar)^3$] for a system with the given values of $I_1 = n\hbar$ and $I_2 = l\hbar$

$$g_{nl} = \int_{-l}^{l} dm = 2l.$$

It corresponds to the statistical weight $g_{nl} = 2l + 1$ in quantum mechanics. Similarly, the next integration over possible l values gives the phase space volume (in units of $(2\pi\hbar)^3$) for a system with the given magnitude of $I_1 = n\hbar$:

$$g_n = \int_0^n g_{nl} dl = n^2$$

The distribution function with fixed I_1 and I_2 values may be obtained by averaging over $m = I_3/\hbar$. It appears the averaged function is not dependent on the θ angle. Finally, we obtain:

$$W_{nl}(r)r^2 dr = \int_0^\pi \sin\theta d\theta \left\{ \frac{1}{g_{nl}} \int_{-l}^{l} W_{nlm}(r,\theta) dm \right\} r^2 dr$$

$$= \frac{r^2 dr}{\left\{ \pi a^2 r \left[\epsilon^2 - (1 - r/a)^2 \right]^{1/2} \right\}}. \qquad (2.101)$$

In a similar way we obtain the distribution function with fixed $I_1 = n\hbar$ value (or energy E_n) by integration over I_2:

$$W_n(r)r^2 dr = \left\{ \frac{1}{g_n} \int_0^n W_{nl}(r) dl \right\} r^2 dr = \frac{2}{\pi} \left[1 - \left(1 - \frac{r}{a} \right)^2 \right]^{1/2} \frac{rdr}{a^2}. \qquad (2.102)$$

The energy conservation law gives the relation between the r and p values:

$$r = 2mZe^2/(p^2 + p_n^2), \qquad p_n^2 = 2m \mid E_n \mid. \qquad (2.103)$$

Using this relation we can rewrite expression (2.102) in the momentum representation:

$$W_n(p)p^2 dp = \frac{32}{\pi} \frac{x^2 dx}{(1 + x^2)^4}, \qquad x = p/p_n. \qquad (2.104)$$

Similarly, in the case of $1 - \epsilon \ll 1$ (when $l \ll n$) we obtain from (2.101) the momentum distribution function with fixed I_1 and I_2 values:

$$W_{nl}(p)p^2 dp = \frac{4}{\pi} \frac{dx}{(1 + x^2)^2}. \qquad (2.105)$$

In conclusion, we consider the "microcanonical" distribution $W_E(\mathbf{p}, \mathbf{r})$ function of the momenta and coordinates with fixed electron energy in the Coulomb field:

$$W_E(\mathbf{p}, \mathbf{r})\frac{d\mathbf{p}d\mathbf{r}}{(2\pi\hbar)^3} = \frac{\delta\left[E - H(\mathbf{p}, \mathbf{r}) \right]}{\rho_{cl}(E)} \frac{d\mathbf{p}d\mathbf{r}}{(2\pi\hbar)^3}, \quad H(\mathbf{p}, \mathbf{r}) = \frac{\mathbf{p}^2}{2m} - \frac{Ze^2}{r}. \qquad (2.106)$$

Here $\rho_{cl}(E)$ is the classical density of states per unit energy interval:

$$\rho_{cl}(E) = \int \delta \left[E - H(\mathbf{p}, \mathbf{r})\right] \frac{d\mathbf{p}d\mathbf{r}}{(2\pi\hbar)^3} = \int \delta \left[E + \frac{m \left(Ze^2\right)^2}{2I_1^2}\right] \frac{d\mathbf{I}d\mathbf{u}}{(2\pi\hbar)^3}$$

$$= \frac{m^{3/2} Z^3 e^6}{\hbar^3 |2E|^{5/2}} = \frac{n^5 \hbar^2}{m Z^2 e^4}. \tag{2.107}$$

It differs from the previously introduced quantum function (1.5) only by the spin factor $g_s = 2$ so that $\rho(E) = 2\rho_{cl}(E)$. The distribution function (2.106) describes the "microcanonical ensemble". It should be noted that distribution functions (2.102, 104) can be derived directly from (2.106) by integration over the $\mathbf{p}$ and $\mathbf{r}$, respectively.

2.4.2 Coulomb Green's Function

The Coulomb Green's function in the closed form was obtained in papers [2.12]. It is a powerful tool in the Rydberg states theory. Here we outline the derivation following papers [2.12]. The Green's function $G(\mathbf{r}, \mathbf{r}', E)$ is the solution of the equation

$$-\frac{\hbar^2}{2m} \left[\Delta + (2k\nu)/r + k^2\right] G(\mathbf{r}, \mathbf{r}', E) = \delta(\mathbf{r} - \mathbf{r}'), \tag{2.108}$$

where

$$\nu = Z/ka_0, \qquad k = \sqrt{2mE}/\hbar.$$

Here E is a complex number but not the eigenvalue of the Hamiltonian. The retarded Green's function defined for real E is obtained from $G(\mathbf{r}, \mathbf{r}', E)$ by taking the limit with E approaching the real axis from above. For $E > 0$ the retarded Green's function consists only of outgoing spherical waves. For $E < 0$ the Green's function decays exponentially as $r \to \infty$.

The general invariance considerations require that $G(\mathbf{r}, \mathbf{r}', E)$ depend on $\mathbf{r}$ and $\mathbf{r}'$ only in the forms r, r', and $\mathfrak{r} = |\mathbf{r} - \mathbf{r}'|$. It is natural to factor out $\mathfrak{r}^{-1}$ from G. Therefore, we introduce $F = -4\pi\mathfrak{r}G$, satisfying the equation

$$\left(\Delta - 2\frac{\mathbf{r} - \mathbf{r}'}{|\mathbf{r} - \mathbf{r}'|^2}\nabla + 2k\nu/r + k^2\right) F(r, r', \mathfrak{r}) = 0, \tag{2.109}$$

with the normalization condition $F(\mathbf{r} = \mathbf{r}') = -2m/\hbar^2$. Now, it is remarkable that $F(r, r', \mathfrak{r})$ is a function of only two variables $x \equiv r+r'+\mathfrak{r}$ and $y \equiv r+r'-\mathfrak{r}$ and is proportional to

$$(\partial/\partial x - \partial/\partial y) f_1(x) f_2(y). \tag{2.110}$$

In view of this result, we rewrite the equation for F in terms of the variables $R \equiv r + r'$ and $\mathfrak{r}$:

$$\left\{ \frac{\partial^2}{\partial \mathfrak{r}^2} + \frac{\partial^2}{\partial R^2} + \frac{\mathfrak{r}^2 + R^2 - 2R\mathfrak{r}r'}{\mathfrak{r}(R - r')} \frac{\partial^2}{\partial \mathfrak{r}\partial R} \right.$$

$$\left. + \frac{\mathfrak{r}^2 - R^2 + 2R\mathfrak{r}r'}{\mathfrak{r}^2(R - r')} \frac{\partial}{\partial R} + \frac{2k\nu}{(R - r')} + k^2 \right\} F(r, r', \mathfrak{r}) = 0.$$

Now, if we let $F(r, r', \mathfrak{r}) = (\partial/\partial \mathfrak{r})D(r, r', \mathfrak{r})$, then the term linear in the differential operators can be eliminated. We find $(\partial/\partial \mathfrak{r})\widehat{M}D(r, r', \mathfrak{r}) = 0$, where

$$\widehat{M} = \frac{\partial^2}{\partial \mathfrak{r}^2} + \frac{\partial^2}{\partial R^2} + \frac{\mathfrak{r}^2 + R^2 - 2R\mathfrak{r}r'}{\mathfrak{r}(R - r')} \frac{\partial^2}{\partial \mathfrak{r}\partial R} + \frac{2k\nu}{(R - r')} + k^2. \quad (2.111)$$

Thus $\widehat{M}D$ is a function of R and r' only, and therefore we can let $\widehat{M}D = 0$. The significance of the transformation to the variables x and y becomes apparent. When $\widehat{M}$ is expressed in terms of these variables, no mixed derivatives occur. The specific feature of the pure Coulomb potential is that $\widehat{M}$ is separable in x and y, and the equation $\widehat{M}D = 0$ may be rewritten in the form

$$\left\{ (x^2 - 2xr')\widehat{O}(x) - (y - 2yr')\widehat{O}(y) \right\} D(x, y, r') = 0,$$
$$\widehat{O}(z) = (\partial^2/\partial z^2 + k^2/4 + k\nu/z). \quad (2.112)$$

The equation $\widehat{O}(z)\mathrm{f}(z)$ is the Whittaker equation. A solution for D is $D = const \cdot [\mathrm{f}_1(x)\mathrm{f}_2(y)]$, where f_1 and f_2 are the Whittaker functions. The choice of the Whittaker functions $W_{\alpha,\beta}(z)$ and $M_{\alpha,\beta}(z)$ flows from the boundary conditions and the choice of the constant is dictated by the normalization requirement $F(\mathbf{r} = \mathbf{r}') = -2m/\hbar^2$. The final expression for the Green's function is

$$G(\mathbf{r}, \mathbf{r}', E) = -\frac{2m}{\hbar^2} \frac{\Gamma(1 - i\nu)}{2\pi} \frac{1}{(x - y)} \frac{1}{ik} \left(\frac{\partial}{\partial y} - \frac{\partial}{\partial x} \right) W_{i\nu,1/2}(-ikx) \cdot M_{i\nu,1/2}(-iky),$$

$$x, y = r + r' \pm \mathfrak{r}. \quad (2.113)$$

In some applications expansion over spherical functions (partial expansion) proves useful. The following expansion:

$$G(\mathbf{r}, \mathbf{r}'; E) = \sum_{l,m} g_l(E; r, r') Y_{lm}^*(\mathbf{r}'/r') Y_{lm}(\mathbf{r}/r) \quad (2.114)$$

may be obtained from partial expansion of the δ-function

$$\delta(\mathbf{r} - \mathbf{r}') = \frac{\delta(r - r')}{rr'} \sum_{lm} Y_{lm}^*(\mathbf{r}'/r') Y_{lm}(\mathbf{r}/r) \quad (2.115)$$

by the standard method. The Schrödinger equation gives the radial equation for g_l:

$$-\left\{ \frac{\hbar^2}{2m} \left[\frac{1}{r^2} \frac{\partial}{\partial r} \left(r^2 \frac{\partial}{\partial r} \right) - \frac{l(l + 1)}{r^2} \right] - U(r) + E \right\} g_l(E; r, r') = \frac{\delta(r - r')}{rr'}. \quad (2.116)$$

If the potential U is the Coulomb one, $U(r) = -Ze^2/r$, then the solution will be

$$g_l(E; r, r') = a_0 \frac{\nu Z}{rr'} \frac{\Gamma(l+1-\nu)}{\Gamma(2l+2)} M_{\nu, l+1/2}(2r_< Z/\nu a_0) W_{\nu, l+1/2}(2r_> Z/\nu a_0),$$

$$\nu = (Z^2 Ry/|E|)^{1/2}. \tag{2.117}$$

The integral representation

$$G(\mathbf{r}, \mathbf{r}'; E) = \frac{1}{2\pi}(rr')^{-1/2} \sum_{l=0}^{\infty} (2l+1) P_l(\hat{\mathbf{r}}\hat{\mathbf{r}}')$$

$$\times \int_0^{\infty} dx \exp\left[-\frac{Z(r+r')}{\nu a_0} \cosh(x)\right] \coth^{2\nu}(x/2) I_{2l+1}\left[\frac{2Z\sqrt{rr'}}{\nu a_0} \sinh(x)\right]$$

$$\tag{2.118}$$

is often useful for analytical investigations.

2.4.3 Density Matrix

a) Coordinate Representation. The density matrix is defined as the "noncoherent" mixing of quantum states. Let a be the total set of the quantum numbers, α is a subset of a, and $a\backslash\alpha$ is complement with respect to a. Then the density matrix of α states is

$$\rho_\alpha(\mathbf{r}, \mathbf{r}') = \sum_{a\backslash\alpha, a'\backslash\alpha} \psi_{a'}^*(\mathbf{r}) \rho_{a'a} \psi_a(\mathbf{r}') = \sum_{a\backslash\alpha, a'\backslash\alpha} |a'\rangle \rho_{a'a}\langle a|, \tag{2.119}$$

where $\rho_{aa'}$ is a positive definite Hermite matrix normalized by the condition:

$$\text{Tr}\{\hat\rho\} = \sum_{a\backslash\alpha, a'\backslash\alpha} \int d\mathbf{r} \psi_{a'}^*(\mathbf{r}, t) \rho_{a'a} \psi_a(\mathbf{r}, t) = \sum_{a\backslash\alpha} \rho_{aa} = 1. \tag{2.120}$$

Below, we briefly discuss the main properties of the density matrix and give explicit expressions in some cases.

In the case of a pure quantum state a_0, the density matrix consists of only one term $|a_0\rangle\langle a_0|$, i.e., $\rho_{aa'} = \delta_{a_0a}\delta_{a_0a'}$. In the general case, the ρ-matrix depends on time according to an equation similar to the Liouville equation in classical mechanics:

$$\frac{d\hat\rho}{dt} = \frac{i}{\hbar}\left[\hat{H}\hat\rho\right] = \frac{i}{\hbar}\left(\hat{H}\hat\rho - \hat\rho\hat{H}\right)$$

$$= \frac{i}{\hbar} \sum_{a\backslash\alpha, a'\backslash\alpha} \left\{\left[\hat{H}\psi_a^*(\mathbf{r})\right] \rho_{aa'} \psi_{a'}(\mathbf{r}') - \psi_a^*(\mathbf{r}) \rho_{a'a} \left[\hat{H}\psi_{a'}(\mathbf{r}')\right]\right\}. \tag{2.121}$$

The ρ-matrix defines the ensemble of the quantum states. The average of a physical value $\hat{O}$ over the ensemble is given by trace of the product:

$$\overline{O}_\alpha = \mathrm{Tr}\left\{\hat{O}\hat{\rho}_\alpha\right\} = \int d\mathbf{r}d\mathbf{r}'O\left(\mathbf{r},\mathbf{r}'\right)\rho_\alpha\left(\mathbf{r},\mathbf{r}'\right),$$

$$O_\alpha\left(\mathbf{r},\mathbf{r}'\right) = \sum_{a\backslash\alpha,a'\backslash\alpha} \psi_{a'}^*\left(\mathbf{r}\right)O_{a'a}\psi_a\left(\mathbf{r}'\right), \quad O_{a'a} = \left\langle a'\left|\hat{O}\right|a\right\rangle.$$

(2.122)

One can see that in the case of a pure state, (2.122) gives the usual diagonal matrix element. Since the trace of a matrix concerning unitary transformations is invariant, the diagonal representation can be considered without loss of generality. In particular, in the diagonal representation (2.122) can be written in the form:

$$\rho_\alpha\left(\mathbf{r},\mathbf{r}'\right) = \sum_{a\backslash\alpha}|a\rangle\rho_a\langle a|, \qquad \overline{O}_\alpha = \mathrm{Tr}\left\{\hat{O}\hat{\rho}_\alpha\right\} = \sum_{a\backslash\alpha}\rho_a\left\langle a|O|a\right\rangle. \tag{2.123}$$

For the coordinate operator $\hat{O} = \delta\left(\mathbf{r} - \mathbf{r}'\right)$ formula (2.123) yields the probability distribution function W_α :

$$W_\alpha\left(\mathbf{r}\right) = \sum_{a\backslash\alpha}\rho_a\psi_a^*\left(\mathbf{r}\right)\psi_a\left(\mathbf{r}\right), \quad \int W_\alpha\left(\mathbf{r}\right)d\mathbf{r} =1. \tag{2.124}$$

The distribution function is reduced to the square of the wave function for the pure state, i.e.,

$$W_\alpha\left(\mathbf{r}\right) = |\psi_\alpha\left(\mathbf{r}\right)|^2. \tag{2.125}$$

In the case of two ensembles, the square of the nondiagonal element, averaged over both ensembles, can be written in terms of the density matrices

$$\sum_{a\backslash\alpha,a'\backslash\alpha'} \rho_{a'}^*\rho_a|\langle a'|O|a\rangle|^2 = \int d\mathbf{r}d\mathbf{r}'\rho_{\alpha'}^*\left(\mathbf{r},\mathbf{r}'\right)O\left(\mathbf{r}\right)O\left(\mathbf{r}'\right)\rho_\alpha\left(\mathbf{r},\mathbf{r}'\right), \tag{2.126}$$

where $\rho_\alpha\left(\mathbf{r},\mathbf{r}'\right)$ and $\rho_{\alpha'}\left(\mathbf{r},\mathbf{r}'\right)$ are the density matrices of the α and α' ensembles, respectively.

Now we give some explicit expressions for the density matrix of the hydrogen-like ions. The ρ-matrix for the nl states (all m sublevels have the same populations) can be written in the form, see (2.39, 41),

$$\rho_{nl}\left(\mathbf{r},\mathbf{r}'\right) = \sum_m R_{nl}(r)Y_{lm}^*(\theta,\varphi)\frac{1}{2l+1}R_{nl}(r')Y_{lm}(\theta',\varphi')$$

$$= R_{nl}(r)R_{nl}(r')\frac{P_l\left(\hat{\mathbf{r}}\hat{\mathbf{r}}'\right)}{4\pi}.$$

(2.127)

According to (2.124), the diagonal element of the $\rho_{nl}\left(\mathbf{r},\mathbf{r}'\right)$ matrix yields the quantum distribution function W_{nl}. Since the obtained expression is not dependent on the θ and φ angles, we find:

$$W_{nl}(r) = \int \rho_{nl}(\mathbf{r}, \mathbf{r}) \, d\Omega = |r\mathcal{R}_{nl}(r)|^2. \tag{2.128}$$

The density matrix for the n-level can be obtained from the spectral expansion of the Green's function:

$$G_E(\mathbf{r}, \mathbf{r'}) = \sum_a \frac{|a\rangle \langle a|}{E - E_a + i0}, \tag{2.129}$$

where sum includes the summation over discrete states and the integral over the continuum. The designation "i0" for an infinitely small imagly value means that if the energy $E > 0$, then the pole $E - E_a$ must be placed in the complex plane above the integration path. Using (2.129), we find for the density matrix

$$g_n\rho_n(\mathbf{r}, \mathbf{r'}) = \sum_a \delta(E_a - E_n) |a\rangle \langle a| = \lim_{E \to E_n} (E - E_n) G_E(\mathbf{r}, \mathbf{r'}). \tag{2.130}$$

Now we use the explicit analytical expression for the Green's function presented above, and take into account that the Γ-function has poles at points $E = E_n$ with residuals $(-1)^{n-1}/(n-1)!$. Then, with the help of the following property of the Whittaker functions:

$$\lim_{\nu \to n} W_{\nu, 1/2} = \Gamma(1 + n) \cdot (-1)^{n+1} M_{n, 1/2},$$

we obtain

$$g_n\rho_n(\mathbf{r}, \mathbf{r'}) = (\pi n a_0^3)^{-1} \frac{\mathfrak{P}_n}{x - y}, \quad \mathfrak{P}_n = \hat{L}\left[M_{n, 1/2}(x/n) M_{n, 1/2}(y/n)\right], \tag{2.131}$$

where $x = Z(r + r' + |\mathbf{r} - \mathbf{r'}|)/a_0$, $y = Z(r + r' - |\mathbf{r} - \mathbf{r'}|)/a_0$,

$$\hat{L} = \frac{\partial}{\partial x} - \frac{\partial}{\partial y}.$$

A similar expression for the density matrix corresponding to the continuum, takes the form

$$g_E\rho_{E+i0}(\mathbf{r}, \mathbf{r'}) = \sum_a \delta(E_a - E + i0) |a\rangle \langle a| = (-1/\pi) \operatorname{Im}\{G_{E+i0}(\mathbf{r}, \mathbf{r'})\},$$

$$g_E = \frac{d}{dE}\left[\frac{4\pi}{3} p^3(E)/(2\pi\hbar)^3\right] = \frac{m}{2\pi^2\hbar^2} k(E). \tag{2.132}$$

Here the outgoing waves and the normalization of the delta-function on energy are assumed, g_E is the density of states per unit volume and energy interval. The use of the explicit analytical expression for the Green's function gives:

$$g_E\rho_{E+i0}(\mathbf{r}, \mathbf{r'}) = \left[(2a_0^3 Ry)(2\pi\chi^2)(1 - e^{-2\pi Z/\chi})\right]^{-1} \frac{\mathfrak{P}_{iZ/\chi}}{x - y}, \quad \chi = ka_0. \tag{2.133}$$

b) Momentum Representation. The density matrix in the momentum representation is introduced in a way similar to the case of the coordinate representation:

$$\rho_\alpha\left(\mathbf{k},\mathbf{k}'\right) = \sum_{a\backslash\alpha,a'\backslash\alpha} G_{a'}^*\left(\mathbf{k}\right)\rho_{a'a}G_a\left(\mathbf{k}'\right) = \sum_{a\backslash\alpha,a'\backslash\alpha} |a'\rangle\rho_{a'a}\langle a|, \tag{2.134}$$

where $\mathbf{k} = \mathbf{p}/\hbar$ is the wave vector. As a result, similar to (2.128) we have for the distribution function of the nl-state:

$$W_{nl}\left(k\right) = \int \rho_{nl}\left(\mathbf{k},\mathbf{k}\right)d\Omega = |g_{nl}(k)k|^2. \tag{2.135}$$

The function is normalized by the condition $\int\limits_0^\infty W_{nl}\left(k\right)dk = 1$. In the case of $l \ll n$, formula (2.135) with the help of (2.95) yields [2.10]

$$W_{nl}\left(k\right) = \frac{4}{\pi}\left(\frac{n_*a_0}{Z}\right)\frac{1 - (-1)^l\cos[2n_*(\beta - \pi)]}{(x^2 + 1)^2},$$

$$x = n_*ka_0/Z, \quad \cos\beta = \frac{x^2 - 1}{x^2 + 1}. \tag{2.136}$$

Averaging over oscillations reduces (2.136) to the following expression:

$$W_{nl}\left(k\right) = |g_{nl}(k)|^2 k^2 = \frac{4}{\pi}\left(\frac{n_*a_0}{Z}\right)\frac{1}{\left[1 + (nka_0/Z)^2\right]^2}, \quad l \ll n, \tag{2.137}$$

which is the same as the pure classical result (2.105) for the distribution function at small orbital momentum.

The general expression for the density matrix of the n-level has been obtained by *Fock* [2.13]. The diagonal element of this matrix (distribution function in the momentum representation) is equal to

$$W_n\left(k\right) = \frac{1}{n^2}\sum_{lm}\int |G_{nlm}(\mathbf{k})|^2\,d\Omega_\mathbf{k} = \frac{1}{n^2}\sum_{l=0}^{n-1}(2l+1)\,|kg_{nl}(k)|^2 =$$

$$= (a_0n/Z)\frac{32\left(nka_0/Z\right)^2}{\pi\left[(nka_0/Z)^2 + 1\right]^4}, \quad \int\limits_0^\infty W_n\left(k\right)dk = 1. \tag{2.138}$$

It should be noted that the expression (2.138) is the same as the classical expression (2.104).

2.4.4 Wigner Function

Let α be the set of the quantum numbers (n, l, m), and $\Delta\alpha$ be the increment set $(\Delta n, \Delta l, \Delta m)$, whereas we assume $|\Delta\alpha| \ll |\alpha|$. The focus of this section is the density matrix or the projection operator $|\alpha\rangle\langle\alpha|$. It extracts only that function from the function space which has quantum number set α relating to some set of the action variable $\mathbf{I} = \alpha \hbar$. Now let us consider a more general density matrix $\widehat{\rho}$ which is defined by the sum of the projection operator with the nearest values of α. In classical mechanics, this sum corresponds to the region of the phase space with the action variable near $\mathbf{I} = \hbar \alpha$. Thus, we have for the averaged projection operator:

$$\frac{(2\pi\hbar)^3}{\Gamma} \cdot \sum_{\alpha' = \alpha - \Delta\alpha}^{\alpha' = \alpha + \Delta\alpha} |\alpha'\rangle\langle\alpha'| = \widehat{\rho} \rightarrow \frac{1}{(2\pi\hbar)^3} \delta\left[\alpha - \mathbf{I}(\mathbf{p}, \mathbf{r})/\hbar\right], \quad (2.139)$$

where Γ is the phase space volume of the box $\alpha \pm \Delta\alpha$, $\Gamma/(2\pi\hbar)^3 = g$ is the number of states. Formula (2.139) establishes the relation of the classical distribution function with the density matrix. A similar relation may be derived also with the help of the function introduced by *Wigner* [2.14]:

$$\mathbf{W}(\mathbf{p}, \mathbf{r}) = (2\pi\hbar)^{-3} \int d\mathbf{r} \exp\left(-\frac{i}{\hbar}\mathbf{p}\cdot\mathbf{r}'\right) \rho(\mathbf{r} - \mathbf{r}'/2, \mathbf{r} + \mathbf{r}'/2). \quad (2.140)$$

The integrals of the Wigner's function over momenta or coordinates are equal to the corresponding diagonal elements of the density matrix. Therefore, the quantum average of any function on momenta or coordinates or the sum of these functions is equal to the corresponding average over the Wigner function. However, it is not valid for a function of the products of momenta and coordinates. Thus, we see that the Wigner function is similar to the classical distribution function, but is not the same. The properties of the Wigner function and its behavior in the limiting case of $\hbar \rightarrow 0$ have been the subject of some papers (see, for example, [2.15]). Here, we just note that the similar function in the action variables is

$$\mathbf{W}(\mathbf{I}, \mathbf{u}) = (2\pi\hbar)^{-3} \int d\mathbf{u}' \exp\left(-\frac{i}{\hbar}\mathbf{I}\cdot\mathbf{u}'\right) \rho(\mathbf{u} - \mathbf{u}'/2, \mathbf{u} + \mathbf{u}'/2) =$$

$$= \frac{1}{g}(2\pi\hbar)^{-3} \int d\mathbf{u}' \exp\left(-\frac{i}{\hbar}\mathbf{I}\cdot\mathbf{u}'\right) \left\{ \frac{1}{(2\pi)^3} \sum_{\alpha' = \alpha - \Delta\alpha}^{\alpha' = \alpha + \Delta\alpha} e^{-i\alpha'\cdot(\mathbf{u} - \mathbf{u}'/2)} \right.$$

$$\left. \cdot e^{-i\alpha'\cdot(\mathbf{u} + \mathbf{u}'/2)} \right\} =$$

$$= \frac{1}{g}(2\pi\hbar)^{-3} \sum_{\alpha' = \alpha - \Delta\alpha}^{\alpha' = \alpha + \Delta\alpha} \delta\left[\alpha - \mathbf{I}(\mathbf{p}, \mathbf{r})/\hbar\right]. \quad (2.141)$$

Hence, for a highly excited state if the quantum averaging can be reduced to averaging over the Wigner function, then it may also be reduced to averaging over the classical distribution function in the phase space.

3. Radiative Transitions and Form Factors

The goal of this chapter is to present the theoretical techniques for calculating the dipole matrix element and form factor for transitions involving highly excited states of atoms and ions. The first one determines the probabilities of radiative transitions, radiative lifetime, as well as, the cross sections of the photoionization and photorecombination processes. On the other hand, the cross sections of collisional processes involving Rydberg atoms can be expressed in terms of the form factor or generalized oscillator strength in many cases of practical importance. We review the pure quantal and semiclassical methods for evaluation of identified matrix elements which are widely used in the modern theory of Rydberg atoms. Our consideration involves the case of hydrogen-like states and those cases when the Rydberg atom can be described within the framework of the one-channel quantum defect theory. The situations in which the interactions between different Rydberg series turn out to be important are out of the scope of this book. The final results are presented in a form convenient for applications.

3.1 Probabilities of Radiative Transitions

3.1.1 General Formulas

Let $\mathbf{a}$ be the total set of quantum numbers; α is a subset of $\mathbf{a}$, whereas $g_\alpha = \sum_{\mathbf{a}\backslash\alpha}$ $g_\mathbf{a}$ specifies its statistical weight; and $\mathbf{a}\backslash\alpha$ is complement with respect to $\mathbf{a}$. The probability of the radiation transition $\alpha \to \alpha'$ is equal to:

$$
A_{\alpha'\alpha} = \frac{4}{3}\frac{\omega^3}{\hbar c^3}\frac{1}{g_\alpha}\sum_{\mathbf{a}'\backslash\alpha'}\sum_{\mathbf{a}\backslash\alpha}|\,\mathbf{D}_{\mathbf{a}'\mathbf{a}}\,|^2 = \frac{4}{3}\frac{\omega^3}{\hbar c^3}\frac{S_{\alpha'\alpha}}{g_\alpha}
$$

$$
= \frac{4}{3}\left(\frac{e^2}{\hbar c}\right)^3\left(\frac{\hbar\omega}{2Ry}\right)^3\frac{1}{g_\alpha}\sum_{\mathbf{a}'\backslash\alpha'}\sum_{\mathbf{a}\backslash\alpha}|\,\mathbf{r}_{\mathbf{a}'\mathbf{a}}/a_0\,|^2\,\tau_0^{-1},
$$

(3.1)

where $\omega = (E_\alpha - E_{\alpha'})/\hbar$ is the transition frequency, $\tau_0 = \hbar^3/me^4$ is the atomic unit of time, $\mathbf{D}_{\mathbf{a}'\mathbf{a}}$, $\mathbf{r}_{\mathbf{a}'\mathbf{a}}$, and $S_{\alpha'\alpha}$ are the matrix elements of the dipole moment, the electron coordinate, and the line-strength, respectively.

In the case of one optical electron, the square of the coordinate matrix element and the line-strength may be expressed through the radial integral:

$$S(n'l',nl) = e^2 \sum_{mm'} |\langle n'l'm' \mid \mathbf{r} \mid nlm \rangle|^2 = e^2 \max(l,l') \left| R_{nl}^{n'l'} \right|^2, \quad l' = l \pm 1,$$

(3.2)

where

$$R_{nl}^{n'l'} = \langle n'l' \mid r \mid nl \rangle = \int_0^\infty \mathcal{R}_{n'l'}(r) r \mathcal{R}_{nl}(r) r^2 dr$$

is the radial integral. There is also a useful relation between the radial integral and the oscillator strength

$$f_{n'l',nl} = \frac{1}{3} \frac{\max(l,l')}{2l+1} \frac{E_{n'l'} - E_{nl}}{Ry} \left| R_{nl}^{n'l'} \right|^2.$$

(3.3)

For a hydrogen-like ion, the explicit expression for the radial integral has been obtained by *Gordon* [3.1]:

$$R_{nl}^{n'l'} = \frac{a_0}{Z} \frac{(-1)^{n'-l}(4nn')^{l+1}(n-n')^{n+n'-2l-2}}{4(2l-1)!(n+n')^{n+n'}} \left[\frac{(n+l)!(n'+l-1)!}{(n'-l)!(n-l-1)!} \right]^{1/2}$$

$$\times \left\{ F\left(-n+l+1, -n'+l; 2l; \frac{-4nn'}{(n-n')^2}\right) - \right.$$

(3.4)

$$\left. - \left(\frac{n-n'}{n+n'}\right)^2 F\left(-n+l-1, -n'+l; 2l; \frac{-4nn'}{(n-n')^2}\right) \right\}.$$

This expression is valid, if $n \neq n'$. In the case of $n = n'$ the radial integral takes the form

$$R_{nl}^{nl-1} = (a_0/Z)(3/2)n(n^2-l^2)^{1/2}.$$

(3.5)

We have the sum rule for the radial integrals

$$\sum_{n'} \left| R_{nl}^{n'l-1} \right|^2 = \sum_{n'} \left| R_{nl}^{n'l+1} \right|^2 = (1/2)(a_0/Z)^2 n^2 (5n^2 + 1 - 3l(l+1)). \quad (3.6)$$

3.1.2 Semiclassical and Asymptotic Approaches

a) **JWKB Method.** The formulas (3.4–6) are valid for hydrogen or hydrogen-like ions. The *Bates* and *Damgaard* [2.9] method based on the regularized Coulomb functions may be used for nonhydrogenic atoms.

The semiclassical approximation for radial matrix elements of hydrogen was developed in [3.2] and was extended to nonhydrogenic atoms [3.3]. The result for the square of the radial integral is

$$\left|R_{nl}^{n'l'}\right|^2 = (a_0/Z)^2 \left|(n_c^2/2\Delta)\left\{[1 - \Delta l\,(l_>/n_c)]J_{\Delta-1}(-x)\right.\right.$$
$$\left.\left. -[1 + \Delta l\,(l_>/n_c)]J_{\Delta+1}(-x) + (2/\pi)\sin(\pi\Delta)(1-\epsilon)\right\}\right|^2,$$

$$\Delta l = l' - l, \quad \Delta = n_*' - n_*, \quad l_> = \max(l, l'),$$

$$n_c = 2n_* n_*'/(n_*' + n_*), \quad \epsilon = \left[1 - (l_>/n_c)^2\right]^{1/2}, \quad x = \epsilon\Delta,$$

$$(3.7)$$

where n_* and n_*' are the effective principal quantum numbers, $E_{nl} = -Z^2 Ry/n_*^2$, and $J_\Delta(x)$ are the Anger functions (see, for example, [2.4]). Instead of $l_>$ in expression (3.7) one can use the mean value of l and l'. Really, the difference in the corresponding values of the radial integral lies in the frame of errors of the semiclassical approach. The advantage of the choice of the $l_>$ value is that formula (3.7) gives the exact quantum result (3.5) in the case of $n = n'$. Formula (3.7) is reduced to a semiclassical formula for hydrogen at integer n_* and n_*', and the Anger functions are reduced to the Bessel functions. If n_* and n_*' are near integer numbers, the difference between the results of (3.7) and those calculated by the Bates and Damgaard method is less than 5% at $n_*, n_*' > 3$. If difference n_*-n_*' is near semi-integer numbers, it may appear relevant. In this case, the Bates and Damgaard method is preferable. If both effective quantum numbers $n_*, n_*' > 6$, the use of the method [2.9] meets some difficulties due to the sign change of the terms in the sums. In this case the semiclassical method (3.7) is justified.

b) **Asymptotic Method.** It was suggested [3.4,5] to clarify the quantum defect dependence of the radial integral explicitly with the help of a series over $\gamma = \Delta l(l/n)$. Let us introduce the function

$$\phi(n_*l, n_*'l \pm 1) = \frac{\langle n_*l|\,r\,|n_*'l \pm 1\rangle}{\langle n_c l|\,r\,|n_c l \pm 1\rangle}, \quad n_c = 2n_* n_*'/(n_*' + n_*) \quad (3.8)$$

and related series expansion

$$\phi(n_*l, n_*'l \pm 1) = \sum_{k=0} g_k(s)\gamma^k, \quad \gamma = \Delta l(l_>/n_c). \quad (3.9)$$

The universal functions $g_k(s)$ include the whole dependence on quantum defect. They have been tabulated in [3.5]. Formula (3.7) allows us to write g_k in terms of the Anger functions. For $k = 1, 2$ the final result is given by

$$g_0(s) = (1/3s)[J_{s-1}(-s) - J_{s+1}(-s)],$$
$$g_1(s) = -(1/3s)[J_{s-1}(-s) + J_{s+1}(-s)]. \quad (3.10)$$

The next functions can be written in terms of g_0 and g_1. We present the series expansion for ϕ up to $k = 3$:

$$\phi(\nu l, \nu'l \pm 1) = g_0(s)[1 + \gamma^2(1 + s/2)] + g_1(s)[\gamma + \gamma^3] - \frac{\sin(\pi s)}{\pi s}\gamma^2. \quad (3.11)$$

It should be noted that there is good agreement between the various methods of calculating the radial integral for the effective quantum numbers n_*, n'_* > 6.

c) **Saddle Point Method.** The method is based on the matrix element in the action variable representation or in the form (2.72). Define function $q(r)$ by

$$\psi_{nl}^{(+)} r \psi_{n'l'}^{(-)} = \exp\left[iq(r)\right]. \tag{3.12}$$

We carry out an integration path through the saddle point r_s at which the first derivative of $q(r)$ is equal to zero:

$$\frac{dq}{dr} = k_{nl}(r_s) - k_{n'l'}(r_s) - \frac{i}{r_s} + \frac{i}{2}\left[\frac{k'_{nl}(r_s)}{k_{nl}(r_s)} - \frac{k'_{n'l'}(r_s)}{k_{n'l'}(r_s)}\right] = 0. \tag{3.13}$$

We used the expression for the phase function in terms of the wave vector (2.66). According to the saddle point method the integral is determined by the close vicinity of the saddle point:

$$R_{nl}^{n'l'} \approx \frac{1}{2} \mathbf{Re}\left\{\psi_{nl}^{(+)}(r_s) r_s \psi_{n'l'}^{(-)}(r_s) \sqrt{\frac{2\pi i}{q''(r_s)}}\right\}. \tag{3.14}$$

Thus, the matrix element (3.14) is a product of the electron coordinate ($r = r_s$), the wave functions evaluated at the saddle point, and a factor which describes the extension or width of the saddle point region. It is easy to see that the two large first terms in (3.13) must nearly cancel each other

$$k_{nl}(r_s) \simeq k_{n'l'}(r_s),$$

therefore, the other terms become important.

Expression (3.14) appears to be remarkably accurate. To illustrate this fact, in Table 3.1 we present the comparison of some radial integrals of (3.14) with the exact quantum ones taken from [2.7].

It should be noted that

− because the saddle point is complex, the matrix element in form (3.14) is not sensitive to the turning point singularity;

Table 3.1. Comparison of the JWKB and quantum radial integrals

Transition	JWKB $R_{nl}^{n'l'}$ (a_0)	Quantum $R_{nl}^{n'l'}$ (a_0)	r_s (a_0)
$2s - 3p$	3.14	3.07	$8.24\exp(-0.86i)$
$2s - 4p$	1.28	1.28	$6.82\exp(-0.90i)$
$2s - 10p$	0.22	0.22	$5.81\exp(-0.93i)$
$2p - 3s$	1.00	0.94	$9.20\exp(-1.06i)$
$2p - 10s$	0.064	0.065	$6.69\exp(-1.15i)$
$2p - 3d$	4.61	4.75	$5.96\exp(-0.85i)$
$2p - 10d$	0.25	0.26	$4.64\exp(-0.98i)$

– to solve (3.13) for the saddle point, the potential is used, but not the wavefunction.

The saddle point method is considered in detail in [2.7].

3.1.3 Summed over Angular Quantum Numbers Line Strength. Kramers Approximation

Radial integrals summed or averaged over all possible orbital quantum numbers are needed for many applications. States with large orbital quantum numbers and small quantum defects make the main contribution, and therefore we may use the hydrogen-like wave functions to calculate the matrix element. The line strength $S(n', n)$

$$S(n', n) = \sum_{ll'} S(n'l', nl) = 3e^2 a_0^2 (Ry/\hbar\omega) \sum_{ll'} (2l + 1) f_{n'l',nl} \quad (3.15)$$

is convenient in this case. The explicit expression for this value was given in [3.6]:

$$S(n', n) = 32 \left(\frac{e}{Z} a_0\right)^2 (nn')^6 \frac{(n - n')^{2(n+n')-3}}{(n + n')^{2(n+n')+4}}$$

$$\times \left\{ \left[F(-n', -n + 1; 1; \frac{-4nn'}{(n - n')^2}) \right]^2 \right.$$

$$\left. - \left[F(-n' + 1, -n; 1; \frac{-4nn'}{(n - n')^2}) \right]^2 \right\}. \quad (3.16)$$

In the semiclassical approximation, the square of the radial integral may be written in the form:

$$S(n', n) = \frac{32}{\pi\sqrt{3}} \left(\frac{e}{Z} a_0\right)^2 \frac{(\varepsilon \cdot \varepsilon')^{3/2}}{(\varepsilon - \varepsilon')^4} \mathcal{G}(\Delta n),$$

$$\mathcal{G}(\Delta n) = \pi\sqrt{3} \mid \Delta n \mid J_{\Delta n}(\Delta n) J'_{\Delta n}(\Delta n), \quad (3.17)$$

where $\varepsilon = 1/n^2$, and $\mathcal{G}(\Delta n)$ is the Gaunt factor. The well-known Kramers approximation corresponds to $\mathcal{G} = 1$. It may be derived by semiclassical methods in the case $n \gg \Delta n \gg 1$. Expression (3.17) for the Gaunt factor may be approximated by the formula

$$\mathcal{G}(\Delta n) \approx (1 - 0.25/ \mid \Delta n \mid).$$

with error less than 2%.

3.2 Photoionization and Photorecombination

3.2.1 General Formulas

The photoionization and photorecombination cross sections for the given nl-state can be written as

$$\sigma_{\text{ph}}(\omega; nl) = \frac{4}{3} \frac{\pi}{(2l+1)} \left(\frac{e^2}{\hbar c}\right) \hbar\omega \left(\frac{d}{dE} S(nl, E)/e^2 a_0^2\right) \pi a_0^2,$$

$$\sigma_{\text{r}}(nl; \omega) = \frac{\pi}{3} \left(\frac{e^2}{\hbar c}\right)^3 \left(\frac{\hbar\omega}{E}\right) \left(\frac{\hbar\omega}{Ry}\right) \hbar\omega \left(\frac{d}{dE} S(nl, E)/e^2 a_0^2\right) \pi a_0^2,$$

$$\frac{d}{dE} S(nl, E) = e^2 \sum_{l'=l\pm1} \max(l, l') R^2(nl, El'),$$

$$R(nl, El') = \int_0^\infty \mathcal{R}_{nl}(r) \mathcal{R}_{El'}(r) r^3 dr, \qquad (3.18)$$

where $\hbar\omega$ and E are the energies of the ionizing photon and electron in continuum, respectively; $\mathcal{R}(r)$ are the radial wave functions of the electron in the discrete and continuous spectra, normalized by the delta-function on energy, and $S(nl, E) = S(E, nl)$ is the line strength for the continuum.

The photoionization cross section σ_{ph} can be also expressed in terms of the continuum oscillator strength $df/d\varepsilon$, where $\varepsilon = |E|/Z^2 Ry$

$$\frac{df}{d\varepsilon} = \frac{1}{4\pi} \left(\frac{\hbar c}{e^2}\right) \left[Z^2 \sigma_{\text{ph}}/\pi a_0^2\right]. \qquad (3.19)$$

Near the ionization limit, the quantity $df/d\varepsilon$ is connected with the usual oscillator strength f for the transition $nl - n'l'$ by the relation:

$$\lim_{\varepsilon\to 0} \frac{df}{d\varepsilon} = \lim_{n'\to\infty} \left[(n_*)^3 f_{n'l',nl}/2\right], \quad n_* = n' - \delta = Z |Ry/E_{n'l'}|^{1/2}, \qquad (3.20)$$

where n_* and $E_{n'l'}$ are the effective quantum number and the energy of level $n'l'$, counted from the ionization limit, and δ is the quantum defect.

3.2.2 Asymptotic Approach

The radial integral for the bound-free transition may be written as [3.7]:

$$R(nl, El') = \frac{a_0}{\sqrt{2Ry}} \frac{\nu^2}{Z^2} C_0 \frac{\exp\left[-2\nu \arctan\left(\frac{n}{\nu}\right)\right]}{(1 - e^{-2\pi\nu})^{1/2}} \left[\frac{4\nu n}{n^2 + \nu^2}\right]^{l_>+1}$$

$$(1 - z)^{(l_>+1-n)/2} \prod_{s=1}^{l'} (s^2 + \nu^2)^{1/2}$$

$$\times \left[F(l_> + 1 - i\nu, -n + l + 1; 2l_>; z) - \frac{1}{1-z} F(l - i\nu, -n + l_> + 1; 2l_>; z)\right],$$

$$l_> = \max(l, l'), \quad \nu^2 = Z^2 Ry/E, \quad z = \frac{4i\nu}{(n+i\nu)^2}, \quad l' = l \pm 1, \qquad (3.21)$$

and the C_0 coefficient is defined by the expression

$$C_0 = \frac{i^{l_> - l}(-1)^{n-l_>}}{4(2l_> + 1)!}\left[\frac{(n+l)!}{(n-l-1)!}\right]^{1/2}. \qquad (3.22)$$

When the electron kinetic energy is much smaller than the binding energy in the final state, i.e. $\nu \ll n$, one can show that the radial integral tends to a constant (see, for example [3.7]):

$$R(nl, El') \underset{\nu \to 0}{\to} \frac{a_0}{\sqrt{2Ry}}\frac{n^2}{Z^2}i^{l-1_>}2^{2l'+4}n^{l_>}e^{-2n}C_0$$

$$\times \begin{cases} \left[F(l-n+1; 2l+2; 4n) - \dfrac{l-n+1}{l+1}F(l-n+2; 2l+3; 4n)\right], & l' = l+1, \\ \left[F(l-n+1; 2l; 4n) - F(l-n-1; 2l; 4n)\right], & l' = l-1. \end{cases}$$
$$(3.23)$$

Assuming the semiclassical approach for Hydrogen (see, for example, [3.8]), the transition strength density is expressed in terms of the Bessel functions:

$$\frac{d}{dE}S(nl, E) = 2(2l+1)n\left(\frac{Ry}{\hbar\omega}\right)^2\left\{[J'_\Delta(\epsilon\Delta)]^2 + (1-\epsilon^2)[J_\Delta(\epsilon\Delta)]^2\right\}e^2a_0^2/Ry,$$

$$\epsilon = [1 - (l+1/2)^2/n^2]^{1/2}, \qquad \Delta = (\hbar\omega)n^3/2Ry. \qquad (3.24)$$

Since $\Delta \gg 1$, the asymptotic expressions of the Bessel functions through the MacDonald functions ($K_{1/3}$ and $K_{2/3}$) may be used. The final result is given by [2.2, 3, 10]:

$$\frac{d}{dE}S(nl, E) = \frac{2(2l+1)}{3\pi^2}\left(\frac{Ry}{\hbar\omega}\right)^2\frac{(l+1/2)^4}{n^3}\left[K_{1/3}^2(\eta) + K_{2/3}^2(\eta)\right]\left(\frac{e^2a_0^2}{Ry}\right),$$

where $\eta = (E/Ry)(l+1/2)^3/6$. The last two expressions are valid in the case of $\eta < 1$.

3.2.3 Kramers Formulas and Gaunt Factor

The photoionization $\sigma_{\rm ph}(nl)$ and photorecombination $\sigma_{\rm r}(nl)$ cross sections are often written in the form:

$$\sigma_{\rm ph}(nl) = \mathcal{G}_{nl}\sigma_{\rm ph}^{\rm Kr}(n),$$
$$\sigma_{\rm r}(nl) = \frac{2l+1}{n^2}\mathcal{G}_{nl}\sigma_{\rm r}^{\rm Kr}(n), \qquad (3.25)$$

where the dimensionless quantity $\mathcal{G}_{nl}$ is termed the bound-free Gaunt factor. $\sigma^{\rm Kr}$ are the Kramers semiclassical cross sections [3.11]. They were obtained assuming the classical consideration of electron scattering in the Coulomb field and the correspondence principle. They relate to the l-averaged cross sections

$$\sigma(n) = \frac{1}{n^2} \sum_l (2l+1)\sigma(nl), \tag{3.26}$$

$$\sigma_{ph}^{Kr}(n) = \frac{64n}{3\sqrt{3}Z^2} \left(\frac{e^2}{\hbar c}\right) \left(\frac{|E_n|}{\hbar\omega}\right)^3 \pi a_0^2, \qquad |E_n| = Z^2 Ry/n^2,$$

$$\sigma_r^{Kr}(n) = \frac{32|E_n|^2}{3\sqrt{3}\hbar\omega\,(\hbar\omega - |E_n|)} \left(\frac{e^2}{\hbar c}\right)^3 \pi a_0^2, \tag{3.27}$$

where $\hbar\omega$ is the photon energy so that $E = \hbar\omega - |E_n|$ is the energy of the electron in the continuous spectrum. The Kramers formulas were derived for the Coulomb field, i.e., for the hydrogen-like ions. However, the formulas written in the form of the right-hand parts of (3.27) can be used in a more general case if one defines $|E_n|$ as the electron binding energy of the atom.

The Kramers photoionization cross section at the threshold $\hbar\omega = |E_n|$ is equal to

$$\sigma_{ph}^{Kr}(n, \hbar\omega = |E_n|) = \frac{64n}{3\sqrt{3}Z^2} \left(\frac{e^2}{\hbar c}\right) \pi a_0^2. \tag{3.28}$$

The Kramers formulas for the photoionization and photorecombination (3.27) cross sections involve the principal dependence on the spectroscopic symbol Z, the binding energy $|E_n|$, and the energy of photoelectron E, but do not describe the transitions from the different l-states. It is possible to find the asymptotic behavior of the Gaunt factor from (3.25) for transitions $nl \to El'$:

$$\mathcal{G}_{nl} = \frac{\exp\left[\frac{4}{\sqrt{\varepsilon}}\left(1 - \frac{\arctan x}{x}\right)\right]}{(1+x^2)^{l+1}} \left(1 - e^{-2\pi/x\sqrt{\varepsilon_0}}\right)^{-1} \prod_{s=1}^{l'} \frac{1 + (xs/n)^2}{1+x^2} \widetilde{\mathcal{G}}_{nl},$$

$$x = (\varepsilon/\varepsilon_n)^{1/2}, \quad x \gg 1, \quad \varepsilon_n = |E_n|/Z^2 Ry, \quad \varepsilon = E/Z^2 Ry, \tag{3.29}$$

where $\widetilde{\mathcal{G}}_{nl}$ is the modified Gaunt factor. The product $\Pi = 1$ at $l' = 0$ and $\widetilde{\mathcal{G}}_{nl} = \mathcal{G}_{nl}$ at $x = 0$. The modified $\widetilde{\mathcal{G}}_{nl}$ factor is practically constant at high energies ε. The photoionization cross sections have the following asymptotic behavior:

$$\sigma_{ph}(nl) \sim \varepsilon^{-(3+l+1/2)}, \qquad \sum_l \sigma_{ph}(nl) \sim \varepsilon^{-7/2}, \qquad \varepsilon \gg \varepsilon_0.$$

3.3 Transition Form Factors

3.3.1 General Formulas

In the momentum transfer representation the cross sections of a number of collisional processes can be expressed in terms of the transition form factor

$$F_{\alpha'\alpha}(\mathbf{Q}) = \frac{1}{g_\alpha} \sum_{a/\alpha, a'/\alpha'} |\langle \alpha'| \exp(i\mathbf{Q} \cdot \mathbf{r}) |\alpha\rangle|^2, \qquad (3.30)$$

where $\mathbf{Q} = \mathbf{k}' - \mathbf{k}$ is the momentum transferred to the Rydberg electron. In particular, this quantity appears in the Born approximation for transitions induced by charged particle collisions or in the impulse approximation for collisions between the Rydberg atoms and neutral particles. The sum in (3.30) is taken over all sublevels of the total α and α' sets of quantum numbers except for α and α', respectively, while g_α is the statistical weight of the initial α-level. It is easy to see, that for inelastic transitions the form factor $F_{\alpha'\alpha}(Q)$ is proportional to $Q^{2\varkappa}$ at small Q ($Q \to 0$), except for the case of $\varkappa = 0$. Here $\varkappa = |l - l'|$ is the minimal possible multipole order, and l, l' are the orbital quantum numbers of the initial and final states. In the case of $\varkappa = 0$ and $l \neq 0$, the inelastic transition form factor $F_{\alpha'\alpha}(Q)$ is proportional to Q^4. The proportional coefficient is defined by the square of the corresponding multipole moment matrix element (dipole moment for the allowed transitions). The asymptotic behavior of the form factor at large values of Q ($Q \to \infty$) is given by $Q^{-2(l+l'+3)}$.

It is also convenient to introduce a quantity which is symmetrical with respect to the initial and final states

$$\mathcal{F}_{\alpha'\alpha}(\mathbf{Q}) = g_\alpha F_{\alpha'\alpha}(\mathbf{Q}) = \sum_{a/\alpha, a'/\alpha'} |\langle \alpha'| \exp(i\mathbf{Q} \cdot \mathbf{r}) |\alpha\rangle|^2. \qquad (3.31)$$

The expression for $F_{\alpha'\alpha}(Q)$ in parabolic coordinates is given in [3.12]. However, these coordinates are inconvenient for many applications. The analytical formulas for the transition form factors of Hydrogen between levels that are low excited have been obtained in [3.13]. The particular case of the transitions between circular orbits ($l = n - 1$ and $l' = n' - 1$) was considered in [3.12]. Application of the general methods of the group theory to the form factor properties is considered in [3.14]. The detailed reference list is given in [3.15].

Below, we consider the form factors summed over all orbital quantum numbers and a special case of transitions with small l. These cases are of great interest for applications.

3.3.2 $n - n'$ Transitions: Quantum Expressions

When we sum over all angular quantum numbers, the form factor becomes

$$\mathcal{F}_{n'n}(Q) = \sum_{lm, l'm'} |\langle n'l'm'| \exp(i\mathbf{Q} \cdot \mathbf{r}) |nlm\rangle|^2. \qquad (3.32)$$

One can see that due to summation over the magnetic quantum numbers, the $\mathcal{F}_{n'n}$ value depends only on the absolute value of Q. The main idea (suggested in [3.16]) consists of summation of the matrix elements before calculating them in explicit form. This becomes possible due to the analytical expression for the Green's function of the Coulomb field and the corresponding formula for the density matrix of the n-level. Using expression (2.131), we can write (3.32) in the form

$$\mathcal{F}_{n'n}(Q) = \frac{1}{\pi^2 (nn')} \int \frac{e^{i\mathbf{Q}\cdot(\mathbf{r}-\mathbf{r}')}\mathfrak{P}_n\mathfrak{P}_{n'}d\mathbf{r}d\mathbf{r}'}{(x-y)^2 a_0^6}, \tag{3.33}$$

where

$$x, y = Z(r + r' \pm |\mathbf{r} - \mathbf{r}'|)/a_0, \quad \mathfrak{P}_n = \left(\frac{\partial}{\partial x} - \frac{\partial}{\partial y}\right)\left[M_{n,1/2}(x/n)M_{n,1/2}(y/n)\right].$$

Let $\mathbf{r} = \mathbf{r} - \mathbf{r}'$, then we integrate over angles, introduce the intermediate variable $\mathbf{r} + \mathbf{r}'$, and reduce the integral in (3.33) to a double integral over x and y:

$$\mathcal{F}_{n'n}(Q) = \frac{1}{24} \int\limits_0^\infty dy \int\limits_y^\infty dx \frac{\sin[Q(x-y)/2]}{Q(x-y)/2} \frac{\mathfrak{P}_n\mathfrak{P}_{n'}}{nn'}(x^2 + y^2 + 4xy). \tag{3.34}$$

Further, we introduce the function $A(Q)$ by the relation:

$$\mathcal{F}_{n'n}(Q) = \frac{1}{Q} \int\limits_0^Q A(Q')dQ'. \tag{3.35}$$

The integrand in (3.34) is symmetrical over the x, y -variables, therefore $A(Q)$ is

$$A(Q) = \frac{1}{48 (nn')} \iint\limits_0^\infty dxdy \cos[Q(x-y)/2](x^2 + y^2 + 4xy)\mathfrak{P}_n\mathfrak{P}_{n'}. \tag{3.36}$$

Now, using the relation between the Whittaker function and the degenerate hypergeometric function

$$M_{n,1/2}(z) = ze^{-z/2}F(1-n;2;z) = e^{-z/2}[F(2-n;1;z) - F(1-n;1;z)],$$

and the known integral

$$I_{n'n} = \int\limits_0^\infty e^{-\lambda x}F(-n;1;px)F(-n';1;p'x)dx$$

$$= \frac{(\lambda-p)^n(\lambda-p')^{n'}}{\lambda^{n+n'+1}}F(-n,-n';1;\frac{pp'}{(\lambda-p)(\lambda-p')}), \tag{3.37}$$

we express $A(Q)$ in terms of the hypergeometric functions [3.16]:

$$A(Q) = \mathbf{Re}\, C_d \left\{ \left[\frac{d}{dQ} I_{n'-1,n-1}(Q) \right] \left[\frac{d}{dQ} I^*_{n',n}(Q) \right] \right.$$

$$- \left[\frac{d}{dQ} I_{n'-1,n}(Q) \right] \left[\frac{d}{dQ} I^*_{n',n-1}(Q) \right] \tag{3.38}$$

$$- \frac{1}{6} \frac{d^2}{dQ^2} \left[I_{n'-1,n-1}(Q) I^*_{n',n}(Q) - I_{n'-1,n}(Q) I^*_{n',n-1}(Q) \right] \right\},$$

$$C_d = (nn')^{-2},$$

where $I_{n'n} \equiv I_{n'n}(Q)$ with $p = 1/n$, $p' = 1/n'$, and $\lambda = (p + p' + iQa_0/Z)/2$.

In a similar way we obtain an analytical expression for the form factor density per unit energy interval, when one of the states belongs to the continuum [3.17]:

$$Ry \frac{d}{dE} \mathcal{F}_{En}(Q) = \frac{1}{Q} \int_0^Q A(Q')dQ', \quad n' = iZ/\varkappa, \quad C_d = -n'/2(Zn)^2. \tag{3.39}$$

3.3.3 $n - n'$ Transitions: Asymptotic Expressions

Now we consider an asymptotic approach to obtain the sum over all angular quantum numbers in the form factor (3.32). It is defined by the function $A(Q)$. The case of $n \gg \Delta n \gg 1$ is of most interest for applications. Here, and below we put $n' > n$, i.e. $\Delta n > 0$. In accordance with (3.37, 38), the $A(Q)$ is dependent on the argument of hypergeometric function

$$z = -\frac{4}{(nn')\{[\Delta n/(nn')]^2 + (Qa_0/Z)^2\}}. \tag{3.40}$$

At finite Q and $n \to \infty$, the z value is proportional to $1/n^2$. However, with increasing n the main contribution to the transition is given by the region of decreasing Q. Therefore, the situation when simultaneously $n \to \infty$ and $Q \to 0$ is of major interest. Again we use the asymptotic Tricomi expansion which is valid at $z > n^{2/3}$:

$$F(-n, -n', 1; z) \to C(-1)^n z^{(n+n')/2} e^{n/2z} J_{\Delta n}(x),$$

$$C = [(n + n' + 1)n]^{-\Delta n/2} \frac{\Gamma(n' + 1)}{2\Gamma(n + 1)} \cdot \left[1 + \frac{\Gamma^2(n + 1)}{\Gamma(n + \Delta n + 1)\Gamma(n - \Delta n + 1)} \right],$$

$$x^2 = 2n(n + n' + 1)/z. \tag{3.41}$$

The following relations follow directly from (3.41) as $n \to \infty$, $\Delta n/n \to 0$:

$$\lim C = 1, \quad \lim \left[x^2 (z/4nn') \right] = 1, \quad \lim \left\{ \left[n^2 + (n')^2 \right] / (2nn') \right\} = 1, \tag{3.42}$$

$$\lim (1 + 1/z)^{-n-n'-1} = e^{-(n+n')/z}.$$

For symmetry we save the difference between n and n', although only the leading term may be justified. The substitution of expressions (3.41, 42) into (3.37, 38) leads to the following result after some rather cumbersome operations:

$$A(Q) = \frac{2\,nn'}{3\,\Delta n} \exp\left\{-(\Delta n)^2\,\eta^2/[2(n+n')]\right\}\left\{a_1(\eta)+a_2(\eta)\right\}\mathfrak{Q}(\Delta n/n),$$

$$a_1(\eta) = \frac{\eta^2-1}{\eta^3}\left(\frac{4}{\eta^2}-1\right)J_{\Delta n}\left[(\Delta n)\,\eta\right]J'_{\Delta n}\left[(\Delta n)\,\eta\right],$$

$$a_2(\eta) = \Delta n\frac{\eta^2-1}{\eta^2}\left\{\frac{\eta^2-1}{\eta^2}\left(1+\frac{12}{\eta^2}\right)[J_{\Delta n}((\Delta n)\eta)]^2\right.$$
$$\left. -\left(1-\frac{12}{\eta^2}\right)[J'_{\Delta n}\{(\Delta n)\,\eta\}]^2\right\},$$

$$\eta = \left[1+\left(\frac{Qa_0nn'}{Z\Delta n}\right)^2\right]^{1/2}.$$

$$\mathfrak{Q}(\Delta n/n) = \left[1+\frac{(\Delta n)^2}{4nn'}\right]^{-2}\exp\left[\frac{(\Delta n)^2}{2(n+n')}\right]. \tag{3.43}$$

Within the framework of the accepted approximation $\Delta n \ll n$, the factor $\mathfrak{Q}(\Delta n/n)$ is equal to unity. However, the $\mathfrak{Q}$ -factor provides that the corresponding generalized oscillator strength

$$f_{n'n}(q) = \left[Z^2\Delta E/(nQa_0)^2\right]\mathcal{F}_{n'n}(Q)$$

reduces in the limiting case $Q \to 0$ to the known semiclassical expression for the oscillator strength:

$$\lim_{Q\to 0} f_{n'n}(Q) = \frac{32}{3n^2}\left(\frac{nn'}{\Delta n(n+n')}\right)^3(\Delta n)\,J_{\Delta n}(\Delta n)\,J_{\Delta n}(\Delta n). \tag{3.44}$$

Figures 3.1, 2 show the comparison of the results of (3.35, 43) with the Born ones [3.8, 17]. For the $\Delta n = 1$ transitions the agreement accounts for about 10% even at $n = 2$ and becomes practically exact for $n \geq 5$. The case of $\Delta n > 1$ is illustrated by the $4 - 7$ and $10 - 13$ transitions. It is seen that (3.35, 43) are satisfactory in the whole relevant region.

In accordance with (3.35, 43) the form factor consists of two terms $\mathcal{F} = \mathcal{F}^{(1)} + \mathcal{F}^{(2)}$ corresponding to a_1 and a_2. Figure 3.3 shows them for the $4 - 7$ transition. Function $\mathcal{F}^{(1)}$ reduces to the usual dipole matrix element at small Q, therefore it corresponds to the dipole interaction. Below, it will be named the "dipole" part. The "dipole" part dominates at small momentum transfers, i.e., it describes large distances r. Function $\mathcal{F}^{(2)}$ tends to zero at small Q and is maximal at $Q \approx Z\Delta n/a_0n^2$. It corresponds to the difference of the initial and final state momenta $Z(1/n - 1/n')/a_0$ at the usual classical momentum transfer and shows that the function $\mathcal{F}^{(2)}$ describes processes of the impulsive kind. This means that in this case the influence of the Coulomb field of

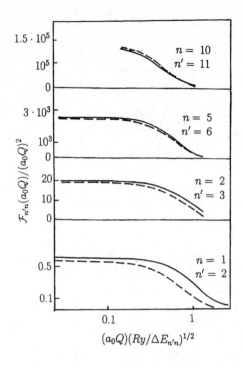

Fig. 3.1. The form factors for the $n - n'$ transitions 1–2, 2–3, 5–6, 10–11. Full curves are the asymptotic expressions; dashed curves are the Born calculations [3.8, 17]

the core on the momentum transfer may be neglected, in contrast to the "dipole" part $\mathcal{F}^{(1)}$. Further, for short, we call $\mathcal{F}^{(2)}$ the "impulsive" part. The comparison of the "dipole" and "impulsive" parts with the form factors of the allowed ($\Delta l = 1$) and forbidden ($\Delta l \neq 1$) transitions is presented in Fig. 3.3. Our interpretation finds the following confirmation: $\mathcal{F}^{(1)}$ and $\mathcal{F}^{(2)}$ correspond qualitatively to the allowed and forbidden transitions, respectively.

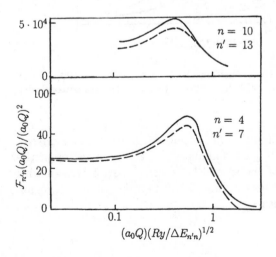

Fig. 3.2. The form factors for the $n - n'$ transitions 4–7, 10–13. Full curves are the asymptotic expressions; dashed curves are the Born calculations [3.8, 17]

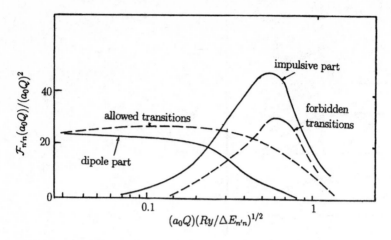

Fig. 3.3. The "dipole" and "impulsive" parts of the 4–7 form factor. Dashed curves are the calculations for the allowed and forbidden transitions [3.8]

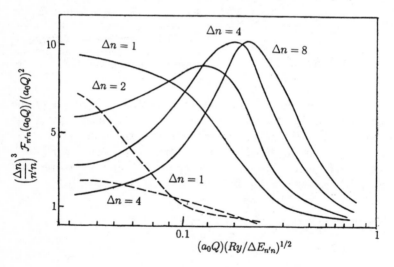

Fig. 3.4. The form factors for the transitions $n \rightarrow n + \Delta n$ ($n = 100$, $\Delta n = 1, 2, 4, 8$). The dashed curves are only the "dipole" parts for the transitions with $\Delta n = 1, 4$

The relative value of the "dipole" and "impulsive" parts depends on Δn. Figure 3.4 shows that the $\mathcal{F}^{(1)}$ part dominates for the form factor of the $\Delta n = 1$ transition. The role of the second "impulsive" part increases with increasing Δn. It is connected with the "impulsive" part describing the region of rather large transfer momenta.

3.3.4 $nl - nl'$ Transitions: Semiclassical Expressions

The symmetric form factor summed over the magnetic quantum numbers is

$$\mathcal{F}_{n'l',nl}(Q) = (2l + 1) F_{n'l',nl}(Q) = \sum_{mm'} |\langle n'l'm'| e^{i\mathbf{Q}\cdot\mathbf{r}} |nlm\rangle|^2 . \tag{3.45}$$

Expanding $e^{i\mathbf{Q}\cdot\mathbf{r}}$ over the spherical waves, we can express the $\mathcal{F}_{n'l',nl}(Q)$ quantity in terms of the $3j$-symbols and the radial integral:

$$\mathcal{F}_{n'l',nl} = \sum_{\varkappa} \mathcal{F}^{(\varkappa)}_{\alpha'\alpha} ,$$

$$\tag{3.46}$$

$$\mathcal{F}^{(\varkappa)}_{n'l',nl} = (2l + 1)(2l' + 1)(2\varkappa + 1) \begin{pmatrix} l & \varkappa & l' \\ 0 & 0 & 0 \end{pmatrix}^2 |\langle n'l'| j_\varkappa(Qr) |nl\rangle|^2 .$$

The main problem is the calculation of the radial integral. In accordance with the correspondence principle, it is given by

$$\langle n'l'| j_\varkappa(Qr)| nl\rangle = \frac{1}{2\pi} \int_0^{2\pi} du e^{-i(\Delta n)u} j_\varkappa \left[\frac{Qn^2}{Z} a_0 (1 - \epsilon \cos \tilde{u}) \right], \tag{3.47}$$

where $u = \tilde{u} - \epsilon \sin \tilde{u}$. From here, we obtain the formula [3.18] for the transition with $\Delta n = 1$ and $l' = l = 0$

$$\langle n+1,0 | j_0(Qr)| n0\rangle = \frac{\sin(x) - x\cos(x)}{(1+x^2)^{1/2}} J_1 \left[(1+x^2)^{1/2}\right], \quad x = Qa_0 n^2/Z. \tag{3.48}$$

We give the expression obtained in [3.19] by means of the asymptotic Tricomi expansion for the hypergeometric functions (2.47) and the Poisson representation for $j_\varkappa(z)$ ($\varkappa = l' - l$):

$$\langle nl'| j_\varkappa(Qr)| nl\rangle = \begin{cases} \Lambda_{2p} j_p(x) J_p(x), & \varkappa = 2p \\ \Lambda_{2p+1} j_p(x) J_{p+1}(x), & \varkappa = 2p + 1 \end{cases}$$

$$\tag{3.49}$$

$$\Lambda_k = \left[\prod_{i=0}^{k-1} \left[1 - ((l - i)/n)^2\right] \right]^{1/2}, \quad x = Qa_0 n^2/Z, \quad p \geq 0.$$

This formula differs from that obtained by *Matsuzawa* [3.20] by the factor Λ_k. The validity region of [3.20] as well as (3.49) is defined by conditions $n \gg 1$, l. The comparison of formulas [3.20] with quantum calculations is presented in [3.21]. The factor Λ_k improves the result, when l is not too small.

3.3.5 Classical Approach

The classical approach is widely used for collisions involving highly excited atoms (see reviews [3.15, 22, 1.3] and references therein). Here we present only the main results for form factors, which can be derived with the use of the classical distribution function method.

We rewrite the form factor expression (3.31) in terms of the density matrix in the momentum representation, and then use (2.139):

$$
\begin{aligned}
\mathcal{F}_{\alpha'\alpha}(Q) &= \sum_{a'\backslash\alpha'}\sum_{a\backslash\alpha} \langle\alpha|\exp(i\mathbf{Q}\cdot\mathbf{r})|\alpha'\rangle\langle\alpha'|\exp(-i\mathbf{Q}\cdot\mathbf{r})|\alpha\rangle \\
&= g_\alpha g_{\alpha'}\langle\rho_{\alpha'}(\mathbf{p})\mid\rho_\alpha(\mathbf{p}-\hbar\mathbf{Q})\rangle \\
&\rightarrow \int\frac{d\mathbf{p}d\mathbf{r}}{(2\pi\hbar)^3}\delta\left[\alpha-\frac{\mathbf{I}(\mathbf{p},\mathbf{r})}{\hbar}\right]\delta\left[\alpha'-\frac{\mathbf{I}'(\mathbf{p}-\hbar\mathbf{Q},\mathbf{r})}{\hbar}\right].
\end{aligned}
\tag{3.50}
$$

Equation expression (3.50) allows us to obtain the results for form factor containing various sums. For example, sum $\mathcal{F}_{n',nl}(Q) = \sum_{m,m'l'}\mathcal{F}_{nlm}^{n'l'm'}(Q)$ over all angular quantum numbers of the final $\alpha' = n'l'm'$ state and over the magnetic quantum numbers of the initial $\alpha = nlm$ state gives:

$$
\begin{aligned}
\mathcal{F}_{n',nl}(Q) &= (n')^2(2l+1)\int\frac{d\mathbf{p}d\mathbf{r}}{(2\pi\hbar)^3}\delta\left[n-I_1(\mathbf{p},\mathbf{r})/\hbar\right] \\
&\times\delta\left[l-I_2(\mathbf{p},\mathbf{r})/\hbar\right]\delta\left[n'-I_1(\mathbf{p}-\hbar\mathbf{Q},\mathbf{r})/\hbar\right].
\end{aligned}
\tag{3.51}
$$

We have substituted the quantum statistical weight $(2l+1)$ instead of the corresponding classical one. We also take into account the double number of states due to the electron spin. After integrating over $\mathbf{r}$, the first two δ-functions give the classical momentum distribution function of the nl-state:

$$
W_{nl}(p) = \frac{p^2}{4\pi}\int\frac{d\mathbf{r}}{a_0^3}\delta\left[n-I_1(\mathbf{p},\mathbf{r})/\hbar\right]\delta\left[l-I_2(\mathbf{p},\mathbf{r})/\hbar\right].
\tag{3.52}
$$

At small values of the orbital angular momentum in the initial state $l \ll n$ this function is reduced to a simple semiclassical expression (2.105) [or equivalent result (2.137)]. The last δ-function in (3.51) describes the states with fixed energy, and may be changed by the microcanonical distribution (2.107). As a result, the form factor can be written as

$$
\begin{aligned}
\mathcal{F}_{n',nl}(Q) &= (2l+1)\frac{2Z^2 Ry}{(n')^3} \\
&\times\int\delta\left(\frac{(\mathbf{p}-\hbar\mathbf{Q})^2}{2m}-\frac{\mathbf{p}^2}{2m}-(E_{n'}-E_{nl})\right)W_{nl}(p)\,dp d\Omega_\mathbf{p}.
\end{aligned}
\tag{3.53}
$$

This formula is really applicable with the quantum distribution function $W_{nl}(p) = |pg_{nl}(p)|^2$. It also can be derived within the framework of the binary encounter approach with the use of the quantum function $g_{nl}(p)$ in the initial state.

Integrating (3.48) over the angles of the $\mathbf{p}$ vector, yields [3.23]

$$F_{n',nl}(Q) = \frac{1}{(2l+1)} \mathcal{F}_{n',nl}(Q) = \frac{Z^2 m Ry}{(n')^3 \hbar Q} \int\limits_{|p_0|}^{\infty} |g_{nl}(p)|^2 \, p \, dp,$$

$$p_0(Q) = \frac{m(E_{n'} - E_{nl}) - \hbar^2 Q^2/2}{\hbar Q}.$$

(3.54)

Summation over all angular quantum numbers of the initial and final states in (3.50), and substitution of the microcanonical distribution also leads to the following result [3.24]:

$$\mathcal{F}_{n',n}(Q) = \frac{4Z^2 (Ry)^2}{(nn')^3} \int \frac{d\mathbf{p} \, d\mathbf{r}}{(2\pi\hbar)^3} \delta\left(\frac{\mathbf{p}^2}{2m} - \frac{Ze^2}{r} - E_n\right)$$

$$\times \delta\left[\frac{(\mathbf{p} - \hbar\mathbf{Q})^2}{2m} - \frac{Ze^2}{r} - E_{n'}\right].$$

(3.55)

Integration of (3.55) yields the expressions [3.17]:

$$\mathcal{F}_{n'n}(Q) = \frac{2^9}{3\pi (nn')^3} \frac{\chi^5}{(\chi^2 + \chi_+^2)^3 (\chi^2 + \chi_-^2)^3},$$

$$\chi = Qa_0/Z, \qquad \chi_\pm = \left|\frac{1}{n} \pm \frac{1}{n'}\right|.$$

(3.56)

or

$$F_{n'n}(Q) = \frac{\mathcal{F}_{n',n}(Q)}{n^2} = \frac{2^9}{3\pi n^5 (n')^3} \frac{\chi^5}{[\chi^4 + 2(\varepsilon + \varepsilon')\chi^2 + (\varepsilon - \varepsilon')^2]^3},$$

$$\varepsilon = 1/n^2, \qquad \varepsilon' = 1/(n')^2.$$

(3.57)

These formulas may also be obtained directly from (3.54) using the Fock's expression (2.138) for $\sum_l (2l+1) \mid g_{nl}(p) \mid^2$. For transition to the continuum $(\alpha \to E)$ the form factor density per unit energy interval is expressed through the form factor of the bound-bound transition $(\alpha \to n')$ presented above by the relation

$$\frac{d}{dE} F_{E\alpha}(Q) = \frac{(n')^3}{2Z^2 Ry} F_{n'\alpha}(Q).$$

(3.58)

It is suggested that the discrete energy $E_{n'} < 0$ be replace by the positive energy E.

3.3.6 Angular Factors for Complex Atoms

The formulas for the radiative transitions and form factors given above can be extended to complex Rydberg atoms or ions which have nonzero angular momentum of the ion core. We consider two of the most often used kinds of descriptions for the total angular function of the Rydberg atom: $L - S$ coupling and $j-j$ coupling. The first one is valid mainly for the neutral atoms if the principal quantum number and orbital momentum of the Rydberg electron are not too large. In the limiting case of the high n values, the $j - j$ coupling becomes predominant.

The general expressions for the oscillator strength, the line-strength and the form factors can be expressed in terms of the corresponding one electron quantities and the angular $Q_{\varkappa}$ factors:

$$F_{\alpha'\alpha} = \sum_{\varkappa} Q_{\varkappa}(\alpha') F^{(\varkappa)}_{n'l',nl} \; , \quad \mathcal{F}_{\alpha'\alpha} = \sum_{\varkappa} Q_{\varkappa}(\alpha') \frac{g_{\alpha}}{g_s(2l+1)} \mathcal{F}^{(\varkappa)}_{n'l',nl} \; ,$$

$$f_{\alpha'\alpha} = Q_1(\alpha') f_{n'l',nl} \; , \quad S_{\alpha'\alpha} = Q_1(\alpha') \frac{g_{\alpha}}{g_s(2l+1)} S^{(\varkappa)}_{n'l',nl} \; ,$$

$$(3.59)$$

where $\varkappa$ is the order of the corresponding multipole ($\varkappa = 1$ for radiative transitions), g_{α} is the statistical weight of the initial state, and $g_s = 2$ is the spin statistical weight. Formulas for the angular $Q_{\varkappa}$ coefficients depend on the kind of coupling.

In the case of $L - S$ coupling, the highly excited atom is defined by the quantum number set $|\alpha\rangle = |nlSLJS_cL_c\rangle$, where SLJ are the spin, the orbital, and total angular momenta of the atom, respectively, S_cL_c are the spin and the orbital momentum of the atomic core. Then, the $Q_{\varkappa}$ factor is given by

$$Q_{\varkappa}(J',J) = \frac{g_c g_s (2l+1)}{2J+1}$$

$$\times |M_{0\varkappa\varkappa}(S'L'J' - SLJ)M_{0\varkappa\varkappa}(L'_c l'L' - L_c lL)M_{000}(S'_c sS' - S_c sS)|^2 \; ,$$

$$(3.60)$$

where $g_c = (2S_c + 1)(2L_c + 1)$ is the statistical weight of the ion core; and the universal M-function and its special cases are

$$M_{qkv}(S'L'J' - SLJ) = [(2J'+1)(2J+1)(2v+1)]^{1/2} \begin{Bmatrix} S' & S & q \\ L' & L & k \\ J' & J & v \end{Bmatrix} ,$$

$$M_{0kv}(S'L'J' - SLJ) = \frac{(-1)^{J+L'+S+k}}{(2S+1)^{1/2}}$$

$$[(2J'+1)(2J+1)]^{1/2} \delta_{kv} \delta_{SS'} \begin{Bmatrix} J & J' & k \\ L' & L & S \end{Bmatrix} ,$$

$$M_{000}(S'L'J' - SLJ) = [(2J+1)/(2L+1)(2S+1)]^{-1/2} \delta_{SS'} \delta_{LL'} \delta_{JJ'} \; .$$

$$(3.61)$$

The M-functions satisfy the sum rule

$$\sum_{J,J',v} M^2_{qkv}(S'L'J' - SLJ) = 1, \qquad \sum_{J,J'} M^2_{0\varkappa\varkappa}(S'L'J' - SLJ) = 1. \qquad (3.62)$$

The advantage of the M-functions is clear representation and convenience. For example, Q-factor of transitions summed over J, J' may be obtained from (3.60) by removing the first M-function and substitution of the corresponding statistical weight of the initial state $(2S+1)(2L+1)$ instead of $(2J+1)$. Similarly, the summation over J, J' and L, L' corresponds to the removal of of the two M-functions and etc.

In the case of $j - j$ coupling, the atom is defined by the set of quantum numbers $|\alpha\rangle = |nlJ_cjJS_cL_c\rangle$, where J_cjJ are the total angular momenta of the atomic core, highly excited electron, and atom, respectively. The corresponding angular factor $Q_\varkappa$ takes the form

$$\begin{aligned} Q_\varkappa &= \frac{g_c g_s (2l+1)}{2J+1} M^2_{0\varkappa\varkappa}(J'_cj'J' - J_cjJ) \\ &\quad M^2_{0\varkappa\varkappa}(sl'j' - slj) M^2_{000}(S'_cL'_cJ'_c - S_cL_cJ_c). \end{aligned} \qquad (3.63)$$

The M-function method and more detailed consideration of the angular Q-factors are given in [3.25].

4. Basic Approaches to Collisions Involving Highly Excited Atoms and Ions

Due to the peculiar properties of highly excited states considered in the preceding chapters, the theoretical description of collisional processes involving Rydberg atoms or ions differs significantly from the case of the atoms or ions in the low excited or ground states. The goal of this chapter is to formulate those physical approaches and general theoretical techniques which will be the basis for describing collisional processes of Rydberg atoms with neutral and charged particles in the following chapters. We give a brief account of various versions of perturbation theory, pure classical and semiclassical methods, the close coupling method, binary-encounter theory and impulse approximation. The results are given in the common form independently of the specific kind of projectile and type of transition. Really, however, the applicability of one or another approach will depend radically on the type of the interaction between the Rydberg atom (ion) and projectile, on the values of quantum numbers and energy transfer as well as, on the relative velocity of colliding partners. We consider here the projectile as a structureless particle, since it is justified for most of the collisional problems considered in this book and makes the main ideas more transparent. However, for the impulse approximation approach we outline the natural generalization of the case when the internal state of projectile is changed as result of a collision.

4.1 Formulation of Problem

4.1.1 Features of Collisions with Neutral and Charged Particles

Before going into a consistent description of the general physical approaches in the collision theory involving Rydberg atoms and ions, we briefly discuss the main characteristic features of elementary collisional processes. In the case of Rydberg atom–neutral atom collisions, the characteristic radii of the interaction potentials of the perturber with both the highly excited electron and the ionic core are, usually, small in comparison with the orbital radius. Therefore, the collisional processes of Rydberg atoms with atomic projectiles (as well as with nonpolar and polar molecules) may often be regarded as two independent processes caused by the scattering of the incident particle on the highly excited (valence) electron and by the scattering on the ionic core. The relative role of the scattering mechanisms is determined by the type of colliding particles and their relative velocity, by the energy defect of transition and by the range of the principal quantum number.

The cross sections of the Rydberg atom-neutral atom collisions can be of the order of the geometrical cross section of a highly excited atom provided the principal quantum number is not too large. It is possible only for the elastic and quasi-elastic processes (i.e., for the processes without change of the highly excited electron energy or with very small energy defect). The values of the cross sections of the inelastic bound-bound and bound-free transitions reveal the strong damping with increasing the energy transferred to the Rydberg atom. Usually, both the excitation and ionization processes with large values of the energy transferred to the highly excited electron result from the perturber-core scattering in contrast to the quasi-elastic and inelastic transitions with small and intermediate energy defects, for which the traditional mechanism of the scattering of the quasi-free electron by a neutral atom is predominant. On the other hand, the large cross sections of the inelastic n-changing transitions and ionization in collisions with molecules are the result of quasi-resonant energy transfer from the internal rotational degrees of freedom of the projectile to the Rydberg electron (or the ion-pair formation in the final channel in collisions with electron-attaching molecules).

For collisions with charged projectiles (electrons or ions) the interaction between the incident particle and the highly excited electron is determined by the long-range Coulomb potential. In collisional processes with small energy transfer (transitions with the change in the orbital angular momentum or transitions for which the change of the principal quantum number is not too large) the Rydberg atom behaves as a whole scatterer. In other words, the interaction of the incident charged particle with the Rydberg electron is appreciably smaller than its interaction with the parent core. Effective cross sections are mainly determined by large distances of the projectile from the Rydberg atom and are very large (comparable with the geometrical area, and in some cases even exceeding it). As a result, the approach based on the dipole approximation for the interaction potential appears to be adequate.

In the opposite case of large energy transfer (transitions with $|\, n - n' \,| \gg 1$ or ionization), the collisional process may be regarded as an encounter of the incident particle with the Rydberg electron without any influence of the parent core. Within the frame of this impulse approach, which is usually called the binary-encounter approach, the cross section is determined by the value of the energy transfer and decreases sharply with energy transfer increasing. General quantitative theory of the Rydberg atom transitions induced by collisions with charged particles should include both mentioned approaches as special cases.

4.1.2 Stationary Problem of Scattering

a) Schrödinger Equation. A general formulation of the scattering problem may be given within the framework of the stationary Schrödinger equation with an appropriate outgoing scattering condition. For collisions involving the Rydberg atom (ion) A and incident structureless particle B, the stationary wave function $\Psi(\mathbf{r}, \mathbf{R})$ of the total system A+B is the solution of the following equation:

$$\mathsf{H}\Psi = \mathsf{E}\Psi, \quad \mathsf{H} = \mathsf{H}_0 + \mathcal{V}, \quad \mathsf{H}_0 = -\frac{\hbar^2}{2\mu}\Delta_{\mathbf{R}} + H_{\mathrm{A}}. \tag{4.1}$$

Here H and E are the total Hamiltonian and energy of the system, respectively; H_0 is the unperturbed Hamiltonian, which consists of the highly excited atom Hamiltonian H_{A} and the relative kinetic energy operator $\left(-\hbar^2\Delta_{\mathbf{R}}/2\mu\right)$ of the colliding particles A+B ($\mu = M_{\mathrm{A}}M_{\mathrm{B}}/(M_{\mathrm{A}} + M_{\mathrm{B}})$ being their reduced mass). We have removed here the Hamiltonian H_{B} of projectile because we consider it as a structureless particle. In the zero order approximation over the projectile-atom interaction $\mathcal{V}$ the solution of the Schrödinger equation

$$\mathsf{H}_0\Psi^{(0)} = \mathsf{E}\Psi^{(0)}, \quad \Psi^{(0)}_{\alpha\mathbf{q}}(\mathbf{r}, \mathbf{R}) = e^{i\mathbf{q}\cdot\mathbf{R}}\psi_\alpha(\mathbf{r}) \tag{4.2}$$

is a product of the plane wave $\psi_{\mathbf{q}}(\mathbf{R}) = \exp(i\mathbf{q}\cdot\mathbf{R})$ describing the relative motion of the colliding particles with the wave vector $\mathbf{q}$, and eigenfunction $\psi_\alpha(\mathbf{r})$ of the Rydberg atom (ion) Hamiltonian

$$H_{\mathrm{A}}\psi_\alpha = E_\alpha\psi_\alpha, \quad H_{\mathrm{A}} = -\frac{\hbar^2}{2\mu_{\mathrm{eA}^+}}\Delta_{\mathbf{r}} + U, \tag{4.3}$$

whereas α and E_α denote the set of quantum numbers characterizing its eigenstate and energy, respectively. Here U is the interaction potential between the Rydberg electron and ion core (the Coulomb-type at large separations $U \to -Ze^2/r$ at $r \to \infty$), and $\mu_{\mathrm{eA}^+} = mM_{\mathrm{A}^+}/(m + M_{\mathrm{A}^+})$ is the reduced mass of $\left(e, \mathrm{A}^+\right)$-pair, i.e. $\mu_{\mathrm{eA}^+} \approx m$. The unperturbed wave function of the colliding particles A+B is normalized by the condition

$$\left\langle \Psi^{(0)}_{\alpha'\mathbf{q}'} \middle| \Psi^{(0)}_{\alpha\mathbf{q}} \right\rangle = \iint \psi^*_{\mathbf{q}'}(\mathbf{R})\psi^*_{\alpha'}(\mathbf{r})\psi_\alpha(\mathbf{r})\psi_{\mathbf{q}}(\mathbf{R})d\mathbf{r}d\mathbf{R}$$

$$= (2\pi)^3 \delta(\mathbf{q}' - \mathbf{q})\delta_{\alpha'\alpha}. \tag{4.4}$$

The radius vector $\mathbf{R}$ of the incident particle B in (4.2) is taken relative to the center of mass of the Rydberg atom A [i.e., to the center of mass of the (e, A^+)-pair], while $\mathbf{r}$ is the valence electron–ion core $\left(e - \mathrm{A}^+\right)$ vector separation.

The exact wave function $\Psi^+_{\alpha\mathbf{q}}(\mathbf{r}, \mathbf{R})$ of system A+B is a solution of the Schrödinger equation (4.1) for the total Hamiltonian $\mathsf{H}=\mathsf{H}_0 + \mathcal{V}$ including the projectile-atom interaction $\mathcal{V}$, which leads to the inelastic transitions between the Rydberg atomic states $|\alpha\rangle \rightarrow |\alpha'\rangle$. This solution satisfies the following boundary condition:

$$\Psi^+_{\alpha q}(\mathbf{r}, \mathbf{R}) \underset{R \to \infty}{\longrightarrow} \Psi^{(0)}_{\alpha q}(\mathbf{r}, \mathbf{R}) + \sum_{\alpha'} f_{\alpha'\alpha}(\mathbf{q}', \mathbf{q}) \frac{e^{iq'R}}{R} \psi_{\alpha'}(\mathbf{r}) , \qquad (4.5)$$

whereas the wave numbers $q \equiv q_\alpha$ and $q' \equiv q_{\alpha'}$ and the atomic energies before and after collision are given by the energy conservation law

$$E = E_\alpha + \frac{\hbar^2 q^2}{2\mu} = E_{\alpha'} + \frac{\hbar^2 (q')^2}{2\mu} , \qquad (4.6)$$

where E is the total energy of system A+B. As is evident from (4.5), at large distances the wave function of this system involves the unperturbed wave function (4.2) and the linear combination of the spherical divergent waves $R^{-1} \exp(iq'R)\psi_{\alpha'}(\mathbf{r})$ corresponding to all possible eigenstates of the Rydberg atom. The coefficient $f_{\alpha'\alpha}(\mathbf{q}', \mathbf{q})$ is the scattering amplitude for the inelastic transition between highly excited states $|\alpha\rangle \rightarrow |\alpha'\rangle$. (It should be noted that for the ionization process the final function $\psi_{\alpha'}$ is a function of the continuous spectrum of Rydberg atom A). As a result, the differential cross section of scattering for the $|\alpha\rangle \rightarrow |\alpha'\rangle$ transition into the interval $d\Omega = \sin\theta d\theta d\varphi$ of solid angles is equal to

$$d\sigma_{\alpha'\alpha}(\mathbf{q}', \mathbf{q}) = \frac{q'}{q} |f_{\alpha'\alpha}(\mathbf{q}', \mathbf{q})|^2 d\Omega . \qquad (4.7)$$

Thus, the Schrödinger equation (4.1) with the boundary condition (4.5) and formula (4.7) give a consistent formulation of the scattering problem. However, direct solution of this equation fails in both the analytical and numerical methods. The next sections will be devoted to various approximated approaches which are widely used in the collision theory involving Rydberg atoms and ions. Here we also present one more general formulation of the scattering problem based on the Lippman-Schwinger equation which turns out to be useful for understanding and analysing a number of approximated methods (e.g., quantum impulse approximation).

b) **Lippman-Schwinger Equation and Green's Resolvent.** By adopting the Lippman-Schwinger formalism, the exact solution of the Schrödinger equation (4.1) with the outgoing scattering condition (4.5) can be rewritten in the two equivalent operator forms (see, for example, [4.1], [4.2])

$$\Psi^+_{\alpha q} = \Psi^{(0)}_{\alpha q} + G^+(E) \mathcal{V}\Psi^{(0)}_{\alpha q} , \qquad G^+(E) = (E - H + i0)^{-1} , \qquad (4.8)$$

$$\Psi^+_{\alpha q} = \Psi^{(0)}_{\alpha q} + G^+_0(E) \mathcal{V}\Psi^+_{\alpha q} , \qquad G^+_0(E) = (E - H_0 + i0)^{-1} . \qquad (4.9)$$

Here the operators $G(E + i0)$ and $G_0(E+i0)$ considered as functions of the complex variable, are the Green's resolvents for the total H and for the unperturbed H_0 Hamiltonians, respectively. To clarify an explicit integral form of the Lippman-Schwinger equation for the exact wave function $\Psi^+_{\alpha q}(\mathbf{r}, \mathbf{R})$ in the coordinate representation, it is convenient to rewrite (4.9) in terms of the unperturbed Green's function of the system

$$G_0^+ \left(\mathbf{R}, \mathbf{r}; \mathbf{R}', \mathbf{r}'; E \right) = -\frac{\mu}{2\pi\hbar^2} \sum_\alpha \frac{\exp\left(iq_\alpha \left| \mathbf{R} - \mathbf{R}' \right|\right)}{\left| \mathbf{R} - \mathbf{R}' \right|} \psi_\alpha^* \left(\mathbf{r}' \right) \psi_\alpha \left(\mathbf{r} \right),$$

$$q_\alpha = \frac{1}{\hbar} 2\mu \left(E - E_\alpha \right)^{1/2},$$

(4.10)

which satisfies the following equation:

$$\left(E + \frac{\hbar^2}{2\mu} \Delta_\mathbf{R} - H_A \right) G_0^+ \left(\mathbf{R}, \mathbf{r}; \mathbf{R}', \mathbf{r}'; E \right) = \delta \left(\mathbf{R} - \mathbf{R}' \right) \delta \left(\mathbf{r} - \mathbf{r}' \right)$$

$$\times G_0^+ \left(\mathbf{R}, \mathbf{r}; \mathbf{R}', \mathbf{r}'; E \right).$$

(4.11)

For the local interaction potential $\mathcal{V}$, the final result takes the form

$$\Psi_{\alpha\mathbf{q}}^+(\mathbf{r}, \mathbf{R}) = \Psi_{\alpha\mathbf{q}}^{(0)}(\mathbf{r}, \mathbf{R}) + \iint G_0^+ \left(\mathbf{r}, \mathbf{R}; \mathbf{r}', \mathbf{R}'; E \right) \mathcal{V} \left(\mathbf{r}', \mathbf{R}' \right) \Psi_{\alpha\mathbf{q}}^+(\mathbf{r}', \mathbf{R}') d\mathbf{r}' d\mathbf{R}'.$$

(4.12)

Using the well-known expression for the Green's function at large separations $R \to \infty$

$$G_0^+ \left(\mathbf{R}, \mathbf{r}; \mathbf{R}', \mathbf{r}'; E \right) = -\frac{\mu}{2\pi\hbar^2} \sum_\alpha \frac{e^{iq_\alpha R}}{R} \psi_\alpha^* \left(\mathbf{r}' \right) \exp\left[-iq_\alpha \left(\mathbf{n}_\mathbf{R} \cdot \mathbf{R}' \right) \right],$$

(4.13)

the basic formula (4.12) for the exact wave function of colliding particles A and B can be reduced to its asymptotic expression (4.5) in terms of the amplitude $f_{\alpha'\alpha}\left(\mathbf{q}', \mathbf{q} \right)$ of inelastic scattering, where $\mathbf{n}_\mathbf{R} = \mathbf{R}/R$ is the unit vector along the radius vector $\mathbf{R}$. Then, we directly derive the final expression for the scattering amplitude in the coordinate representation

$$f_{\alpha'\alpha}\left(\mathbf{q}', \mathbf{q} \right) = -\frac{\mu}{2\pi\hbar^2} \left\langle \Psi_{\alpha'\mathbf{q}'}^{(0)} \left| \mathcal{V} \right| \Psi_{\alpha\mathbf{q}}^+ \right\rangle$$

$$= -\frac{\mu}{2\pi\hbar^2} \iint d\mathbf{r} d\mathbf{R} \exp\left(-i\mathbf{q}' \cdot \mathbf{R} \right) \psi_{\alpha'}^* \left(\mathbf{r} \right) \mathcal{V} \left(\mathbf{r}, \mathbf{R} \right) \Psi_{\alpha\mathbf{q}}^+(\mathbf{r}, \mathbf{R}).$$

(4.14)

A number of general results of stationary scattering theory may also be formulated in terms of the so-called scattering T-operator (or T-matrix), which transforms the unperturbed wave function $\Psi_{\alpha\mathbf{q}}^{(0)}$ of the total system to the exact solution $\Psi_{\alpha\mathbf{q}}^+$ of the scattering problem. In accordance with the basic equations (4.8, 9) this operator has the form

$$\mathsf{T}(E) \Psi_{\alpha\mathbf{q}}^{(0)} = \mathcal{V}\Psi_{\alpha\mathbf{q}}^+, \qquad \mathsf{T}(E) = \mathcal{V} + \mathcal{V}G^+(E)\mathcal{V}.$$

(4.15)

The Lippman-Schwinger equation for the T(E)-operator and for the Green's G(E)-resolvent

$$T(E) = \mathcal{V} + \mathcal{V}G_0^+(E)\,T(E), \qquad G^+(E) = G_0^+(E) + G_0^+(E)\,\mathcal{V}G^+(E), \qquad (4.16)$$

directly proceeds from (4.8–15) if we take into account the following operator equality $G(E)\mathcal{V} = G_0(E)T(E)$. As is apparent from (4.14, 15) the scattering amplitude (4.6) for the inelastic $\alpha \to \alpha'$ transition can be expressed in terms of the T-matrix elements on the energy shell [see (4.7)] taken over the unperturbed wave functions (4.2) of the system. This relationship is given by

$$f_{\alpha'\alpha}(\mathbf{q}',\mathbf{q}) = -\frac{\mu}{2\pi\hbar^2} T_{\alpha'\alpha}(\mathbf{q}',\mathbf{q}) \equiv -\frac{\mu}{2\pi\hbar^2} \left\langle \Psi_{\alpha'\mathbf{q}'}^{(0)} \,|\, T(E) \,|\, \Psi_{\alpha\mathbf{q}}^{(0)} \right\rangle, \qquad (4.17)$$

if the wave function $\Psi_{\alpha\mathbf{q}}^{(0)}$ is normalized by the condition (4.4). By introducing standard Dirac's designations for the eigenvectors of the Rydberg atom and for the plane wave of scattering

$$|\alpha\rangle \equiv |\psi_\alpha(\mathbf{r})\rangle, \qquad |\mathbf{q}\rangle \equiv |\psi_\mathbf{q}(\mathbf{R})\rangle = |\exp(i\mathbf{q}\cdot\mathbf{R})\rangle \qquad (4.18)$$

we can represent the T–matrix element in the following form:

$$T_{\alpha'\alpha}(\mathbf{q}',\mathbf{q}) = \langle \mathbf{q}',\alpha' \,|\, T(E) \,|\, \mathbf{q},\alpha\rangle = \langle \mathbf{q}',\alpha' \,|\, \mathcal{V} \,|\, \Psi_{\alpha\mathbf{q}}^+\rangle, \qquad (4.19)$$

which is especially convenient for further analysis. The second equation in (4.19) is in full agreement with (4.14, 15).

Using the well-known expansion of the Green's resolvent over the eigenfunctions of the Rydberg atom $|\alpha\rangle = \psi_\alpha(\mathbf{r})$ and over the plane waves of scattering $|\mathbf{q}\rangle = \psi_\mathbf{q}(\mathbf{R})$

$$G_0^+(E) = \sum_\alpha \int \frac{d\mathbf{q}}{(2\pi)^3} \frac{|\mathbf{q},\alpha\rangle\langle\mathbf{q},\alpha|}{E - E_\alpha - \hbar^2 q^2/2\mu + i0}, \qquad (4.20)$$

the Lippman-Schwinger operator equation (4.16) can be rewritten in the following integral form for the transition matrix elements:

$$\langle \mathbf{q}',f \,|\, T(E) \,|\, \mathbf{q},i\rangle = \langle \mathbf{q}',f \,|\, \mathcal{V} \,|\, \mathbf{q},i\rangle$$
$$+ \sum_\alpha \int \frac{d\mathbf{q}''}{(2\pi)^3} \frac{\langle \mathbf{q}',f \,|\, \mathcal{V} \,|\, \mathbf{q}'',\alpha\rangle \langle \mathbf{q}'',\alpha \,|\, T(E) \,|\, \mathbf{q},i\rangle}{E - E_\alpha - \hbar^2(q'')^2/2\mu + i0}. \qquad (4.21)$$

The square of the transition matrix element of the scattering T–operator directly yields the probability or the cross section of the collision process. For example, the differential cross section of the $|\,i,\mathbf{q}\rangle \to |\,f,\mathbf{q}'\rangle$ transition into the interval of the wave vectors $d\mathbf{q}'$ is expressed in terms of the T-matrix on the energy shell (4.7) in accordance with the following equation:

$$d\sigma_{fi}(\mathbf{q}',\mathbf{q}) = \frac{2\pi}{\hbar}\left(\frac{\mu}{\hbar q}\right) |T_{fi}(\mathbf{q}',\mathbf{q})|^2 \,\delta(E_f + \mathcal{E}' - E_i - \mathcal{E})\,\frac{d\mathbf{q}'}{(2\pi)^3}, \qquad (4.22)$$

$$\mathbf{q} = \mu\mathbf{V}/\hbar, \qquad \mathcal{E} = \hbar^2 q^2/2\mu.$$

Here $\mathcal{E}$ and $\mathcal{E}'$ are the kinetic energies and $\mathbf{V}$ and $\mathbf{V}'$ are the relative velocities of the incident particle with respect to the Rydberg atom before and after collision. Integration of (4.22) over all possible values of the final wave numbers q' and over the solid scattering angles $d\Omega_{\mathbf{q'q}} = \sin\theta d\theta d\varphi$ leads to the basic formula

$$\sigma_{fi}(q) = \left(\frac{\mu}{2\pi\hbar^2}\right)^2 \frac{q'}{q} \int |\mathsf{T}_{fi}(\mathbf{q'},\mathbf{q})|^2 d\Omega_{\mathbf{q'q}} \tag{4.23}$$

for the integral cross section σ_{fi} of the bound-bound $i \to f$ transition between Rydberg states. For the bound-free $i \to f$ transition the σ_{fi} quantity on the left-hand side of (4.23) should be replaced, as usual, by the differential ionization cross section $d\sigma_{fi}/dE_f$ per unit energy interval of the ejected electron, provided its final wave function ψ_f of the continuous spectrum is normalized on the δ-function of energy.

The form of (4.21) is such that we may easily establish the relationship between the imaginary part of the transition matrix element for elastic scattering in the forward direction $\theta_{\mathbf{q'q}}$ and the total cross section $\sigma_i^{\mathrm{tot}} = \sum_f \sigma_{fi}$ of all inelastic and elastic processes. Indeed, using (2.94) for the integration path and (4.19) and assuming $(f = i)$ and $(\mathbf{q'} = \mathbf{q})$ in (4.21), we have

$$\mathrm{Im}\left\{\mathsf{T}_{ii}(\mathbf{q},\mathbf{q})\right\} = -\frac{\hbar^2 q}{2\mu}\sigma_i^{\mathrm{tot}}(q), \quad \mathrm{Im}\left\{f_{ii}(q,\theta_{\mathbf{q'q}}=0)\right\} = \frac{q}{4\pi}\sigma_i^{\mathrm{tot}}(q). \tag{4.24}$$

The second relation in (4.24) was rewritten in terms of the usual scattering amplitude (4.17) in accordance with (4.17). This is the so-called optical theorem, which reflects the basic analytical properties of the scattering amplitude.

c) Rydberg Atom-Projectile Interaction and Mass Disparity Relation. The most important phenomena in collisions involving the Rydberg atom (ion) and structureless neutral or charged projectile may be described in terms of the local interaction potential $\mathcal{V} \equiv \mathcal{V}_{\mathrm{AB}}$, which is the sum of both the interaction between the incident particle B and the highly excited electron $\mathcal{V}_{\mathrm{eB}}$ and the ion core interaction $\mathcal{V}_{\mathrm{A+B}}$. In other words, we shall assume the total potential interaction to be of the form

$$\mathcal{V} = \mathcal{V}_{\mathrm{eB}}(\mathbf{r}_{\mathrm{eB}}) + \mathcal{V}_{\mathrm{A+B}}(\mathbf{R}_{\mathrm{A+B}}). \tag{4.25}$$

Neglecting, as usual, the additional small polarization interaction between the ion core and projectile which results from the induced electric field originated by the Rydberg electron. This effect is fully absent in the case of an electron projectile.

The radius vectors $\mathbf{r}_{\mathrm{eB}}$ and $\mathbf{R}_{\mathrm{A+B}}$ of projectile B relative to the Rydberg electron e and to the ion core $\mathrm{A^+}$ can be expressed in terms of its position vector $\mathbf{R}$ relative to the $(\mathrm{e}, \mathrm{A^+})$ center of mass, and the vector $\mathbf{r}$ of the electron-core separation, which are involved in the general equations of scattering theory

$$\mathbf{r}_{\mathrm{eB}} = \mathbf{R} - (M_{\mathrm{A^+}}/M_{\mathrm{A}})\mathbf{r}, \quad \mathbf{R}_{\mathrm{A+B}} = \mathbf{R} + (m/M_{\mathrm{A}})\mathbf{r}, \tag{4.26}$$

where $M_A = m + M_{A^+}$ is the total mass of the Rydberg atom. Since the electron mass is small compared to the mass of the Rydberg atom ($m \ll M_A$) it is natural to expand the core-projectile interaction in a power series of the ($m\mathbf{r}/M_A$) ratio. As a result, we have

$$V_{A+B}(\mathbf{R}_{A+B}) = V_{A+B}(\mathbf{R}) + (m/M_A)\mathbf{r} \cdot \nabla_{\mathbf{R}} V_{A+B}(\mathbf{R}) + \dots \quad (4.27)$$

and, hence, the matrix element of the core-perturber interaction over the Rydberg atom wave functions takes the form

$$\langle \alpha' | V_{A+B} | \alpha \rangle = V_{A+B}(\mathbf{R}) \delta_{\alpha'\alpha} + (m/M_A) \langle \psi_{\alpha'}(\mathbf{r}) | \mathbf{r} | \psi_\alpha(\mathbf{r}) \rangle \cdot \nabla_{\mathbf{R}} V_{A+B}(\mathbf{R}) + \dots$$

$$(4.28)$$

The second term in (4.27) corresponds to the inertial force $\mathbf{F}_e = md^2(\mathbf{R}_{A^+})/dt^2$ acting on the Rydberg electron in the noninertial frame of reference moving with the ion core A^+. Since this term contains a factor $(m/M_A) \ll 1$ its contribution to the total transition amplitude is small and usually can be neglected. Therefore, for most collision problems of practical importance, we can neglect the so-called noninertial effects in collisions of Rydberg atoms with neutral or charged particles. (The situations, in which these effects become important, will be discussed in Chap. 7).

Thus, neglecting the weak noninertial effects, we can count off the projectile position vector $\mathbf{R}$ from the center of mass of the ion core A^+. In this approximation the total interaction of the Rydberg atom with the projectile can be written as

$$V \approx V_{eB}(\mathbf{R} - \mathbf{r}) + V_{A+B}(\mathbf{R}), \quad (4.29)$$

whereas the potential V_{eB} of the electron-projectile interaction depends only on the difference of $\mathbf{r}_{eB} = \mathbf{R} - \mathbf{r}$, while the core-projectile interaction does not contain the dependence on the Rydberg electron coordinates. Hence, the matrix elements $\langle \psi_{\alpha'}(\mathbf{r}) | V_{A+B}(\mathbf{R}) | \psi_\alpha(\mathbf{r}) \rangle$ of the core-projectile interaction are rigorously equal to zero (if $\alpha' \neq \alpha$) due to the orthogonality of the Rydberg atom wave functions. As a result, in collisions of the Rydberg atom with the structureless particles, the cross section of the inelastic transition ($\alpha \neq \alpha'$) is mainly determined by the interaction V_{eB} between the Rydberg electron and projectile. The core-perturber interaction V_{A+B} is certainly important in the process of elastic scattering ($\alpha' = \alpha$) and electron Rydberg ion collision since the potential V_{A+B} changes the trajectory of the incident particle.

4.2 Born Approximation: Momentum Representation

We start the consideration of approximate theoretical methods for describing the inelastic transitions between the Rydberg states of an atom (ion) in its collision with a projectile from the simplest quantum approach called the

Born approximation. This approach is based on the perturbation theory over the interaction potential. Within the frame of first order perturbation theory the solution of (4.1) with boundary condition (4.5) can be written in explicit form. The asymptote of this solution directly yields the transition amplitude $f_{\alpha'\alpha}(\mathbf{q}', \mathbf{q})$. The final result for the transition cross section is given by

$$\sigma_{\alpha'\alpha}(q) = \frac{q'}{q}\left(\frac{\mu}{2\pi\hbar^2}\right)^2 \frac{1}{g_\alpha} \sum_{a\backslash\alpha} \sum_{a'\backslash\alpha'} \int d\Omega_{\mathbf{q}'\mathbf{q}}$$

$$\times \left|\iint d\mathbf{R}d\mathbf{r}\exp(-i\mathbf{q}'\cdot\mathbf{R})\psi_{\alpha'}^*(\mathbf{r})\mathcal{V}(\mathbf{r},\mathbf{R})\psi_\alpha(\mathbf{r})\exp(i\mathbf{q}\cdot\mathbf{R})\right|^2 ,$$

(4.30)

where $\mu = M_A M_B/(M_A + M_B)$ is the reduced mass of the colliding particles; and the momentum wave numbers q and q' before and after collision in (4.30) are connected by the law of energy conservation (4.6). This formula is written taking into account the degeneracy of the initial and final levels. The same expression can also be obtained from the well-known Fermi's gold rule. The details of both derivations are presented in any available monograph on quantum mechanics (e.g., [2.2]). It should be emphasized that within the framework of general scattering theory the Born approximation directly follows from (4.6, 17) with the use of only the first term $T^B = \mathcal{V}$ on the right-hand side of the Lippman-Schwinger equation (4.16).

Integration over the solid angle of scattering $d\Omega$ in the basic formula (4.30) can be changed by integration over the value of the momentum transfer $\mathbf{Q} = \mathbf{q}'-\mathbf{q}$ in the limits from $Q_{min} = |q' - q|$ up to $Q_{max} = q'+q$ in accordance with the relations

$$Q^2 = q^2 + (q')^2 - 2qq'\cos\theta , \qquad QdQ = qq'\sin\theta d\theta , \qquad (4.31)$$

in which the q' value is determined from the law of energy conservation. Further, we take into account the previous discussion about the nonvanishing contribution of only the Rydberg electron-projectile interaction $\mathcal{V}_{eB}$ to the cross section of inelastic transitions ($\alpha' \neq \alpha$). Since it depends only on the difference $\mathbf{r}_{eB} = \mathbf{R} - \mathbf{r}$, we can rewrite expression (4.30) in the following general form:

$$\sigma_{\alpha'\alpha} = \frac{2\pi}{q^2}\left(\frac{\mu}{\mu_{eB}}\right)^2 \int_{Q_{min}}^{Q_{max}} |f_{eB}(Q)|^2 F_{\alpha'\alpha}(Q)QdQ , \qquad (4.32)$$

if we introduce the Born amplitude $f_{eB}(Q)$ for scattering of an electron by projectile B and the form factor for transition between highly excited states $|\alpha\rangle$ and $|\alpha'\rangle$

$$f_{eB}(Q) = -\frac{\mu_{eB}}{2\pi\hbar^2}\int \mathcal{V}_{eB}(\mathbf{r}_{eB})\exp(-i\mathbf{Q}\cdot\mathbf{r}_{eB})d\mathbf{r}_{eB} ,$$

(4.33)

$$F_{\alpha'\alpha}(Q) = \frac{1}{g_\alpha} \sum_{a \backslash \alpha} \sum_{a' \backslash \alpha'} \left| \int \psi_{\alpha'}^*(\mathbf{r}) \exp(i\mathbf{Q} \cdot \mathbf{r}) \psi_\alpha(\mathbf{r}) d\mathbf{r} \right|^2 ,$$

where $\mu_{eB} = mM_B/(m+M_B)$ is the reduced mass of electron e and projectile B. It is evident that $\mu_{eB} \approx m$ for collisions of the Rydberg atom with the heavy particles (atom, molecule or ion), while $\mu_{eB} = m/2$ in its collision with an electron. Formulas (4.32–33) are especially convenient for specific applications and have an obvious physical meaning according to which the transition cross section can be expressed in terms of the Fourier components of the wave functions product and the Fourier component of the interaction corresponding to the same value of the momentum transfer. They open a direct way for further generalizations. In particular, the expression for the cross section in terms of the form factor (4.32) can be valid with exact f_{eB} amplitude, when the Born approximation for the electron-projectile scattering does not hold.

4.3 Time-Dependent Approach: Impact-Parameter Representation

The general quantum problem of collision between the highly excited atom and projectile is significantly simplified if their relative motion can be considered as classical motion along a given trajectory. Such classical description is valid provided the de Broglie wavelength $1/q$ of the projectile is much smaller than the characteristic radius R_{int} of the interaction potential ($qR_{int} \gg 1$) and the energy $\Delta\mathcal{E} = \hbar^2 \left[(q')^2 - q^2 \right]/2\mu$ transferred to the translational motion of the colliding particles is small as compared to the kinetic energy $\mathcal{E} = \hbar^2 q^2/2\mu$ of the projectile ($\Delta\mathcal{E} \ll \mathcal{E}$). The first condition shows that the action function $S \sim \hbar q R_{int}$ is much greater than $\hbar$ and hence clearly demonstrates the classical character of motion in accordance with the basic correspondence principle [2.2]. The second one means that the classical description of the projectile motion along a given trajectory is certainly inapplicable for large values of transition energy defects and, in particular, near the threshold of the process $\mathcal{E} \sim \Delta\mathcal{E}$, where the quantum effects in scattering becomes important. For practical estimates it is convenient to rewrite condition $\Delta\mathcal{E} \ll \mathcal{E}$ in terms of the projectile-Rydberg atom relative velocity V, orbital electron velocity $v_n \sim v_0/n$ and its binding energy $|E_n| = Z^2 Ry/n^2$

$$V \gg v_n \left(\frac{m}{\mu} \frac{\Delta\mathcal{E}}{|E_n|} \right)^{1/2} . \tag{4.34}$$

One can see that it has the same form for collisions with neutral and charged projectiles. However, the applicability range of the first condition $qR_{int} \gg 1$ turns out to be quite different for collisions with neutral and charged projectiles and depends on the principal quantum number, relative velocity of colliding particles, transition energy defect, etc.

For instance, for quasi-elastic or weakly inelastic collisions with charged particles $R_{int} \sim a_0 n^2/Z$ and, hence, condition $qR_{int} \gg 1$ does not lead to any new restrictions in addition to (4.34). Moreover, the projectile motion relative to the center of mass of the Rydberg atom is usually taken to be classical following the straight-line trajectory. In the case of a multicharged ion ($\mathcal{Z}$ is its charge) this means that the energy of the projectile must be greater than the characteristic energy of the Coulomb interaction $V_{A+B} = \mathcal{Z}Ze^2/R$ in the region of separations $R \sim r_n = a_0 n^2/Z$. This additional restriction on the projectile velocity can be written as

$$V \gg v_n \left(\frac{\mathcal{Z}m}{\mu} \right)^{1/2} . \tag{4.35}$$

It is important to stress that for collisions with the heavy charged particles the identified conditions are usually fulfilled in most situations of practical importance. However, for the electron-Rydberg ion collisions there is a wide range of the incident electron velocities in which the last condition ($V \gg v_n$) does not hold and the approximation of rectilinear trajectory becomes invalid. Then, the distortion effects should be taken into account, for example, within the framework of the Coulomb-Born approximation.

For collision with a neutral projectile (atom or molecule) its motion relative to the ion core of the Rydberg atom may be considered as classical if $(\mu V/\hbar) R_{int} \gg 1$, where μ is the reduced mass of atom and projectile. Since the characteristic radius R_{int} of the core-projectile interaction is usually about or greater than a few atomic values a_0 this means that the classical description of projectile motion would certainly be valid in the range of about or greater than thermal energies. The quantum character of projectile motion is realized only at small relative velocities (subthermal energies), particularly so, for collisions of atoms with small reduced mass. A more detailed discussion concerning the applicability of the classical and quasi-classical descriptions of the atom-atom and atom-molecule collisions is given in [2.2].

4.3.1 Close Coupled Equations for Transition Amplitudes

Given the classical character of a projectile-Rydberg atom's relative motion along a given trajectory $\mathbf{R}(t)$, one of the most efficient theoretical method for calculation of transition probabilities and cross sections is based on the time-dependent impact-parameter approach. According to this approach the projectile B affects the Rydberg atom A by means of perturbation interaction $V[\mathbf{r}, \mathbf{R}(t)]$, which contains an implicit dependence on time t through its position vector $\mathbf{R}(t)$ relative to the center of mass of the ion core A^+ and highly excited electron e. (For most problems the radius vector of the projectile is taken with respect to the nuclei of A^+). This allows us to remove the kinetic energy operator from the basic time-dependent Schrödinger equation for the total quantum wave function of the system. As a result, within the framework of the impact-parameter time-dependent approach the wave function $\psi\,[\mathbf{r}, \mathbf{R}(t)]$ of the colliding particles is determined from the following equation:

$$i\hbar\frac{\partial}{\partial t}\psi\,[\mathbf{r}, \mathbf{R}(t)] = \{H_A + V\,[\mathbf{r}, \mathbf{R}(t)]\}\,\psi\,[\mathbf{r}, \mathbf{R}(t)]\,. \tag{4.36}$$

Since the main contribution to the cross sections of Rydberg atom-neutral (charged) particle collisions is determined by the large separations R between the ionic core A^+ and projectile B (i.e., by large values of the impact parameter ρ), their relative motion is usually taken to be rectilinear, so that

$$\mathbf{R}(t) = \boldsymbol{\rho} + \mathbf{V}t. \tag{4.37}$$

The solution of the Schrödinger equation (4.36) can be obtained by Dirac's method of variation of constants according to which the time-dependent wave function $\psi\,[\mathbf{r}, \mathbf{R}(t)]$ of the colliding particles is expanded in the eigenfunctions $\psi_\alpha\,(\mathbf{r})$ of the unperturbed Rydberg atom Hamiltonian H_A

$$\psi\,[\mathbf{r}, \mathbf{R}(t)] = \sum_\alpha \mathfrak{a}_\alpha\,(t)\,\psi_\alpha\,(\mathbf{r})\,. \tag{4.38}$$

Substitution of this expression into the time-dependent Schrödinger equation (4.36) leads to a system of close coupled differential equations for transition amplitudes $\mathfrak{a}_\alpha\,(t)$

$$i\hbar\frac{d\mathfrak{a}_\alpha}{dt} = E_\alpha\mathfrak{a}_\alpha + \sum_{\alpha'} V_{\alpha\alpha'}\,(\mathbf{R}(t))\,\mathfrak{a}_{\alpha'}\,. \tag{4.39}$$

Here E_α is the eigenenergy of the Rydberg atom Hamiltonian H_A, and

$$V_{\alpha'\alpha}\,[\mathbf{R}(t)] = \langle\alpha'\,|V\,[\mathbf{r}, \mathbf{R}(t)]|\,\alpha\rangle = \int \psi_{\alpha'}^*\,(\mathbf{r})\,V\,[\mathbf{r}, \mathbf{R}(t)]\,\psi_\alpha\,(\mathbf{r})\,d\mathbf{r} \tag{4.40}$$

is the matrix element of the interaction potential, which depends on time only through the position vector of the projectile $\mathbf{R}(t)$. Standard replacement of variables

$$\mathfrak{a}_\alpha(t) = a_\alpha(t)\exp(-iE_\alpha t/\hbar), \tag{4.41}$$

transforms this system to the form

$$i\hbar\frac{da_\alpha}{dt} = \sum_{\alpha'} V_{\alpha\alpha'}\,[\mathbf{R}(t)]\exp(i\omega_{\alpha'\alpha}t)a_{\alpha'}, \qquad \omega_{\alpha'\alpha} = (E_{\alpha'} - E_\alpha)/\hbar, \tag{4.42}$$

which is called the system of the close coupled equations in the interaction representation.

A solution of (4.42) satisfying the initial conditions

$$\lim_{t\to-\infty} a_\alpha^{(i)}(\rho, V; t) = \delta_{i\alpha}\,,$$

directly yields the result for the probability W_{fi} of the $i \to f$ transition and for the corresponding cross section after integration over the impact parameters

$$W_{fi}(\rho, V) = \lim_{t \to \infty} \left| a_f^{(i)}(\rho, V; t) \right|^2, \qquad \sigma_{fi}(V) = 2\pi \int_0^\infty W_{fi}(\rho, V)\rho d\rho. \qquad (4.43)$$

This formula can be rewritten in terms of the S-matrix of scattering in the impact-parameter representation. The solution set of differential equations (4.42) forms its matrix elements

$$S_{fi}(\rho, V) = a_f^{(i)}(\rho, V; t = \infty), \qquad \sigma_{fi}(V) = 2\pi \int_0^\infty |S_{fi}(\rho, V)|^2 \rho d\rho. \qquad (4.44)$$

The S-matrix is unitary $SS^\dagger = I$ (where I is the unit matrix) so that its elements satisfies the relation $\sum_f |S_{fi}(\rho, V)|^2 = 1$. Hence, the total cross section σ_i^{tot} of elastic σ_i^{el} and inelastic $\sigma_i^{in} = \sum_{f \neq i} \sigma_{fi}$ scattering

$$\sigma_i^{tot} = \sigma_i^{el} + \sigma_i^{in} \qquad (4.45)$$

can be written as

$$\sigma_i^{tot}(V) = 4\pi \int_0^\infty \left(1 - \mathbf{Re}\left\{S_{ii}(\rho, V)\right\}\right) \rho d\rho,$$

$$\sigma_i^{el}(V) = 2\pi \int_0^\infty |1 - S_{ii}(\rho, V)|^2 \rho d\rho, \qquad (4.46)$$

$$\sigma_i^{in}(V) = 2\pi \int_0^\infty \left(1 - |S_{ii}(\rho, V)|^2\right) \rho d\rho.$$

Note also that the S-matrix of scattering is expressed through the T-matrix, introduced previously in Sect. 4.1.2, by the following relation $S = I + T$.

It is important to stress that the system (4.42) [or (4.39)] consists of an infinite number of differential equations. Usually, they are cut off and are considered in some finite dimension space of states. The choice of the Rydberg levels with different magnitudes of the principal quantum number involves in the description of collision dynamics depends on the specific problem, accuracy, on the range of n, and transition frequencies. However, an important feature of any process involving the Rydberg atom or ion is a great number of closely spaced quasi-degenerate sublevels.

4.3.2 Normalized Perturbation Theory

In the range of weak coupling, when the transition probability is sufficiently small $W_{fi} \ll 1$, it can be calculated using the basic formula of first-order perturbation theory. This formula follows directly from the system of close coupling equations (4.42) by substitution of the relation $a_\alpha^{(i)}(t = -\infty) = \delta_{i\alpha}$ into its right-hand side

$$W_{fi}(\rho, V) = \left| a_f^{(i)}(\rho, V; t = +\infty) \right|^2,$$

(4.47)

$$a_f^{(i)}(\rho, V; t = \infty) = -\frac{i}{\hbar} \int_{-\infty}^{\infty} V_{fi}[\mathbf{R}(t)] \exp(i\omega_{fi}t) \, dt,$$

where $a_f^{(i)}(\rho, V; t = \infty)$ is the transition amplitude in the impact-parameter representation. These results often identified as the amplitude and probability are also called Born's formulas similar to the corresponding formulas in the preceding section.

If the interaction potential is not small, the transition probability calculated according to (4.47) may become greater than unity and, hence, a standard version of perturbation theory breaks down. In order to restore the unitarity property for the transition probability and to reasonably describe the behavior of the cross section in the range of close coupling states *Seaton* [4.3] suggested a normalized version of perturbation theory. According to [4.3] one possible way for normalization of the transition probability is to use the following simple relation for its actual value:

$$W_{fi}(\rho, V) = \frac{W_{fi}^B(\rho, V)}{1 + W_{fi}^B(\rho, V)} \rightarrow \begin{cases} 1, & W_{fi}^B \gg 1 \\ W_{fi}^B(\rho, V), & W_{fi}^B \ll 1 \end{cases}$$

(4.48)

in terms of the Born's W_{fi}^B defined by (4.47). There is also another simple way for normalization of the transition probability in the range of close coupling which turns out to be effective in practical calculations of processes involving Rydberg atoms. It consists of separation of the whole range of impact parameters into two regions $(0 \leq \rho \leq \rho_0)$ and $(\rho_0 < \rho)$ with qualitatively different behavior of the transition probability

$$W_{fi}(\rho, V) = \begin{cases} W_{fi}^B(\rho, V) & \rho \geq \rho_0, \\ c & \rho < \rho_0. \end{cases}$$

(4.49)

At large impact parameters $\rho > \rho_0$ the coupling between Rydberg states is weak, and the probability can be calculated using formula (4.47) of the Born approximation, i.e., $W_{fi} = W_{fi}^B$. However, at small $\rho < \rho_0$, due to strong coupling between Rydberg states, the first-order perturbation theory leads to overestimated values for the transition probability and it is normalized to a constant c of the order of unity. The magnitude of the impact parameter ρ_0, which separates the region of weak coupling from that of close coupling is to be found from the relation $W_{fi}^B(\rho_0) = c$. As a result, the cross section of the $i \rightarrow f$ transition is given by

$$\sigma_{fi}(V) = c\pi\rho_0^2 + 2\pi \int_{\rho_0}^{\infty} W_{fi}^B(\rho, V)\rho d\rho.$$

(4.50)

The choice of the normalizing constant c contains some ambiguities and is determined by the specific problem.

On the whole the recipes (4.48, 49) for normalization of transition probabilities can be rigorously justified if the contribution of the close coupling region (where $W_{fi}^B > 1$) to the cross section is not too large. Nevertheless they provide quite a reasonable explanation of the cross section behavior of the inelastic and quasi-elastic collisions involving Rydberg atoms in a wide range of the principal quantum number and the energy transfer. Another sophisticated method of generalizing the perturbation theory is based on the K-matrix of scattering. As was mentioned above, an exact set of solutions of the close coupled differential equations (4.39) [or (4.42)] forms the unitary S-matrix of scattering. The values of the transition amplitudes a_{fi} calculated within the framework of first order perturbation theory by formula (4.47) exceed unity in the range of strong interaction but they form the elements of some Hermitian matrix. Then, according to the unitarized perturbation theory in its general form [4.3], the normalized transition probability W_{fi} can be expressed in terms of some unitary S-matrix of scattering. The unitarity of this matrix will be achieved if we define it by means of the following relations:

$$W_{fi} = |S_{fi}|^2, \qquad S = \frac{1 + iK}{1 - iK}, \qquad K_{fi} = \frac{1}{2} a_{fi} \qquad (4.51)$$

with the amplitudes a_{fi} calculated in the Born approximation (4.47). Some times one can also use another expression for the normalized S-matrix

$$S = \exp(2iK). \qquad (4.52)$$

It is important to stress that in the case of weak interaction, the elements of the identified S-matrices (4.54, 52) are reduced to the usual Born's expressions (4.47), obtained in first-order perturbation theory. It is evident that the normalization recipes of (4.54, 52) based on the K-matrix are somewhat better than (4.48, 49). Actually, they incorporate not only the effects of normalization but also pretend to describe approximately the region of strong interaction without direct solution of the close coupling equations. However, this approach needs the evaluation of all matrix elements of the interaction potential and, hence, it turns out to be sufficiently complicated for calculation of the processes involving a great number of Rydberg states.

4.3.3 Connection with Momentum Transfer Representation

With the use of relation (4.31) the basic quantum formula (4.30) of the Born approximation for the transition cross section can be rewritten in terms of the transition amplitude $a_{fi}(Q, V)$ in the momentum transfer representation

$$\sigma_{fi}(q) = \frac{1}{(2\pi)^2} \int\limits_0^{2\pi} d\varphi \int\limits_{Q_{min}}^{Q_{max}} |a_{fi}(Q, V)|^2 Q dQ, \qquad (4.53)$$

$$a_{fi}(Q, V) = -\frac{i}{\hbar V} \int \int d\mathbf{R} d\mathbf{r} \exp(-i\mathbf{Q} \cdot \mathbf{R}) \psi_f^*(\mathbf{r}) V(\mathbf{r}, \mathbf{R}) \psi_i(\mathbf{r}). \qquad (4.54)$$

Here we discuss the relationship between this amplitude $a_{fi}(Q, V)$ and the first-order transition amplitude $a_{fi}(\rho, V; t = \infty)$ in the impact-parameter representation. We shall show below that the amplitudes of (4.42, 54) are connected to each other by means of the two-dimensional Fourier transform provided the energy transferred to the Rydberg atom is not too large (i.e., if $\Delta E_{fi} \ll \mathcal{E} = \hbar^2 q^2 / 2\mu$). Then, the wave numbers of the projectile-Rydberg atom relative motion before and after collision can be approximately presented as

$$q' = \left[q^2 - 2\mu(E_f - E_i)/\hbar^2\right]^{1/2} \approx q - \omega/V, \qquad \omega = (E_f - E_i)/\hbar.$$

As a result of decomposing the $\mathbf{Q}$ and $\mathbf{R}$ vectors into parallel and perpendicular components of the initial wave vector $\mathbf{q}$ for the projectile-Rydberg atom relative motion we obtain

$$\mathbf{Q} \cdot \mathbf{R} = Q_\parallel R_\parallel + \mathbf{Q}_\perp \cdot \boldsymbol{\rho}, \quad d\mathbf{R} = d\boldsymbol{\rho} dR_\parallel, \quad Q_\parallel = q - q' \cos\theta.$$

As apparent from the condition $\Delta E_{fi} \ll \hbar^2 q^2 / 2\mu$ and the energy conservation law the $Q_\parallel$ value is reduced to ω/V, i.e.

$$Q_\parallel \xrightarrow[\mu \to \infty]{} \omega/V.$$

Thus, the integration over the momentum transfer in (4.53) taken over a sphere transforms to a plane in distance ω/V that formally corresponds to the limiting case of a large reduced mass of colliding partners $\mu \to \infty$. Further, we take into account that $R_\parallel = Vt$ and the region of small scattering angles θ (tending to zero) becomes predominant in this case. Then, expression (4.54) can be rewritten as

$$a_{fi}(Q, V) = -\frac{i}{\hbar} \int d\boldsymbol{\rho} e^{-i\mathbf{Q}_\perp \cdot \boldsymbol{\rho}} \int_{-\infty}^{\infty} \frac{dR_\parallel}{V} \exp(i\omega_{fi} t) \int d\mathbf{r} \psi_f^*(\mathbf{r}) V(\mathbf{r}, \mathbf{R}) \psi_i(\mathbf{r}).$$

$$(4.55)$$

Finally, using formulas (4.47, 55) we derive the required relationships for the direct and inverse transforms of the transition amplitudes

$$a_{fi}(Q, V) = \int d\boldsymbol{\rho} e^{-i\mathbf{Q}_\perp \cdot \boldsymbol{\rho}} a_{fi}(\boldsymbol{\rho}, V),$$

$$a_{fi}(\boldsymbol{\rho}, V) = \frac{1}{2\pi} \int d\mathbf{Q}_\perp e^{i\mathbf{Q}_\perp \cdot \boldsymbol{\rho}} a_{fi}(Q, V). \qquad (4.56)$$

According to the Parseval equality these formulas lead to the following expression for the transition cross section:

$$\sigma_{fi}(q) = \left(\frac{1}{2\pi}\right)^2 \int d\mathbf{Q}_\perp |a_{fi}(Q, V)|^2 = \int d\boldsymbol{\rho} |a_{fi}(\boldsymbol{\rho}, V)|^2. \qquad (4.57)$$

It should be emphasized that relations (4.56, 57) can actually be proved for any order of perturbation theory [4.4]. Therefore, they are also valid for exact amplitudes and cross sections if the mass of the incident particle is large enough.

The relation (4.57) opens a possibility to use the methods developed for calculation of the transition probability in the impact parameter representation, in order to evaluate the cross sections in the momentum transfer representation. In particular, formula (4.48) for the normalized transition probability in the impact parameter representation can be modified for the momentum transfer representation [4.5]. It consists of replacing the Born form factor by the "normalized" one in accordance with the following relationship:

$$F_N(Q) = \frac{F_B(Q)}{\left[1 + 4(v_0/V)^2 |F_B(Q)|^2\right]^{1/2}}. \qquad (4.58)$$

4.4 Semiclassical Approach in Action Variables

The classical description of motion is based on the concept of trajectory. A similar basic role in quantum mechanics is the concept of the eigenstate of a system. It is well known, that the eigenstate can not be reduced to any classical trajectory in the limiting case of $\hbar \to 0$. There are two alternative approaches to establish the correspondence between classical and quantum values. The first one is based on the consideration of the linear combinations of the eigenstates (such as the wave packets), which correspond to the classical trajectory as $\hbar \to 0$. This approach was widely used in optics and atomic physics by Glauber (see, for example, [4.2]). Another approach deals with the set of the classical trajectories described by the action function. The second approach allows us to use the classical mechanics technique for calculations of quantum values, and we shall follow it in this section. We shall use the impact-parameter representation and, therefore, the motion of the incident particle is assumed to be pure classical.

4.4.1 Classical Perturbation Theory

The classical perturbation theory technique for the atomic problem had been developed in the twenties before the final formulation of nonrelativistic quantum mechanics had been completed. It is based on the action-angle variables approach, which was used for the description of the unperturbed Rydberg atom in Chap. 2. Here this approach will be aimed for a consistent formulation of the collision problem. Let the motion of the incident particle be assumed as given. Then the projectile influence on the Rydberg atom (ion) may be described with the help of the additional interaction term $\mathcal{V}(\mathbf{u}, \mathbf{I}, t)$ to the atomic Hamiltonian, where $\mathbf{I} = (I_1, I_2, I_3)$ and $\mathbf{u} = (u_1, u_2, u_3)$ are the three-dimensional vectors of the highly excited electron action variables and the corresponding angle variables, introduced in Sect. 2.1.2. As a result, the corresponding Hamilton-Jacobi equation is given by

$$\frac{\partial S_{tot}}{\partial t} - H_0 \left(\frac{\partial S_{tot}}{\partial u_1}\right) - V\left(\mathbf{u}, \frac{\partial S_{tot}}{\partial \mathbf{u}}, t\right) = 0, \tag{4.59}$$

$$S^{(0)}(\mathbf{u}, t) = \mathbf{I} \cdot \mathbf{u} - E_n t, \quad \mathbf{I} \cdot \mathbf{u} = \sum_{i=1}^{3} I_i u_i$$

with the boundary condition $\lim_{t \to -\infty} S_{tot}(\mathbf{u}, t) = S^{(0)}(\mathbf{u}, t)$. Here H_0 and $S^{(0)}(\mathbf{u}, t)$ are the Hamilton function and the action function describing the unperturbed Rydberg atom with energy E_n and action variable $I_1 = n\hbar$. The classical frequency of the Rydberg electron motion $\omega_n = \partial H_0 / \partial I_1$ is equal to $2Z^2 Ry/\hbar n^3$ for the hydrogen-like atom or ion.

Further, we present the total action function of this system as the sum $S_{tot} = S^{(0)} + S$ of its unperturbed part $S^{(0)}$ and the action increment S induced by the projectile-atom interaction. If the collisional process results in a small enough action increment $S \ll S^{(0)}$, then (4.59) is reduced to a rather simple form

$$\frac{\partial S}{\partial t} - \omega_n \frac{\partial S}{\partial u_1} = V\left(\mathbf{u}, \frac{\partial S^{(0)}}{\partial \mathbf{u}}, t\right) \tag{4.60}$$

with the accuracy of the first order terms and under the following boundary condition $S(u_i, t \to -\infty) = 0$. The solution of (4.60) was obtained in [4.6] by the Fourier transform method

$$S(\mathbf{u}, t) = \sum_{\boldsymbol{\kappa}} A_{\boldsymbol{\kappa}}(t) \exp[i(\boldsymbol{\kappa} \cdot \mathbf{u} + \kappa_1 \omega_n t)],$$

$$A_{\boldsymbol{\kappa}} = \int_{-\infty}^{t} V_{\boldsymbol{\kappa}}(t') \exp(-i\kappa_1 \omega_n t') dt', \tag{4.61}$$

$$V_{\boldsymbol{\kappa}}(t) = (2\pi)^{-3} \int_{0}^{2\pi} d\mathbf{u} V\left(u_i, t, \frac{\partial S^{(0)}}{\partial u_i}\right) \exp(-i\boldsymbol{\kappa} \cdot \mathbf{u}).$$

Here, for brevity we denote the set of the indices $(\kappa_1, \kappa_2, \kappa_3)$ by the three dimensional vector, which is determined by the difference of the quantum numbers $\boldsymbol{\kappa} = (n' - n, l' - l, m' - m)$. The first index is indentified since the Hamilton function H_0 of electron in the Coulomb field is dependent on the $I_1 = n\hbar$ variable alone. The classical perturbation theory should reflect this fact in explicit form. *Richards* [4.7] suggested another form for the solution of the equation (4.60)

$$S(\mathbf{u}, t) = \int_{-\infty}^{t} V(u_1 + \omega_n(t - t'), u_2, u_3; t') dt'. \tag{4.62}$$

Since any function $f(u_1 + \omega t, u_2, u_3)$ is the solution of the homogeneous equation

$$\frac{\partial f}{\partial t} - \omega \frac{\partial f}{\partial u_1} = 0,$$

it is easy to check that formula (4.62) is the solution of (4.60).

If we neglect the term $\omega_n(t - t')$, formula (4.62) reduces to the well-known "sudden" approximation

$$S(\mathbf{u}, t) = \int_{-\infty}^{t} \mathcal{V}(\mathbf{u}; t') dt'. \tag{4.63}$$

Formulas (4.61, 62) give the solution of the scattering problem in terms of the action function and the corresponding action-angle variables. It is convenient to use the quantum results for the cross section calculations and for the evaluation of integrals in (4.61, 62). It should be noted that in this section we have used the so-called unperturbed action variables. In the next order of the perturbation theory these variables must be overspecified taking into account the corrections (4.61, 62). However, the calculation difficulties already increase dramatically in the second order and high-order corrections to the results presented above have not yet been found.

4.4.2 Relation Between Classical and Quantum Values

We proceed from the semiclassical approximation for the atomic wave function $\psi_\alpha = \exp(iS_\alpha/\hbar)$ according to which the probability of the $\alpha \to \alpha'$ transition is given by

$$W_{\alpha'\alpha} = \lim_{t \to 0} \left| \left\langle \psi_{\alpha'}^{(0)}(t) \mid \psi_\alpha(t) \right\rangle \right|^2 = \lim_{t \to 0} \left| \left\langle \exp\left(\frac{i}{\hbar} S_{\alpha'}^{(0)}\right) \right| \exp\left[\frac{i}{\hbar}\left(S_\alpha^{(0)} + S\right)\right] \right\rangle \right|^2 \tag{4.64}$$

Here the superscript "0" corresponds to the unperturbed atom, i.e. $\psi_\alpha(t) \to \psi_\alpha^{(0)}$ at $t \to -\infty$. In the semiclassical approximation the $\left(S_\alpha^{(0)} + S\right)$ quantity is the action function defined by the Hamilton-Jacobi equation and by the initial conditions (4.60). The use of (4.59) for the unperturbed action function $S^{(0)}$, gives the following expression for the transition amplitude:

$$a_{\alpha'\alpha} = (1/2\pi)^3 \int\int\int_0^{2\pi} d\mathbf{u} \exp\left\{-i\left[\boldsymbol{\kappa} \cdot \mathbf{u} - \frac{1}{\hbar} S(\alpha, \mathbf{u})\right]\right\}. \tag{4.65}$$

As was mentioned above, the $\boldsymbol{\kappa} = \alpha' - \alpha$ vector in (4.65) denotes the three-dimensional difference of the quantum numbers $\boldsymbol{\kappa} = (n' - n, l' - l, m' - m)$, while $\alpha = (n, l, m)$ is a set of the Rydberg atom quantum numbers;

and $S(\mathbf{u}) = \lim\limits_{t \to \infty} S(\mathbf{u}, t)$ is the total increment of the action function due to the collision. Formula (4.65) yields the probability of the transition between atomic eigenstates in terms of the action function increment which is the result of the solution of the classical problem

$$S(\alpha, \mathbf{u}) = \sum_{\kappa} A_{\kappa} \exp(-i\kappa \cdot \mathbf{u}), \tag{4.66}$$

where the classical amplitudes A_{κ} are given by formulas (4.61).

a) Strong Interaction. Now we consider expression (4.65) for strong interactions, when the action increment S is large in comparison with the Plank constant $\hbar$. First, we analyze the simplest one-dimensional case for the action variables corresponding to the principal quantum number n. We use here the designation κ and u instead of component $\kappa_1 = \Delta n$ and u_1, while other variables u_2 and u_3 will be considered as parameters. We also assume that the well-known validity conditions of the method of the steepest descents are satisfied for the $S/\hbar$ function. Under these conditions one can obtain from (4.65) the following relation:

$$\frac{1}{2\pi} \int\limits_{0}^{2\pi} du \exp\left[-i\left(\kappa u - S/\hbar\right)\right]$$

$$\approx \frac{1}{2\pi} \sum_{j} \exp\left\{i\left[-\kappa u_j + S(u_j)/\hbar + \pi/4\right]\right\} \left|\frac{1}{2\pi\hbar}\frac{\partial^2 S}{\partial u^2}\right|_{u=u_j}^{-1/2}. \tag{4.67}$$

The points of steady phase u_j are given by equation $\partial S/\partial u = \hbar\kappa$. The square of the absolute value (4.67) may be determined without interference terms, since $S/\hbar \gg 1$. (Note that in most cases of practical importance there is only one point of the steady phase). Therefore, the probability $W_{n'n}$ of the $n \to n'$ transition can be written as

$$W_{n'n} = 2\pi\hbar \sum_{j} \left|2\pi\frac{\partial\Lambda}{\partial u}\right|_{\Lambda=2\pi\hbar\kappa}^{-1}, \qquad \Lambda = 2\pi\frac{\partial S}{\partial u} = 2\pi\left[I(u) - n\hbar\right]. \tag{4.68}$$

This expression gives a quantity $W_{n'n}$ as a function of the collision parameters (e.g. the impact parameter ρ). Integration of (4.68) over $2\pi\rho d\rho$ leads to the following expression for the transition cross section:

$$\sigma_{n'n} = 2\pi \int\limits_{0}^{\infty} W_{n'n}(\rho)\rho d\rho = 2\pi\hbar \int\limits_{0}^{\infty} \rho d\rho \left|\frac{\partial\Lambda}{\partial u}\right|_{\Lambda=2\pi\hbar\kappa}^{-1}. \tag{4.69}$$

Each given value of the impact parameter ρ in (4.69) corresponds to some definite value of u, for which the condition $\Lambda = 2\pi\hbar\kappa$ is fulfilled, i.e., the variables ρ and u are connected by the relation $I(u) - n\hbar = \hbar\kappa$. Therefore, we can replace the integration over the impact parameter ρ by integration over the u variable

$$\sigma_{n'n} = 2\pi \int_0^{2\pi} du \rho(u, \Lambda) \left| \frac{d\rho(\Delta u)}{d\Lambda} \right|_{\Lambda = 2\pi\hbar\kappa} \tag{4.70}$$

Let us denote $\Delta E = E - E_n$. Then the use of expression (4.70) and formulas of classical mechanics from Sect. 2.2, yields

$$\left. \left| \frac{d\rho}{d\Lambda} \right| \right|_{\Lambda = 2\pi\hbar\kappa} = \frac{1}{2\pi} \left| \frac{d\rho}{d(\Delta E)} \frac{d(\Delta E)}{dI} \right|_{\Delta E = |E_{n'} - E_n|}$$

$$= \frac{\omega_{n'}}{2\pi} \left| \frac{d\rho}{d(\Delta E)} \right|_{\Delta E = |E_{n'} - E_n|}. \tag{4.71}$$

Further, we can introduce the quantity $d\sigma/dE$ by the relation

$$\int_0^{2\pi} \frac{du}{2\pi} 2\pi\rho \left| \frac{d\rho}{d(\Delta E)} \right| = \frac{d\sigma}{d(\Delta E)}, \tag{4.72}$$

which can be considered as the classical differential cross section for the energy transfer averaged over the "initial phase". Then, with the help of (4.70), the integral cross section of the inelastic $n \to n'$ transition can be presented as

$$\sigma_{n'n} = \hbar \left(\frac{d\sigma}{d(\Delta I)} \right)_{\Delta I = \hbar(\Delta n)} = \hbar\omega_{n'} \left(\frac{d\sigma}{d(\Delta E)} \right)_{\Delta E = E_{n'} - E_n}. \tag{4.73}$$

Here $\omega_{n'}$ is the frequency of the classical electron motion with energy $E_{n'}$ which is determined by the energy difference between two neighboring Rydberg levels $\hbar\omega_{n'} \approx E_{n'} - E_{n'-1} \approx E_{n'+1} - E_{n'}$.

Formula (4.73) establishes the correspondence between the classical cross section for the energy transfer and the usual quantum expression for the cross section of the $n \to n'$ transition. It directly follows from the aforementioned analysis and shows that the cross section of inelastic transition $\sigma_{n'n}$ may be expressed in terms of the classical cross section for the energy transfer $d\sigma/d(\Delta E)$, if the resulting action increment in the collision process is large as compared to the Plank constant, i.e., $S \gg \hbar$.

It may be worthwhile to point out that in the three-dimensional case, the general relationship between quantum and classical cross sections can be written as

$$\sigma_{nlm}^{n'l'm'} = \hbar^3 \left(\frac{d^3\sigma}{d(\Delta I_1) d(\Delta I_2) d(\Delta I_3)} \right)_{\Delta I = \hbar\kappa}. \tag{4.74}$$

One can see that quantum cross section is determined by the classical differential cross section at the point

$$\{\Delta I_1, \Delta I_2, \Delta I_3\} = \{\hbar(n' - n), \hbar(l' - l), \hbar(m' - m)\}.$$

Integration of this expression over the possible action transfer ΔI_2, and ΔI_3 is equivalent to the summation of the quantum cross section over the magnetic and orbital quantum numbers and we arrive at the formula (4.73).

b) Weak Interaction. Now we consider the opposite case, when the action increment is small enough $S \ll \hbar$. Then, expression (4.65) may be significantly simplified. We expand $\exp(-iS/\hbar)$ in a series over the $S/\hbar$ value and keep only the first nonvanishing term

$$a_{\alpha'\alpha} = (-i/\hbar)A_\kappa = (-i/\hbar) \int_{-\infty}^{\infty} V_\kappa(t) \exp(i\Delta n \omega_n t) dt. \tag{4.75}$$

In the limit of $\hbar \to 0$ the matrix element $V_{\alpha+\kappa,\alpha}$ reduces to the corresponding Fourier component V_κ, so that formula (4.75) is the same as the first order result of quantum perturbation theory. Therefore, the A_κ coefficient may be determined from the quantum calculations of the transition amplitudes in the framework of perturbation theory.

The probability $W_{n'n}$ of the $n \to n'$ transition summed over the angular quantum numbers of the final states and averaged over the initial states can be obtained from (4.65). Using the relation

$$\sum_k \exp\left[ik\,(u - u')\right] = 2\pi\delta\,(u - u'), \tag{4.76}$$

we obtain

$$n^2 W_{n'n} = \sum_{lm} \sum_{l'm'} |\, a_{\alpha'\alpha}\,|^2$$

$$= \sum_{lm} \frac{1}{(2\pi)^4} \int_0^{2\pi} du_2 \int_0^{2\pi} du_3 \left| \int_0^{2\pi} du_1 \exp\left[-i(\Delta n)u_1 - \frac{i}{\hbar}S(\alpha, \mathbf{u})\right] \right|. \tag{4.77}$$

It is important to note that this result will reduced to the one-dimensional expression

$$W_{n'n} = \left| \frac{1}{2\pi} \int_0^{2\pi} du \exp\left[-i\,(\Delta n)\,u - \frac{i}{\hbar}S(n, u)\right] \right|^2 \tag{4.78}$$

if the increment of the action function S is independent of the u_2 and u_3 variables and, consequently, is independent of the angular quantum numbers l and m. Note also that in this formula we have omitted the index "1" from the u_1 components of the $\mathbf{u}$ vector denoting them simply as u.

c) Averaged Value of Energy Transfer Square. A number of general relations for transition probabilities and cross sections may be obtained from the semiclassical expression (4.65) using simple the formula (4.76). In particular, as is apparent from this property, formula (4.65) satisfies the following normalizing relation:

$$\sum_{\alpha'} W_{\alpha'\alpha} = \frac{1}{(2\pi)^3} \int\!\!\!\int\!\!\!\int_0^{2\pi} d\mathbf{u} \int\!\!\!\int\!\!\!\int_0^{2\pi} d\mathbf{u}' \delta(\mathbf{u} - \mathbf{u}') \exp\left[\frac{i}{\hbar}(S(\mathbf{u}) - S(\mathbf{u}'))\right] = 1.$$

$$(4.79)$$

Since the period of the action function $S(\mathbf{u})$ is equal to 2π, then the integration by parts yields the relation

$$\int_0^{2\pi} \kappa^s \exp\left[i\left(\kappa u + S/\hbar\right)\right] du = i^s \int_0^{2\pi} du \exp\left(i\kappa u\right) \frac{d\left[\exp\left(iS/\hbar\right)\right]}{du}$$

$$= (-\hbar)^{-s} \int_0^{2\pi} \left[\left(\frac{dS}{du}\right)^s + \frac{d}{du}\mathfrak{P}\left(\frac{dS}{du}\right)\right] \exp\left[i\left(\kappa u + S/\hbar\right)\right] du,$$

where $\mathfrak{P}(dS/du)$ is the polynomial over dS/du. We multiply this equation by $\mathfrak{a}_{\alpha\alpha'}$ and sum over $k = \Delta n$. As a result, we have

$$\sum_k k^s W_{n,n+k} = (-\hbar)^{-s}(2\pi)^{-3} \int\!\!\!\int\!\!\!\int_0^{2\pi} d\mathbf{u} \left[\left(\frac{dS}{du_1}\right)^s\right]. \qquad (4.80)$$

Put, for example, $s = 2$, then from (4.80) we obtain the relation for the average square of the energy transfer

$$\sum_{\Delta n} (\Delta E_{n,n+\Delta n})^2 W_{n,n+\Delta n} = \left(\frac{2Z^2 Ry}{n^3}\right)^2 \sum_k (\Delta n)^2 W_{n,n+\Delta n}$$

$$= \left(\frac{\Delta E_{n,n+1}}{\hbar}\right)^2 \int_0^{2\pi} \frac{du}{2\pi} \left[\left(\frac{dS}{du_1}\right)^2\right]. \qquad (4.81)$$

This formula clearly demonstrates that the quantum average square of the energy transfer is the same as a purely classical average one. A similar statement may be derived from (4.80) for any classical value (e.g., for the average square of the angular momentum transfer, for the projection of the angular momentum, etc.).

With the help of the Fourier series expansion of the action increment (4.66), we obtain from (4.81)

$$\sum_k (\Delta E_{n,n+k})^2 W_{n,n+k} = \sum_k (A_k/\hbar)^2 (\Delta E_{n,n+k})^2. \qquad (4.82)$$

Since $(A_k/\hbar)^2$ is the Born transition probability, a surprising statement follows from formula (4.82), that is "the Born average of the square of the energy transfer is equal to the average of the square of the energy transfer in the framework of the classical perturbation theory in spite of the significant difference in their region of applicability". In this approximation the average energy transfer is equal to zero. Similar statements about angular momentum or projection of the angular momentum are also valid. However this is not so for powers greater than 2.

4.4.3 Model of Equidistant Levels and Correspondence Principle for S-Matrix

The model of equidistant levels for the description of transitions between highly excited states induced by electron collisions was developed in [4.8] for the case of the "one-dimensional" Rydberg atom characterized only by the principal quantum number n. It is based on the analogy between a highly excited atom and the highly excited harmonic oscillator. In the framework of the usual close coupling method, the transition amplitudes are described by equations

$$i\hbar \, \dot{a}_n = E_n a_n + \sum_{n'} V_{nn'}(\rho, t) a_{n'} , \quad a_n \xrightarrow[t \to -\infty]{} \delta_{nn_0} , \tag{4.83}$$

where $V_{n'n}(\rho, t)$ is the matrix element of the Rydberg atom interaction with an external particle in the impact parameter representation. Shifting the point of account of quantum numbers to n_0 and introducing $a_k = a_{n-n_0} \exp\left(-iE_{n_0}t/\hbar\right)$ we rewrite (4.83) in the form

$$i\hbar \, \dot{a}_k = (E_{n_0+k} - E_{n_0}) a_k + \sum_{k'} V_{n_0+k,n_0+k'}(\rho, t) a_{k'} . \tag{4.84}$$

In the case of $n \gg 1$, the spectrum of the highly excited atom energy levels is almost equidistant, i.e., $E_{n+k} \approx E_n + k(\hbar\omega_n)$, where $\hbar\omega_n = 2Z^2 Ry/n^3$. The matrix element of interaction is dependent essentially on the quantum number difference k, while the n-dependence is assumed to be weak enough, so that one can approximately put $V_{n,n+k} \approx V_{n,n-k} \approx V_k$. Under this condition the system of the close coupled differential equations (4.84) can be rewritten as

$$i\hbar \frac{da_k}{dt} = k\hbar\omega_{n_0} a_k + \sum_{k'} V_{k'}(\rho, t) a_{k+k'} . \tag{4.85}$$

The solution of the infinite system (4.85) is given in [4.8] with the help of the generating $G(t, \varphi)$ function

$$G(t, \varphi) = \sum_{\kappa} a_\kappa(t) e^{i\kappa\varphi}. \tag{4.86}$$

As is evident from (4.85, 86) this function satisfies the following equation:

$$i\hbar \frac{\partial}{\partial t} G(t, \varphi) - i\hbar \frac{\partial}{\partial \varphi} G(t, \varphi) = V(t, \varphi) G(t, \varphi), \tag{4.87}$$

$$V(t, \varphi) = \sum_k V_k(t) e^{-ik\varphi}.$$

Use of the substitution

$$G(t, \varphi) = \exp\left[-\frac{i}{\hbar} S(t, \varphi)\right] \tag{4.88}$$

reduces (4.87) to the basic equation (4.60) of the semiclassical perturbation theory [4.6], [4.9] for the action increment, whereas the phase variable u corresponds to the quantity φ.

The amplitude a_k in (4.84) can be considered as the S$-$matrix element S_{n_0+k,n_0}. According to the *correspondence principle* it corresponds to the k-Fourier component of the scattering operator over the action variables S_k, and the matrix element $V_{n,n+k}$ corresponds to the k-Fourier component of the interaction potential V_k. This means that in the framework of the *correspondence principle*, system (4.84) is reduced to

$$i\hbar \frac{d}{dt} S_k = k\hbar\omega_{n_0} S_k + \sum_{k'} V_{k'}(\rho, t) S_{k+k'} , \tag{4.89}$$

which is different from (4.85) only by designations.

Equivalence of the all three approaches (semiclassical perturbation theory, system of the equations based on the *correspondence principle*, and the model of equidistant levels) for the problem of collisions between the Rydberg atom and charged particles was established in [4.7]. It is well known that the semiclassical description of a highly excited atom provides the equidistant spectrum of energy levels and suggests the matrix element dependence on the quantum number differences alone. Equivalence of (4.87) and those of the semiclassical perturbation theory shows that the semiclassical description is applicable, if the energy spectrum of the system is approximately equidistant.

It may be worthwhile to point out that the theoretical approach based on the model of equidistant levels incorporates the results for a highly excited harmonic oscillator obtained early by *Feynmann* [4.10] and others [4.11, 12].

4.5 Impulse Approximation Approach

The impulse approximation was originally proposed by *Fermi* [4.13] and Chew and coworkers [4.14] in nuclear physics. In particular, it was widely used for describing high-energy neutron scattering by complex nuclei (see [4.1] for more details). The quantum impulse approximation and its semiclassical version (the so-called binary-encounter approach in the momentum representation) found wide employment in the collision theory involving Rydberg atoms [4.15], [1.3]. It is also well known that the pure classical impulse-approximation approach has been used to describe energy transfer processes in collisions of a Rydberg atom with neutral and charged particles. According to [4.1] there are two classes of interactions of the combine system for which the impulse approximation would be valid: they are the weak binding and quasi-classical binding. The first one corresponds to the case when the kinetic energy $\mathcal{E} = \mu V^2/2$ of the incident particle B relative to the Rydberg atom appreciably exceeds the binding energy $|E_n| = Ry/n_*^2$ of the outer electron e with its parent core A^+. This is the standard condition for the applicability of the impulse approximation. The impulse approximation approach may also be used for the cross section calculations if the weak binding condition is broken down but the collision time of the projectile with the Rydberg electron is small compared to the characteristic period of the electron orbital motion. This second case corresponds to the so-called quasi-classical binding condition, when the interaction potential between the Rydberg electron and its ionic core assumes a very small variation ΔU during the scattering of the incident particle on the highly excited electron. Therefore, the ionic core A^+ does not affect the process of the encounter and its role reduces only to the generation of the momentum space wave functions in the initial and final Rydberg states.

4.5.1 Quantum Impulse Approximation

The general quantum treatment of the impulse approximation for the description of collisional transitions may be given in terms of the scattering $T(E)$-operator. Let us consider the Rydberg atom A as a weakly bound two-particle system consisting of the valence electron e and its ionic core A^+ with a large orbital radius. In accordance with the basic idea of the impulse approximation our goal here is to represent the scattering amplitude from a complex system (Rydberg atom) as a superposition of scattering amplitudes from "free" independent scatterers (highly excited electron and ion core) which have the same momentum distribution as the initially bound particles. This means that we may neglect the Coulomb type electron-core interaction U during the scattering of the incident particle on both the highly excited electron e and the ion core A^+. At the same time the bound character of the Rydberg electron motion around its ion core appears in this approach due to the use of exact atomic wave functions in the initial $G_i(\kappa)$ and final $G_f(\kappa')$ states.

To derive the basic formulas of the quantum impulse approximation it is convenient to rewrite the total Hamiltonian of system A+B in the form

$$H = K_{tot} + \mathcal{V} + U, \qquad \mathcal{V} = \mathcal{V}_{eB} + \mathcal{V}_{A^+B} , \qquad (4.90)$$

$$K_{tot} = \frac{\hbar^2 q^2}{2\mu} + \frac{\hbar^2 \kappa^2}{2\mu_{eA^+}} , \qquad \mu \equiv \mathfrak{M}_B = \frac{M_B(m + M_{A^+})}{M_{A^+} + m + M_B} , \qquad (4.91)$$

which allows us to explicitly distinguish the electron-ion core interaction potential U, the total kinetic energy operator K_{tot} of three particles $[(e, A^+), B]$, and the two-body operators $\mathcal{V}_{eB}$ and $\mathcal{V}_{A^+B}$ for the electron-projectile and core-projectile interaction potentials. Here the first term corresponds to the kinetic energy of the projectile (B) motion relative to the center of mass of the (e, A^+)-pair, while the second term is the kinetic energy operator of electron motion relative to the ion core. Further, we present the Green's resolvent $G^+(E) = (E - H + i0)^{-1}$ as an expansion in a power series of U and retain only its first term

$$G^+(E) \approx (E - K_{tot} - \mathcal{V} + i0)^{-1}, \tag{4.92}$$

which involves the two-body operators $\mathcal{V}_{eB}$ and $\mathcal{V}_{A+B}$ for the electron-projectile and core-projectile interaction potentials, and the term K_{tot}.

The next step consists of separation of the variables describing the projectile B motion relative to the highly excited electron and its motion relative to the ion core. Starting from (4.92) and the basic expression for the scattering T-operator (4.15), one can present it in the form

$$T \approx T^{imp} = T_{eB} + T_{A+B}, \tag{4.93}$$

following the well-known work by *Chew* and *Goldberger* [4.14]. Thus, in the impulse approximation the scattering operator contains only the scattering T_{eB} and T_{A+B} operators

$$T_{eB}(E) = \mathcal{V}_{eB} + \mathcal{V}_{eB} \left(\frac{1}{E - K_{tot} - \mathcal{V}_{eB} + i0} \right) \mathcal{V}_{eB}, \tag{4.94}$$

$$T_{A+B}(E) = \mathcal{V}_{A+B} + \mathcal{V}_{A+B} \left(\frac{1}{E - K_{tot} - \mathcal{V}_{A+B} + i0} \right) \mathcal{V}_{A+B} \tag{4.95}$$

involving the two-body (electron-perturber) $\mathcal{V}_{eB}$ and (core-perturber) $\mathcal{V}_{A+B}$ interactions, respectively. The removed terms in (4.93) describe the non-impulsive corrections to the impulse approximation and the effects of multiple scattering. As is evident from (4.93), in the impulse approximation the contributions of the perturber-electron $(B-e)$ and perturber-core $(B-A^+)$ scattering to the transition matrix (4.19) and cross section can be calculated independently:

$$T_{fi}^{imp}(\mathbf{q}', \mathbf{q}) = \langle \mathbf{q}', f | T_{eB}(E) | \mathbf{q}, i \rangle + \langle \mathbf{q}', f | T_{A+B}(E) | \mathbf{q}, i \rangle, \tag{4.96}$$

$$\sigma_{fi}^{imp} = \sigma_{fi}(e - B) + \sigma_{fi}(A^+ - B). \tag{4.97}$$

It is worthwhile to remember that these matrix elements are taken over the plane waves $|\mathbf{q}\rangle = |\exp(i\mathbf{q} \cdot \mathbf{R})\rangle$ describing the projectile motion relative to the center of mass of the Rydberg atom, i.e., we work in the total center of mass system of the three particles $[B-(e, A^+)]$.

a) Electron-Projectile Scattering. Let us first consider the major contribution to the transition amplitude induced by the electron-perturber scattering mechanism. The core-perturber contribution will be analyzed in Chap. 7. The fundamental step in deriving the basic formula of the impulse approximation is to use the expansion of the Rydberg atom wave functions in the initial $|i\rangle$ and final $|f\rangle$ states over the plane waves $\exp(i\boldsymbol{\kappa} \cdot \mathbf{r})$ describing the relative electron-core motion with the wave vector $\boldsymbol{\kappa}$

$$|\alpha\rangle = (2\pi)^{-3/2} \int G_\alpha(\kappa) |\kappa\rangle \, d\kappa , \qquad |\kappa\rangle = \exp(i\kappa \cdot \mathbf{r}) , \qquad |\alpha\rangle = |\psi_\alpha(\mathbf{r})\rangle \quad (4.98)$$

Here $\kappa = \mu_{eA+} \mathbf{v}_{eA+}/\hbar$, and $\mathbf{v}_{eA+}$ is the relative velocity, $\mu_{eA+} = mM_{A+}/(m+M_{A+})$ being the reduced mass of the (e, A^+)-pair, and the expansion coefficient $G_\alpha(\kappa)$ corresponds to the momentum space wave function introduced by the relation (2.83). Then, the transition matrix element takes the form

$$\langle \mathbf{q}', f | T_{eB}(E) | \mathbf{q}, i \rangle = (2\pi)^{-3} \int\int G_f^*(\kappa') G_i(\kappa) \langle \mathbf{q}' | \langle \kappa' | T_{eB}(E) | \kappa \rangle | \mathbf{q} \rangle \, d\kappa d\kappa'.$$

$$(4.99)$$

Further we note that the momentum $\hbar \mathbf{q}_{A+}$ of the ion core A^+ and its kinetic energy $(\hbar^2 \mathbf{q}_{A+}^2/2M_{A+})$ in the (total) center of mass system do not changed during the electron-perturber scattering in accordance with the basic assumption of the impulse approximation (i.e., A^+ is only a spectator in the e-B encounter). Moreover, the total kinetic energy operator of system (4.91) can be rewritten in the equivalent form

$$K_{tot} = \frac{\hbar^2 \mathbf{q}_{A+}^2}{2\mathfrak{M}_{A+}} + \frac{\hbar^2 \mathbf{k}^2}{2\mu_{eB}} , \qquad \mathfrak{M}_{A+} = \frac{M_{A+}(m+M_B)}{M_{A+}+m+M_B}.$$

Here $\hbar \mathbf{k} = \mu_{eB} \mathbf{v}_{eB}$ is the electron momentum relative to the projectile B, and $\mu_{eB} = mM_B/(m+M_B)$ being their reduced mass. Thus, the transition matrix element of the T_{eB}-operator (4.94), taken over the plane waves, involves the delta function

$$\langle \mathbf{q}' | \langle \kappa' | T_{eB}(E) | \kappa \rangle | \mathbf{q} \rangle = (2\pi)^3 \delta(\mathbf{q}'_{A+} - \mathbf{q}_{A+}) \langle \mathbf{k}' | t_{eB}(\epsilon = \hbar^2 \mathbf{k}^2/2\mu_{eB}) | \mathbf{k} \rangle .$$

$$(4.100)$$

Expression (4.100) corresponds to the reduction of the three-body matrix elements in terms of the two-body scattering operator for electron-perturber scattering.

The relation between the wave vectors $(\mathbf{q}, \kappa)$ and $(\mathbf{q}_{A+}, \mathbf{k})$

$$\mathbf{k} = (\mu_{eB}/m)\kappa - (\mu_{eB}/\mu)\mathbf{q}, \qquad \mathbf{v}_{eB} = (\mu_{eA+}/m)\mathbf{v}_{eA+} - \mathbf{V}, \qquad (4.101)$$

$$\mathbf{q}_{A+} = -\kappa - \frac{M_{A+}}{m+M_{A+}}\mathbf{q}, \qquad (4.102)$$

allows us to rewrite the delta-function as

$$\delta(\mathbf{q}'_{A+} - \mathbf{q}_{A+}) = \delta\left(\kappa - \kappa' + \frac{M_{A+}}{m+M_{A+}}\mathbf{Q}\right), \qquad (4.103)$$

where $\hbar\mathbf{Q} = \hbar(\mathbf{q} - \mathbf{q}')$ is the momentum transfer vector in the collision of the projectile with the Rydberg atom in the total center of mass of the three particles $\left(B - (e, A^+)\right)$, and $\mathbf{V} = \hbar\mathbf{q}/\mu$ is the relative velocity of the projectile with respect to the center of mass of the (e, A^+)-pair. Then, substituting relations (4.100, 103) into (4.99) and performing the integration over κ', we obtain the final formula of the impulse approximation for the contribution of the electron-perturber scattering to the transition matrix element between the Rydberg atomic states $i \to f$

$$
\begin{aligned}
T_{fi}^{\mathrm{eB}}(\mathbf{q}',\mathbf{q}) &= \langle G_f(\kappa')|\, t_{\mathrm{eB}}(\mathbf{k}',\mathbf{k};\epsilon)\,|G_i(\kappa)\rangle \\
&= \int G_f^*(\kappa') t_{\mathrm{eB}}(\mathbf{k}',\mathbf{k};\epsilon) G_i(\kappa) d\kappa.
\end{aligned}
\tag{4.104}
$$

The matrix element $t_{\mathrm{eB}}(\mathbf{k}',\mathbf{k};\epsilon)$ of the two-body operator for scattering of the free electron by a perturbing particle B in the final formula (4.104) has the form

$$
t_{\mathrm{eB}}(\mathbf{k}',\mathbf{k};\epsilon) = \left\langle \mathbf{k}' \left| \mathcal{V}_{\mathrm{eB}} + \mathcal{V}_{\mathrm{eB}} \left(\frac{1}{\epsilon - \mathrm{K}_{\mathrm{eB}} - \mathcal{V}_{\mathrm{eB}} + \mathrm{i}0} \right) \mathcal{V}_{\mathrm{eB}} \right| \mathbf{k} \right\rangle .
\tag{4.105}
$$

It is taken over the plane waves $(|\mathbf{k}\rangle = |\exp(\mathrm{i}\mathbf{k}\mathbf{r}_{\mathrm{eB}})\rangle$ describing the electron-perturber relative motion in the system of their center of mass. The relative momentum transfers in (4.104) are given by the relations

$$
\mathbf{k}' - \mathbf{k} = (\mu_{\mathrm{eB}}/\mu)\,[(M_{\mathrm{A}^+}/M_{\mathrm{A}}) + (m/\mu)]\,\mathbf{Q} ,
\tag{4.106}
$$

$$
\kappa' = \kappa + \mathbf{K}, \qquad \mathbf{K} = (M_{\mathrm{A}^+}/M_{\mathrm{A}})\mathbf{Q},
\tag{4.107}
$$

where $\hbar\mathbf{Q} = \hbar(\mathbf{q} - \mathbf{q}')$ is the momentum transfer vector for collision of projectile B with the Rydberg atom A^* ($M_{\mathrm{A}} = m + M_{\mathrm{A}^+}$ is its mass). Note also that in contrast to (4.94) the two-body scattering t_{eB}-operator (4.105) in the final formula of the impulse approximation contains only the kinetic energy operator $\mathrm{K}_{\mathrm{eB}} = \hbar^2\mathbf{k}^2/2\mu_{\mathrm{eB}}$ for the electron-perturber relative motion and the interaction potential $\mathcal{V}_{\mathrm{eB}}$ of the (e, B)-pair.

Expression (4.104) for the contribution of the electron-perturber scattering may be rewritten in the following equivalent form [4.15], [1.3]:

$$
\begin{aligned}
f_{fi}^{\mathrm{eB}}(\mathbf{q}',\mathbf{q}) &= \frac{\mu}{\mu_{\mathrm{eB}}}\,\langle G_f(\kappa')|\, f_{\mathrm{eB}}(\mathbf{k}',\mathbf{k})\,|G_i(\kappa)\rangle \\
&= \frac{\mu}{\mu_{\mathrm{eB}}}\int G_f^*(\kappa') f_{\mathrm{eB}}(\mathbf{k}',\mathbf{k}) G_i(\kappa) d\kappa.
\end{aligned}
\tag{4.108}
$$

Here $f_{fi}(\mathbf{q}',\mathbf{q})$ is the standard inelastic scattering amplitude (4.17) of the Rydberg atom A^* by projectile B associated with the corresponding matrix element $T_{fi}(\mathbf{q}',\mathbf{q})$ of the transition operator on the energy shell [see (4.17)]; and $f_{\mathrm{eB}}(\mathbf{k}',\mathbf{k})$ denotes the two-particle (electron-perturber) scattering amplitude defined by the relation

$$
\begin{aligned}
f_{\mathrm{eB}}(\mathbf{k}',\mathbf{k}) &= -\frac{\mu_{\mathrm{eB}}}{2\pi\hbar^2} t_{\mathrm{eB}}(\mathbf{k}',\mathbf{k};\epsilon) \\
&= -\frac{\mu_{\mathrm{eB}}}{2\pi\hbar^2}\int d\mathbf{r}_{\mathrm{eB}} \exp(-\mathrm{i}\mathbf{k}'\cdot\mathbf{r}_{\mathrm{eB}})\, \mathcal{V}_{\mathrm{eB}} \psi_{\mathbf{k}}^+(\mathbf{r}_{\mathrm{eB}}).
\end{aligned}
\tag{4.109}
$$

Note that in the impulse approximation, the electron-perturber scattering amplitude $f_{\mathrm{eB}}(\mathbf{k}',\mathbf{k})$ in the basic equation (4.108) (or the matrix element

$t_{eB}(\mathbf{k}', \mathbf{k}; \epsilon)$ of the scattering $t_{eB}(\epsilon)$-operator) should be taken, generally, both on $(k = k')$ and off the energy shell. The final expression of the quantum impulse approximation for the contribution of the perturber-quasi-free electron scattering to the cross section $\sigma_{fi}(q)$ of the $i \to f$ transition can be presented as

$$\sigma_{fi}(q) = \left(\frac{\mu}{\mu_{eB}}\right)^2 \frac{q'}{q} \int |\langle G_f(\kappa')| f_{eB}(\mathbf{k}', \mathbf{k}) |G_i(\kappa)\rangle|^2 \, d\Omega_{\mathbf{q}'\mathbf{q}} \,, (4.110)$$

where $\mathbf{q}' = \mathbf{q} - \mathbf{Q}$. It is worthwhile to point out that the shift of the Rydberg atom wave function in momentum space $\kappa \to \kappa'$ is performed by the translational operator

$$G_\alpha(\kappa + \mathbf{K}) = \exp(i\mathbf{K} \cdot \hat{\mathbf{r}})G_\alpha(\kappa), \tag{4.111}$$

where $\hat{\mathbf{r}} = i\partial/\partial\kappa$ is the radius vector operator of the valence electron.

In the most interesting case of collisions between the Rydberg atom and the heavy projectile (atom, ion, or molecule with mass $M_B \gg m$) one can take into account that $\mu_{eB} \approx m$, $\mu_{eA^+} \approx m$ and $M_A = m + M_{A^+} \approx M_{A^+}$ so that the relationships for the relative momenta and velocities of three particles can be rewritten as

$$\mathbf{k} \approx \kappa - (m/\mu)\mathbf{q}, \qquad \mathbf{v}_{eB} \approx \mathbf{v}_{eA^+} - \mathbf{V}, \quad \kappa' - \kappa \approx \mathbf{q} - \mathbf{q}' \quad (4.112)$$

i.e., $\mathbf{K} \approx \mathbf{Q}$.

For collisions of the Rydberg atom with an electron $\mu_{eB} = m/2$, and we have

$$\kappa' - \kappa \approx \mathbf{Q}, \quad \mathbf{v}_{eB} \approx \mathbf{v}_{eA^+} - \mathbf{V}, \quad \mathbf{k}' - \mathbf{k} = \mathbf{q} - \mathbf{q}' = \mathbf{Q} \,. \tag{4.113}$$

It is important to stress that the error of the identified relations is of the order of the mass ratio (m/M_A) and certainly is less than the one of collision theory. The next approximations are connected with the separate analysis of fast and slow collisions that will be considered below.

b) Extension to the Projectile Internal Degrees of Freedom. All results presented above describe the case of collisions between the Rydberg atom and neutral or charged particle when the projectile state is not changed during the scattering process. Actually, however, the final formulas of this section are also applicable in the case when transitions between highly excited atomic states are accompanied by the inelastic excitation (de-excitation) of the projectile. Then, the basic formula of the impulse approximation for the transition amplitude $f_{\alpha'\alpha}^{\beta'\beta}(\mathbf{q}', \mathbf{q})$ of the process

$$A(\alpha) + B(\beta) \to A(\alpha') + B(\beta')$$

can be rewritten as

$$f_{\alpha'\alpha}^{\beta'\beta}(\mathbf{q}', \mathbf{q}) = \frac{\mu}{\mu_{eB}} \langle G_{\alpha'}(\kappa')| f_{eB}^{\beta'\beta}(\mathbf{k}', \mathbf{k}) |G_\alpha(\kappa)\rangle,$$

$$f_{eB}^{\beta'\beta}(\mathbf{k}', \mathbf{k}) = -\frac{\mu_{eB}}{2\pi\hbar^2} \langle \mathbf{k}', \beta'| t_{eB}(\epsilon) |\beta, \mathbf{k}\rangle \,.$$

$$\tag{4.114}$$

Here $f_{eB}^{\beta'\beta}(\mathbf{k}',\mathbf{k})$ is the amplitude for the scattering of the free electron by a perturber

$$e + B(\beta) \rightarrow e + B(\beta')$$

and β is the set of quantum numbers characterizing its internal state (e.g., the vibrational-rotational states of the perturbing molecule $\beta = v, j, j_z$). As in the case of pure elastic quasi-free electron-perturber scattering the two-body matrix elements $t_{eB}^{\beta'\beta}(\mathbf{k}',\mathbf{k};\epsilon)$ should be taken in (4.114) both on and off $[E_\beta + \hbar^2\mathbf{k}^2/2\mu_{eB} \neq E_{\beta'} + \hbar^2(\mathbf{k}')^2/2\mu_{eB}]$ the energy shell.

4.5.2 Binary Encounter Approach

In the following we present general formulas of binary encounter theory for the cross sections of the bound-bound $nl \rightarrow n'$ [4.16] and bound-free $nl \rightarrow E$ [4.17] transitions summed over possible quantum numbers l' and m' in the final state, and averaged over the magnetic m-sublevels of the initial nl-level. Consider first the process of direct ionization $A(nl) + B(\beta) \rightarrow A^+ + B(\beta') + e$. We shall proceed from the basic equation (4.22) of scattering theory for the differential cross section $d\sigma_{\alpha'\alpha}^{\beta'\beta}(\mathbf{q}',\mathbf{q})$ of the inelastic transition $|\alpha,\beta,\mathbf{q}\rangle \rightarrow |\alpha',\beta',\mathbf{q}'\rangle$ and the quantum impulse approximation (4.114). Integration of this equation over the solid angles $d\Omega_{\mathbf{q}'\mathbf{q}}$ of scattering and its summation over all final states of the Rydberg atom in the discrete spectrum and integration in the continuous spectrum will lead to the following expression for the differential cross section $d\sigma_{nl}(q',q)/d\mathcal{E}'$ per unit energy interval $d\mathcal{E}' = \hbar^2 q' dq'/\mu$ of the colliding A and B particles:

$$\frac{d\sigma_{\alpha'\alpha}^{\beta'\beta}(q',q)}{d\mathcal{E}'} = \frac{\mu_{q'}^2}{\mu_{eB}^2 q(2l+1)} \sum_m \int d\Omega_{\mathbf{q}'\mathbf{q}} \sum_{\alpha'} \left\langle G_\alpha(\boldsymbol{\kappa}) \left| \left[f_{eB}^{\beta'\beta}(\mathbf{k}',\mathbf{k}) \right]^* \right| G_{\alpha'}(\boldsymbol{\kappa}) \right\rangle$$

$$\times \left\langle G_{\alpha'}(\boldsymbol{\kappa}') \left| f_{eB}^{\beta'\beta}(\mathbf{k}',\mathbf{k}) \right| G_\alpha(\boldsymbol{\kappa}) \right\rangle \delta(E_{\alpha'} - E_\alpha + \Delta\mathcal{E}),$$

$$\Delta\mathcal{E} = E_{\beta'} - E_\beta + \mathcal{E}' - \mathcal{E}. \tag{4.115}$$

Now we use the Fourier transform for the δ-function

$$\delta(E_{\alpha'} - E_\alpha + \Delta\mathcal{E}) = \frac{1}{2\pi\hbar} \int\limits_{-\infty}^{\infty} \exp\left[\frac{\mathrm{i}}{\hbar}(E_{\alpha'} - E_\alpha + \Delta\mathcal{E})t\right] dt$$

and the following relations for the Heisenberg evolution operators:

$$\exp\left(-\frac{\mathrm{i}}{\hbar}H_A t\right)|G_\alpha(\boldsymbol{\kappa})\rangle = \exp\left(-\frac{\mathrm{i}}{\hbar}E_\alpha t\right)|G_\alpha(\boldsymbol{\kappa})\rangle. \tag{4.116}$$

As a result, using the completeness property $\sum_{\alpha'}|G_{\alpha'}\rangle\langle G_{\alpha'}| = \mathbb{1}$ for the Rydberg atom wave functions, expression (4.115) can be rewritten as

$$\frac{d\sigma_{nl}(q',q)}{d\mathcal{E}'} = \frac{\mu_{q'}^2}{\mu_{eB}^2 q(2l+1)} \sum_m \int d\Omega_{q'q} \langle G_{nlm}(\boldsymbol{\kappa})| \hat{\mathcal{O}} |G_{nlm}(\boldsymbol{\kappa})\rangle \,,$$

$$\hat{\mathcal{O}} = \frac{1}{2\pi\hbar} \int\limits_{-\infty}^{\infty} dt \exp\left(\frac{i}{\hbar}\Delta\mathfrak{E}t\right) \left[f_{eB}^{\beta'\beta}(\mathbf{k}',\mathbf{k})\right]^*$$

$$\times \exp\left(\frac{i}{\hbar}H_A't\right) f_{eB}^{\beta'\beta}(\mathbf{k}',\mathbf{k}) \exp\left(-\frac{i}{\hbar}H_A t\right). \quad (4.117)$$

Thus, in the impulse approximation the differential cross section $d\sigma_{nl}(q',q)/d\mathcal{E}'$ of ionization can be expressed in terms of some operator $\hat{\mathcal{O}}$, averaged over the initial $|i\rangle \equiv |nlm\rangle$ state of the Rydberg atom. This fact is in agreement with the general results of the theory of quasi-free scattering on a system of weakly bound particles (see [4.1]). Here $H_A = H_A(\boldsymbol{\kappa}, \hat{\mathbf{r}})$ and $H_A' = H_A(\boldsymbol{\kappa}', \hat{\mathbf{r}})$ are the Hamiltonians of the Rydberg atom A^* in the momentum representation for the initial $\boldsymbol{\kappa}$ and final $\boldsymbol{\kappa}' = \boldsymbol{\kappa} + \mathbf{K}$ electron momenta:

$$H_A \equiv H_A(\boldsymbol{\kappa}, \hat{\mathbf{r}}) = \hbar^2\kappa^2/2\mu_{eA^+} + \hat{U}_{eA^+} \,. \quad (4.118)$$

$\hat{U}_{eA^+}(\hat{\mathbf{r}})$ is the potential energy operator of electron-core interaction, and $\hat{\mathbf{r}} = i\partial/\partial\boldsymbol{\kappa}$ is the radius vector operator of the valence electron.

Within the framework of the impulse approximation the evolution operator $\exp[-(i/\hbar)H_A t]$ in (4.117) commutes with the two-particle t_{eB} operator of the electron-perturber scattering. Then, using the Baker-Campbell-Hausdorf formula for expanding in series the $\exp[(i/\hbar)H_A't] \exp[-(i/\hbar)H_A t]$ -operator, we obtain

$$\exp[(i/\hbar)H_A't] \exp[-(i/\hbar)H_A t] = \exp\left[\frac{i}{\hbar}(H_A' - H_A)t + [H_A', H_A]\frac{t^2}{2\hbar^2} + \ldots\right]$$

$$= \exp\left\{\frac{i\hbar t\left[(\kappa')^2 - \kappa^2\right]}{2\mu_{eA^+}} + \frac{it^2}{2}\left(\frac{\mathbf{K}}{\mu_{eA^+}}\right)\hat{\mathbf{F}}_{eA^+} + \ldots\right\}. \quad (4.119)$$

Here $\hat{\mathbf{F}}_{eA^+}$ is the operator of the force acting on the outer electron of the Rydberg atom A^* by the ionic core A^+.

As follows from the comparison of the first and second terms in (4.119), all terms (nonlinear in time t), corresponding to the change of the potential energy U_{eA^+} during the interaction of the colliding e and B particles, can be neglected if the following condition $\hbar K_{eB} \gg \tau_{eB}F_{eA^+}$ is to be satisfied. Here $\hbar K_{eB}$ is the characteristic momentum transferred to the outer electron e in its binary collision with the perturbing particle B, and $(\tau_{eB}F_{eA^+})$ is the impulse of the force $F_{eA^+} = e^2/r_{eA^+}^2$ acting on this electron by the ionic core A^+ during the collision time τ_{eB} of e and B particles. It should be noted that this condition is one of the validity criteria of the impulse approximation and the quasi-free electron model (see below Sect. 5.4.8). Thus within the framework of the binary-encounter theory in the impulse treatment, only the first (linear in time t) term of the series expansion should be retained in (4.119). This term is determined by the change of the kinetic energy of the quasi-free electron e in its collision with the perturber B.

Then, substituting (4.119) into (4.117) and performing the integration over dt, we obtain the following general expression of binary-encounter theory for the differential cross section $d\sigma_{E,nl}(\mathcal{E})/dE$ of ionization per unit energy interval of the ejected electron [4.17]:

$$\frac{d\sigma_{E,nl}^{\beta'\beta}(\mathcal{E})}{dE} = \frac{\mu^2 q'}{\mu_{eB}^2 q(2l+1)} \sum_m \int d\Omega_{\mathbf{q'q}} \int d\kappa \, |G_{nlm}(\kappa)|^2 \left| f_{eB}^{\beta'\beta}(\mathbf{k'},\mathbf{k}) \right|^2$$
$$\times \delta \left[\hbar^2 \left((\kappa')^2 - \kappa^2\right)/2\mu_{eA^+} + E_{nl} - E \right]. \tag{4.120}$$

Here $E_{nl} = -Ry/(n-\delta_l)^2 < 0$ is the energy of the Rydberg electron in the initial discrete state and $E > 0$ is the kinetic energy of the ejected electron. The final kinetic energy $\mathcal{E}' = \hbar^2 q'^2/2\mu$ of colliding A^* and B particles is determined from the law of energy conservation for the direct ionization process

$$\mathcal{E} + E_\beta + E_{nl} = \mathcal{E}' + E_{\beta'} + E, \qquad \mathcal{E} = \hbar^2 q^2/2\mu, \qquad \mathcal{E}' = \hbar^2 q'^2/2\mu.$$

The total ionization cross section $\sigma_{nl}^{\beta'\beta}(\mathcal{E})$ is determined from (4.120) by integrating over all possible values of the ejected electron energy

$$\sigma_{nl}^{\beta'\beta}(\mathcal{E}) = \int_0^{E_{max}} [d\sigma_{E,nl}(\mathcal{E})/dE] \, dE, \qquad E_{max} = \mathcal{E} - (|E_{nl}| - \Delta E_{\beta\beta'}), \tag{4.121}$$

where $\Delta E_{\beta\beta'} = E_\beta - E_{\beta'}$ is the change of the internal energy of the perturbing particle B. As follows from (4.121), direct ionization of the Rydberg atom is possible, when the kinetic energy $\mathcal{E} = \hbar^2 q^2/2\mu$ of colliding A^* and B particles relative motion satisfies the condition $\mathcal{E}_{min} \leq \mathcal{E}$ (where $\mathcal{E}_{min} = \max(0, |E_{nl}| - \Delta E_{\beta\beta'})$).

The general equation (4.120) of the binary-encounter theory relates the ionization cross section with the differential cross section $d\sigma_{eB}^{\beta'\beta}/d\Omega_{\mathbf{k'k}} = \left| f_{eB}^{\beta'\beta}(\mathbf{k'},\mathbf{k}) \right|^2$ for electron-perturber scattering and with the momentum distribution function $|G_{nlm}(\kappa)|^2$ of the Rydberg electron in the initial atomic $|i\rangle = |nlm\rangle$ state. Thus, there is a definite analogy between this equation, obtained [4.17] in the impulse approximation directly from the quantum scattering theory, and the basic equation of semiclassical theory [4.18. 19]. However, an important feature of the semiclassical equation (4.120) for the ionization cross section is due to the presence of the delta function, which plays the role of a microcanonical distribution. This δ-function brings out from the entire momentum space only those momentum values which correspond to the classical energy transfer $\Delta E_{fi} = |E_{nl}| + E = \hbar^2(\kappa'^2 - \kappa^2)/2\mu_{eA+}$ for a given bound-free $nl \to E$ transition.

General equation (4.120) of the binary-encounter theory may also be used in the case of the bound-bound $nl \to n'$-transition $A(nl) + B(\beta) \to A(n') + B(\beta')$, if we additionally introduce the quasi-continuum approximation for the hydrogen-like degenerate n' levels of the Rydberg atom [4.16]. Then, in order to obtain the cross section $\sigma_{n',nl}^{\beta'\beta}(\mathcal{E})$ of the inelastic (or quasi-elastic) $nl \to n'$ transition, the differential cross section $d\sigma_{E,nl}^{\beta'\beta}(\mathcal{E})/dE$ per unit energy interval of the quasi-continuous spectrum $E_{n'}$ should be multiplied by the factor $|dE_{n'}/dn'| = 2Ry/(n')^3$, while the final energy $E > 0$ on the right-hand side of (4.120) should be replaced by $E_{n'} < 0$, i.e.,

$$\frac{d\sigma_{E,nl}^{\beta'\beta}(\mathcal{E})}{dE} \to \sigma_{n',nl}^{\beta'\beta} \left|\frac{dE_{n'}}{dn'}\right|^{-1} = \sigma_{n',nl}^{\beta'\beta}\frac{(n')^3}{2Z^2Ry}, \qquad E \to E_{n'} . \tag{4.122}$$

The law of energy conservation for the identified process can be rewritten as

$$\frac{\hbar^2 q^2}{2\mu} + E_\beta - \frac{Ry}{(n-\delta_l)^2} = \frac{\hbar^2(q')^2}{2\mu} + E_{\beta'} - \frac{Ry}{(n')^2}. \tag{4.123}$$

It should be noted that a particular form of the general equation (4.120) is very convenient to perform some physical simplifications and for further analysis of the different special cases. We shall show later that it can be reduced to a rather simple form for the most interesting cases of slow $V \ll v_0/n$ and fast $V \gg v_0/n$ collisions between the Rydberg atom and neutral perturbing particle.

5. Collisions of Rydberg Atom with Neutral Particles: Weak-Coupling Models

In the present chapter we start the consideration of collisions involving the Rydberg atoms and neutral particles. Within the framework of the weak-coupling approximations we present general techniques for calculation of transition probabilities and cross sections between highly excited states and ionization induced by scattering of the quasi-free Rydberg electron by a neutral target. Theoretical description of these collisions will be based on two approaches, outlined in the preceding chapter. The first one is the semiclassical time-dependent approach in the impact-parameter representation combined with the first-order perturbation theory. It deals directly with the potential of the electron-perturber interaction. To derive analytic formulas for probabilities and cross sections and to demonstrate their dependencies on the main physical parameters this interaction will be described by the Fermi pseudopotential of the zero-range. The second approach based on the impulse-approximation will allow us to express the cross section of the Rydberg atom-neutral collisions in terms of the amplitude $f_{eB}(\mathbf{k'}, \mathbf{k})$, differential $d\sigma_{eB}(k, \theta)/d\Omega$ or total $\sigma_{eB}(k)$ cross sections for scattering of the free electron by the perturbing atom or molecule. It will provide the general treatment of the Rydberg atom-neutral collisions within the framework of the quasi-free electron model and will be valid for any arbitrary form of the electron-perturber scattering amplitude.

5.1 Quasi-free Electron Model

Most of the available methods for the description of collisions involving Rydberg atom A^* and the atoms or molecules in the ground or low excited states are based on the key assumption that processes caused by scattering of neutral projectile B on the highly-excited electron and on the parent core A^+ may be treated independently. Thus, the resultant three-body problem is rather simplified because the perturbing atom or molecule does not interact simultaneously with both the outer electron and ion core. This is true for a wide range of the principal quantum numbers n due to the orbital radius $r_n \sim n^2 a_0$ of the Rydberg atom being large as compared to the characteristic r_{eB} and R_{A+B} dimensions of both the electron-perturber and core-perturber interaction regions. Then, neglecting the interference effects and multiple scattering, the resultant cross section of the Rydberg atom–neutral collision is determined by an additive contribution of the electron-perturber and core-perturber scattering mechanisms in accord with (4.97). In many situations the contribution of the scattering of a highly excited electron on the perturbing particle turns out to be predominant. Then, theoretical analysis of collisional processes involving highly excited atom and neutral particle is based on the quasi-free-electron model proposed first by *Fermi* [5.1] in order to explain the spectral line shift behavior of Rydberg atomic series in a buffer gas. This model has been formulated in its final form as a result of intensive investigations of elastic, quasi-elastic, inelastic and ionizing collisions between the Rydberg atoms and neutral projectiles (see reviews [5.2, 4.15, 18, 1.3]).

The main idea of the quasi-free-electron model is that the collision of the Rydberg atom A^* with atom or molecule B is treated as an elastic or inelastic binary encounter between the valence (slow and free) electron and neutral projectile. The parent core A^+ is only a "spectator" in this encounter, so that its interaction with the neutral perturber is entirely neglected. However, it is responsible for the generation of the electron momentum distribution function in the given Rydberg atomic state. The momentum distribution function is determined (see Sect. 2.5) by the interaction $U(r)$ of electron with the parent core A^+ (i.e., primarily by their Coulomb interaction at large distances r).

Only sufficiently small electron energies $\epsilon = \hbar^2 k^2 / 2m \ll Ry$ and the wave numbers $ka_0 \ll 1$ are responsible for thermal collisions between the Rydberg atoms and neutral particles. The characteristic range of k, which makes the main contribution to one or another process, depends essentially on both the principal quantum number n and the energy ΔE transferred to the highly excited electron in the binary e–B encounter. In pure elastic and quasi-elastic collisions (when $\Delta E = 0$) the characteristic values of k are about the mean orbital wave number $k_n \sim 1/na_0$ of the Rydberg electron. This means that $ka_0 \sim k_n a_0 \approx 0.2$, 0.1, and 0.03 and, hence, the averaged Rydberg electron energies are equal to $\epsilon_n \approx Ry/n^2 \sim 0.5$, 0.1, and 0.01 eV for $n = 5$, 10, and 30, respectively. At $n \sim 100$ (when $k_n a_0 \sim 10^{-2}$) the milli- and submilli-electron-volt energy diapason is becoming important, while for very high values of $n \sim 1000$ the averaged kinetic energy of Rydberg electron corresponds to the micro-electron-volt energy region. It is important to stress, that the inelastic n-changing excitation (de-excitation) and ionization processes with large energy ΔE transferred to the translational motion of colliding atoms are determined by much greater values of $k > |\Delta E| / 2\hbar V$ than the mean orbital momentum $1/(na_0)$. Nevertheless, the characteristic momenta of k in such transitions are also satisfied by the condition $kr_0 \ll 1$ if the principal quantum number is not too small. Here r_0 is the effective radius of the short-range interaction of the electron with the target atom or molecule B, whose value is usually a few atomic units. As follows from the analysis of different elementary processes involving Rydberg atoms and neutral targets, commonly of practical use are only the energies $\epsilon = \hbar^2 k^2/2m$, which do not exceed $0.1 - 1$ eV $(ka_0 < 0.1\text{--}0.3)$.

5.2 Scattering of Ultra-Slow Electrons by Atoms and Molecules

The electron scattering by atomic and molecular targets has been the subject of intensive theoretical and experimental research for many years (see [4.2, 5.3–7] and references therein). Since the de Broglie wavelength $1/k$ of a ultra-low-energy electron is large compared to the effective range r_0 of the electron-atom or electron-molecule short-range interaction, one of the most adequate methods for describing the electron scattering in the sub-thermal energy region is based on the effective range theory. This theory gives an appropriate expansion of the scattering amplitude in the vicinity of zero energy $(0 \leq k \ll 1/r_0)$. It is mainly determined by the type of long-range interaction between the electron and target atom or molecule (i.e., by the polarization, quadrupole or dipole potentials). The short-range interaction $(r < r_0)$ is usually described by means of several parameters such as the scattering length L and the effective radius r_0. It is well known that the major terms of the elastic electron-atom scattering amplitude and cross section as $k \rightarrow 0$ are determined by the contribution of the s-wave alone and can be expressed in terms of one parameter $(f_{\mathrm{eB}} = -L$ and $\sigma_{\mathrm{eB}}^{\mathrm{el}} = 4\pi L^2$, where L is the scattering length).

Until recently, this simplest description of electron scattering at ultra-low energies was used in the most part of theoretical works on collisions involving Rydberg atoms and neutral atomic targets. However, detailed studies showed that the scattering length approximation fails to reliably describe even thermal collisions of highly excited atoms with the ground-state heavy-rare gas atoms, particularly so, for the inelastic transitions with large energy transfer. It also does not hold for collisions with the neon atom having an anomalously low value of the scattering length. Moreover, it certainly becomes inapplicable for collisions of Rydberg atoms with strongly polarizable targets (e.g., alkali-metal atoms). The effective range of the long-range (polarization) part of the electron interaction with these target systems is characterized by the Weisskopf radius $r_{\mathrm{W}} = (\pi\alpha/4ka_0)^{1/3}$. Therefore, at large values of atomic polarizability α it may become comparable to the electron wavelength $1/k$ even for submilli-electron-volt energies. A similar situation occurs for electron scattering by polar molecules having a substantial permanent dipole moment, for which the effective range of the long range interaction turns out to be particularly large. This leads to the appearance of anomalously large values of the elastic electron-atom scattering as well as rotationally elastic and inelastic processes in electron-molecule scattering at very low energies. They are primarily the result of the presence of real discrete or virtual levels of very weakly bound negative atomic or molecular ions or the shape resonances on the quasi-discrete (quasi-stationary) levels. These phenomena and their revealings in collisions involving Rydberg atoms were the subject of intensive research in recent years (see [1.3, 5.8] and references therein).

In this section we shall briefly discuss the behavior of the scattering amplitude and the differential cross section of ultra-slow electron scattering by the rare gas and alkali-metal atoms as well as by the polar and nonpolar molecules. We shall present only some of the final formulas, which are necessary for the next analysis of the Rydberg atom-neutral collisions.

5.2.1 Electron-Atom Scattering

The basic expressions for the elastic scattering amplitude and for the differential cross section averaged over the two possible $S_+ = s_B + 1/2$ and $S_- = | s_B - 1/2 |$ values of the total spin S of the B + e system are given by [2.2]

$$f_{eB}^{(S)}(k, \theta) = \sum_\ell (2\ell + 1) f_\ell^{(S)}(k) P_\ell(\cos \theta),$$

$$f_\ell^{(S)} = \left[k \cot\left(\eta_\ell^{(S)} \right) - ik \right]^{-1}, \tag{5.1}$$

$$\frac{d\sigma_{eB}}{d\Omega} = |f_{eB}(k, \theta)|^2$$

$$= \sum_{S=S_+, S_-} C(S) \left| f_{eB}^{(S)}(k, \theta) \right|^2. \tag{5.2}$$

Here $f_\ell^{(S)}(k)$ and $\eta_\ell^{(S)}(k)$ are the amplitude and the phase shift of the partial wave with the orbital momentum ℓ; θ is the scattering angle; $C(S) = (2S + 1)/[2(2s_B + 1)]$ is the spin factor, s_B is the spin of the target B. Note that $S_+ = S_- = 1/2$ and, hence, $C(S_+) = C(S_-) = 1/2$ for the ground-state rare gas atoms, for which $s_B = 0$. At the same time, there are the triplet $[S_+ = 1, C(S_+) = 3/4]$ and singlet $[S_- = 0, C(S_-) = 1/4]$ waves for electron scattering by the ground-state alkali-metal atoms.

a) **Potential Scattering by Rare Gas Atoms.** In the case of potential scattering of a slow electron by the atom B with sufficiently small polarizability α (e.g., the rare gas atoms) the modified effective range theory has been formulated by *O'Malley* et al. [5.9, 10]. In this theory the scattering partial phase shifts $\eta_\ell(k)$ are given by the expressions:

$$\frac{\tan[\eta_0(k)]}{k} = -L - \frac{\pi\alpha}{3a_0} k - \frac{4\alpha L}{3a_0} k^2 \ln(ka_0) + D_0 k^2 + F_0 k^3 + O(k^4),$$

$$\frac{\tan[\eta_\ell(k)]}{k} = \frac{\pi\alpha k}{a_0(2\ell + 3)(2\ell + 1)(2\ell - 1)} + D_\ell k^2 + O(k^3), \qquad \ell \geq 1. \tag{5.3}$$

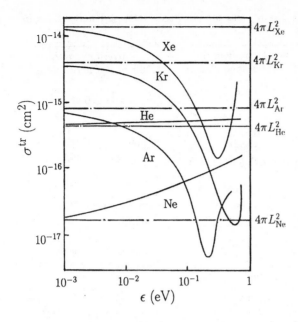

Fig. 5.1. The energy dependence of the momentum transfer cross sections $\sigma^{tr}(\epsilon)$ for elastic electron scattering by the rare gas atoms. Full curves are the results [5.11] obtained from the electron-cyclotron-resonance absorption spectra using the modified effective range theory [5.10]. Dashed curves correspond to scattering length approximation $\sigma^{tr}(\epsilon) = 4\pi L^2$

One can see, that the values of η_ℓ strongly decrease with an increase of the orbital momentum ℓ for electron–target-atom relative motion. The low-energy expansions for the amplitude and differential cross section of electron-atom scattering are as follows:

$$\mathbf{Re}\{f_{eB}(k,\theta)\} = -L - \frac{\pi\alpha}{2a_0}k\sin\left(\frac{\theta}{2}\right) - \frac{4\alpha L}{3a_0}k^2\ln(ka_0)$$
$$-bk^2 + O\left(k^3\right), \tag{5.4}$$

$$|f_{eB}(k,\theta)|^2 = L^2 + \frac{\pi\alpha L}{a_0}k\sin\left(\frac{\theta}{2}\right) + \frac{8\alpha L^2}{3a_0}k^2\ln(ka_0)$$
$$+Bk^2 + O\left(k^3\right). \tag{5.5}$$

Here L is the scattering length of the target atom B; D_ℓ, F_0, b and B are the constant coefficients ($D_\ell > 0$ for $\ell = 0$, $D_\ell < 0$ for $\ell = 1$ and all D_ℓ values with $\ell \geq 2$ are usually taken to be $D_\ell = 0$). The energy range in which these expressions are valid is limited by the condition $\alpha k^2/a_0 \ll 1$. The specific feature of expansions (5.4, 5) is the presence of the second linear k term and the third logarithmic term, which appear due to the long-range polarization interaction $V^{l.r.} = -\overset{.}{\alpha}e^2/2r^4$. The second term, which is proportional to the momentum transfer $Q = |\mathbf{k}' - \mathbf{k}| = 2k\sin(\theta/2)$ in the electron-atom collision, vanishes in the forward direction $\theta = 0$.

The theory [5.9, 10] provides a successful explanation of all phenomena to occur in electron-rare gas atom scattering at small energies. For He, Ne, Ar, Kr, and Xe a detailed analysis of the phase shifts and the cross sections was made in [5.10–13]. The results for the momentum transfer cross sections, ob-

Table 5.1. Polarizabilities α (in a_0^3) and the electron scattering lengths L (in a_0) of the ground-state rare gas atoms

Atom	Polarizability	Scattering length Effective Range Theory	Scattering length Shift of Rydberg Levels
He	$1.383^{a)}$	$1.19^{d)}$ $1.17^{f)}$	$1.11 \pm 0.2^{g)}$
Ne	$2.67^{b)}$	$0.214^{e)}$ $0.204^{f)}$ $0.2065^{b)}$	$0.192 \pm 0.003^{g)}$ $0.24 \pm 0.01^{h)}$
Ar	$11.08^{a,c)}$	$-1.70^{d)}$ $-1.55^{f)}$ $-1.593^{c)}$	$-1.4 \pm 0.1^{g)}$
Kr	$16.74^{a,c)}$	$-3.70^{d)}$ $-3.5^{f)}$ $-3.478^{c)}$	$-3.05 \pm 0.06^{g)}$ $-4.0 \pm 0.3^{h)}$
Xe	$27.29^{c)}$	$-6.50^{d)}$ $-6.5^{f)}$ $-6.527^{c)}$	$-6.1 \pm 0.3^{g)}$

$^{a)}$ *Radzig* and *Smirnov* [5.16] – reference data for atomic polarizability. $^{b)}$ *Gulley* et al. [5.12], $^{c)}$ *Weyhreter* et al. [5.13], $^{d)}$ *O'Malley* [5.10], $^{e)}$ *O'Malley* and *Crompton* [5.17], $^{f)}$ *Golovanivsky* and *Kabilan* [5.11] – extrapolation of experimental data on electron-atom scattering to zero energy using the modified effective range theory. $^{g)}$ *Heber* et al. [5.14], $^{h)}$ *Thompson* et al. [5.15] – measurements of pressure shift of high-Rydberg levels in the rare gases.

tained in [5.11] by numerical simulation of the Electron-Cyclotron-Resonance experimental curves, are presented in Fig. 5.1. It can be seen that the scattering length approximation (in which the elastic scattering cross section is independent of the electron energy $\sigma_{eB}^{el} = 4\pi L^2 = const$ and is determined by the contribution of the partial s-wave) may be reliably used in a wide energy region $\epsilon < 1$ eV for the helium atom alone. For the heavy rare gas atoms Ar, Kr, and Xe there are the deep Ramsauer-Townsend minima, which appear in the range of $\epsilon \approx 0.1$–1 eV due to negative values of the scattering lengths. For a neon atom, having an anomalously low value of the scattering length, the momentum transfer cross section σ_{eB}^{tr} grows with increase of electron momenta and considerably exceeds the value of $4\pi L^2$ in the range above 0.01 eV. It should be noted that the magnitudes of the scattering lengths found by extrapolation of electron-atom cross sections to zero energy using the modified effective range theory [5.9, 10] are in quite reasonable agreement with the data obtained in the experiments [5.14, 15] on the pressure shift of the high-Rydberg atomic series by rare gas atoms (see Table 5.1).

b) Resonance Scattering by Alkali-Metal Atoms. For highly polarizable alkali-metal atoms, the theory [5.9, 10] fails to describe the scattering
amplitude in the energy region above a few meV. An important feature of
electron scattering by the ground state alkali atom B is the presence of the
shape ^3P-resonance on the quasi-discrete triplet level ($E_r > 0$) of the corresponding negative ion B^-. This leads to enormous high values of the cross
sections in the range of about 10–100 meV depending on the polarizability
of the alkali-metal atom (see Fig. 5.2). The appearance of the shape resonance for the p-scattering wave is the result of the potential barrier in the
effective electron-atom interaction, which includes both the polarization and
centrifugal terms $V_{\text{eff}}(r) = -\alpha e^2/2r^4 + \hbar^2\ell(\ell+1)/2mr^2$ (where $\ell = 1$). At
high polarizability α this barrier occurs at large distances $r_{\text{max}} = (\alpha/a_0)^{1/2}$
and has small eight $V_{\text{eff}}^{\text{max}} = Ry(a_0^3/\alpha)$ so that possible quasi-discrete levels
may exist only in the ultra-low energy region $E_r < V_{\text{eff}}^{\text{max}}$ (i.e., below about
30–80 meV for alkali atoms, for which the typical values of α are given in
Table 5.2).

The very presence of the ^3P-shape resonances proceeds from the close-
coupling [5.23, 3] and variational [5.20] calculations of the partial phase shifts
for electron scattering by Li, Na, K, and Rb as well as from the effective
range theory [5.18, 19]. Recent relativistic R-matrix calculations [5.22] taking into account the electron correlation effects showed that there is also the
quasi-discrete ^3P-level of the negative Cs^--ion (It is split by the spin-orbit
interaction into three fine-structure components with total angular momentum values $J = 0$, 1 and 2, for which the shape resonance positions are
$E_r = 1.78$, 5.56, and 12.76 meV, respectively). There is also a direct experiment [5.25] on measuring the elastic electron scattering cross sections,
which clearly demonstrates that the ^3P-resonances occur for Na and Rb at

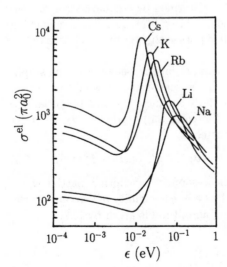

Fig. 5.2. The energy dependence of
total cross sections $\sigma^{\text{el}}(\epsilon)$ for elastic
electron scattering by the alkali-metal
atoms. Full curves are the results of the
modified effective range theory for K,
Rb, and Cs [5.19] and variational calculations for Li, and Na [5.20]

Table 5.2. Polarizabilities α (in a_0^3); singlet L_- and triplet L_+ electron scattering lengths (in a_0); positions E_r (in meV) and widths Γ_r (in meV) of the ^3P-resonances for the ground-state alkali-metal atoms

Atom	α	L_-	L_+	E_r	Γ_r
Li	$162^{f)}$	$3.65^{e)}$	$-5.66^{e)}$	$59^{a)}$	$77^{a)}$
	$165^{g)}$			$60^{b)}$	$57^{b)}$
Na	$162^{f)}$	$4.23^{e)}$	$-5.91^{e)}$	$83^{a)}$	$188^{a)}$
	$166^{g)}$			$83^{b)}$	$85^{b)}$
K	$303^{a)}$	$0.57^{a)}$	$-15.4^{a)}$	$19^{a)}$	$16^{a)}$
	$287^{f)}$	$0.55^{e)}$	$-15.0^{e)}$	$20^{c)}$	$21^{c)}$
Rb	$328^{a)}$	$2.03^{a)}$	$-16.9^{a)}$	$23^{a)}$	$25^{a)}$
	$310^{f)}$			$28^{c)}$	$31^{c)}$
Cs	$402^{a)}$	$-2.4^{a)}$	$-22.7^{a)}$	$12.6^{a)}$	$9.1^{a)}$
	$385^{f)}$	$-4.04^{e)}$	$-25.3^{e)}$	$9.14^{d)}$	$6.04^{d)}$

[a] *Fabrikant* [5.18, 19] – effective-range theory.
[b] *Sinfailam* and *Nesbet* [5.20] – variational calculations.
[c] *Lebedev* and *Marchenko* [5.21] – impulse-approximation results detected from experiments on impact broadening of high-Rydberg levels in alkali vapors.
[d] *Thumm* and *Norcross* [5.22] – the J-averaged data for the resonance position and width obtained from relativistic R-matrix calculations.
[e] *Karule* [5.23] – two-state close coupling calculations of scattering lengths.
[f] *Radzig* and *Smirnov* [5.16], [g] *Dalgarno* [5.24] – data for atomic polarizabilities.

the low-energy range. A resonance feature of electron scattering by the heavy alkali-metal atoms K, Rb, and Cs is also confirmed by comparison of the calculated cross sections for impact broadening [4.16, 5.21, 19, 26] and quenching [4.16] of Rydberg atomic levels in alkali vapors with experimental data [5.27–29] (see below Sects. 6.6.2, 9.3).

Consider the influence of the ^3P-resonance on the behavior of the amplitude and differential cross section for elastic scattering. It would be convenient to separate the contributions of the resonance f_{eB}^r and potential f_{eB}^P scattering in the basic expression for the total amplitude $f(\epsilon, \theta)$ of triplet $(S_r = S_+ = 1)$ and singlet $(S_- = 0)$ scattering

$$f_+(\epsilon, \theta) = f_+^P(\epsilon, \theta) + (2\ell_r + 1)f_+^r(\epsilon)P_{\ell_r}(\cos \theta), \qquad f_-(\epsilon, \theta) \equiv f_-^P(\epsilon, \theta), \quad (5.6)$$

where ℓ_r is the orbital moment of the resonance scattering partial wave. With the quasi-discrete level $(E_r > 0)$ being in the vicinity of zero energy the specific form of the resonance scattering amplitude is given by [2.2]

$$f_+^r(\epsilon) = -\frac{\hbar}{\sqrt{2m}} \frac{\gamma \epsilon^{\ell_r}}{\left(\epsilon - \epsilon_0 + i\gamma \epsilon^{\ell_r + 1/2}\right)} \,. \qquad (5.7)$$

Here ϵ_0 and γ are the dimensioned parameters (ϵ_0, $\gamma > 0$) defining the position of maximum E_r in electron-atom scattering $\sigma^r_{max} = \sigma^r_{el}(E_r) \approx 9\pi/k^2_r$ and the width Γ_r of the ^3P-resonance ($E_r \sim \epsilon_0$ and $\Gamma_r/2 \sim \gamma\epsilon_0^{\ell_r+1/2}$).

In the case of ^3P-resonance (when $\ell_r = 1$, and spin factor $C(S_+) = 3/4$) its contribution to the total differential scattering cross section averaged over spins (5.2) can be described by the simple formula [4.16]

$$\frac{d\sigma^r_{eB}}{d\Omega} = \frac{27}{4}\left|f^r_+(\epsilon)\right|^2\cos^2\theta, \qquad \left|f^r_+(\epsilon)\right|^2 = \frac{\hbar^2}{2m}\frac{\gamma^2\epsilon^2}{(\epsilon - \epsilon_0)^2 + \gamma^2\epsilon^3}. \qquad (5.8)$$

Then, the real and imaginary parts of the resonance scattering amplitude $f^r_{eB}(\epsilon, \theta) = (2\ell_r + 1)f^r_+(\epsilon)\cos\theta$ are given by

$$\mathbf{Re}\left\{f^r_+(\epsilon)\right\} = \frac{\sqrt{2m}(\epsilon_0 - \epsilon)}{\hbar\gamma\epsilon}\left|f^r_+(\epsilon)\right|^2, \quad \mathbf{Im}\left\{f^r_+(\epsilon)\right\} = \frac{\sqrt{2m\epsilon}}{\hbar}\left|f^r_+(\epsilon)\right|^2. \ (5.9)$$

Comparison of the simple analytic expression (5.8) with the results [5.19, 20] of numerical calculations shows that it provides quite reasonable quantitative description of the ^3P-resonance curve in the energy region $\epsilon \sim E_r$ for all alkali atoms.

As is evident from Fig. 5.2, the ^3P-resonance gives a major contribution to the electron-alkali atom scattering in a wide range of energies $|\epsilon - E_r| \sim \Gamma_r$ ($E_r \sim \Gamma_r \sim 10$–100 meV, see Table 5.2). However, in the close vicinity of zero energy $\epsilon \ll E_r$ the main role is played by the potential scattering, which can be described (at $\alpha k^2/a_0 \ll 1$) by the standard expression (5.5) of the effective range theory [5.9, 10] averaged over spins:

$$\frac{d\sigma^P_{eB}}{d\Omega} = L_1^2 + \frac{\pi\alpha}{a_0}L_2 k\sin\left(\frac{\theta}{2}\right) + \frac{8\alpha}{3a_0}L_1^2 k^2\ln(ka_0) + Bk^2 + O\left(k^3\right). \qquad (5.10)$$

Here L_1 and L_2 are the effective scattering lengths defined by the relations

$$L_1^2 = \frac{3}{4}L_+^2 + \frac{1}{4}L_-^2, \qquad L_2 = \frac{3}{4}L_+ + \frac{1}{4}L_-. \qquad (5.11)$$

L_+ and L_- are the usual triplet and singlet scattering lengths, respectively. Since the value of L_2 in (5.11) is negative (see Table 5.2), the Ramsauer-Townsend minimum is observed for any alkali atom in the region between 1–20 meV. It is important to stress that available experimental data of *Heinke* et al. [5.27] for the averaged over spins scattering lengths of the heavy alkali-metal atoms $L_2^{Rb} = -13.7\,a_0$ and $L_2^{Cs} = -18.3\,a_0$, obtained from measurements of the pressure shift of high-Rydberg levels by Rb and Cs, are in good agreement with corresponding results ($L_2^{Rb} = -12.2\,a_0$ and $L_2^{Cs} = -17.6\,a_0$) found from the effective range theory [5.19].

c) Real and Virtual States of Weakly Bound Negative Ions. The behavior of the amplitude for elastic electron scattering by a neutral target B is also appreciably changed compared to the simplest law $f_{eB} = -L = const$ if there is a discrete or virtual level of a corresponding negative ion B$^-$ with energy closed to zero. If this level corresponds to the electron momentum

ℓ, then the low-energy expansion for the contribution of the partial ℓ-wave to the scattering amplitude can be written as

$$f_{eB}^{(\ell)}(\epsilon, \theta) \approx (2\ell + 1) \frac{(-1)^{\ell+1} |E_0|^\ell}{\beta (\epsilon + |E_0|)} P_\ell (\cos \theta), \qquad \epsilon \sim |E_0| \qquad (5.12)$$

in accordance with general theory [2.2]. Here $|E_0| = \hbar^2 \varkappa^2 / 2m$ is the energy of a real ($\varkappa > 0$) or virtual ($\varkappa < 0$) bound state ($E_0 < 0$), and β is the parameter of low energy expansion. In the most interesting case of s-level ($\ell = 0$) the scattering amplitude is an isotropic function and the coresponding total cross section is given by $\sigma_{eB} = 4\pi / (k^2 + \varkappa^2)$ at $\epsilon \sim |E_0|$. Therefore, the presence of a bound or virtual level of a weakly bound negative ion (e.g., supported by the long-range interaction) leads to the scattering cross section, which turns out to be significantly greater than the value $4\pi r_0^2$ determined by the characteristic radius r_0 of the short-range interaction.

It is important to stress that in recent years considerable experimental and theoretical efforts have been devoted to the study of very weakly bound energy ^2P-levels of negative Ca^-, Sr^-, and Ba^--ions [5.30–32]. Their role in collisional processes involving Rydberg atoms attracts growing attention (for example, charge transfer process: $Ca(4snl) + Ca(4s^2) \rightarrow Ca^+(4s) + Ca^-(4s^2 4p)$, see [5.33]).

5.2.2 Electron-Molecule Scattering

The main features of the electron-molecule scattering are due to the long-range dipole or quadrupole interaction and to the presence of rotational degrees of freedom. For instance, the rotational excitation and de-excitation of a molecule play a very important role at small energies and give a substantial contribution to the total and momentum transfer cross sections.

a) **Nonpolar Molecules.** Consider first the scattering of low-energy electrons ($k \ll 1/r_0$) by a nonpolar diatomic molecule B in the $^1\Sigma$-state, for which a long-range potential is given by the expression

$$\mathcal{V}^{l.r.}(r, \Theta) = -\left(\frac{eQ}{r^3} + \frac{\alpha_2 e^2}{2r^4} \right) P_2(\cos \Theta) - \frac{\alpha_0 e^2}{2r^4} + O(1/r^5) . \quad (5.13)$$

Here Θ is the angle of the radius vector of the incident electron **r** relative to the vector of the internuclear **N**-axis; Q is the quadrupole moment. The symmetrical α_0 and orientation-dependent α_2 polarizabilities are determined by the main magnitudes $\alpha_\parallel$ and $\alpha_\perp$ of the polarizability tensor, i.e., $\alpha_0 = (\alpha_\parallel + 2\alpha_\perp)/3$ and $\alpha_2 = 2(\alpha_\parallel - \alpha_\perp)/3$. For the potentials with the r^{-n} asymptotic dependence, the contribution to the scattering amplitude f_{eB}, which arises from the large distances (when $|\mathcal{V}| \ll \hbar^2/mr^2$), can be calculated in the perturbation theory, if $n > 2$ [2.2]. However, the validity condition of perturbation theory for the short-range interaction may fail to hold. The low-energy expansion of the scattering amplitude has been found by *O'Malley*

[5.10] in the case when the electron-point quadrupole interaction is a leading term at large distances. The final expressions for the differential cross sections $d\sigma_{eB}^{j'j}/d\Omega = (k'/k)\left|f^{j'j}(\mathbf{k}',\mathbf{k})\right|^2$ of pure elastic $j \to j$ scattering and for the rotational $j \to j \pm 2$ transitions can be written as

$$\frac{d\sigma_{eB}^{jj}}{d\Omega} = L^2 + \frac{\pi\alpha_0}{2a_0}QL + \frac{j(j+1)}{(2j+3)(2j-1)}$$
$$\left[\frac{4a_0^2}{45}\left(\frac{Q}{ea_0^2}\right)^2 + \frac{\pi\alpha_2 Q}{60}\left(\frac{Q}{ea_0^2}\right)\right], \tag{5.14}$$

$$\frac{d\sigma_{eB}^{j'j}}{d\Omega} = \frac{j_>(j_>-1)}{(j+j_>+1)(j+j_>-1)}\frac{k'}{k}\left[\frac{2a_0^2}{15}\left(\frac{Q}{ea_0^2}\right)^2 + \frac{\pi\alpha_2 Q}{40}\left(\frac{Q}{ea_0^2}\right)\right]. \tag{5.15}$$

Here $Q = |\mathbf{k}' - \mathbf{k}|$ is the momentum transfer; $(k')^2 = k^2 - 2m(E_{j'} - E_j)/\hbar^2$; j and $j' = j \pm 2$ are the rotational quantum numbers of the molecule before and after collision (E_j and $E_{j'}$ are its internal rotational energies), $j_> = \max(j,j')$, and $L = -\lim_{k\to 0}\left(k^{-1}\tan[\eta_0(k)]\right)$ is the standard scattering length. As is apparent from (5.14), in the presence of the quadrupole interaction the contribution to the cross sections of elastic scattering as $k \to 0$ gives not only the S-wave, but also all of the highest partial scattering waves.

b) Polar Molecules. In this case the asymptotic form of the interaction potential is given by

$$V^{l.r.}(r,\Theta) = -\frac{eD}{r^2}\cos\Theta - \left(\frac{eQ}{r^3} + \frac{\alpha_2 e^2}{2r^4}\right)P_2(\cos\Theta) - \frac{\alpha_0 e^2}{2r^4} + O(1/r^5). \tag{5.16}$$

The main features of electron-polar molecule scattering are due to the presence of a leading dipole term. It is predominant for rotational excitation (de-excitation) processes, provided that the dipole moment $\mathcal{D}$ is not too small. The quadrupole and polarization interactions in (5.16) give significant contribution to the scattering amplitude if the Qk/ea_0 and $\alpha_0 k^2/a_0$ parameters become of the same order of magnitude as $\mathcal{D}/ea_0$. Because of sufficiently slow fall (as r^{-2}) of dipole potential in the asymptotic region of r, the distant collisions between the low-energy electron and polar molecule are, on the whole, more important in rotational $j \to j \pm 1$ transitions than the close collisions. Thus, a simple description of such processes can be made on the basis of first order perturbation theory with the use of point-dipole interaction. As it has been found by *Massey* [5.34], in the Born approximation the differential $d\sigma_{eB}^{j'j}/d\Omega$ cross section of the rotational excitation (de-excitation) of the polar molecule is given by

$$\frac{d\sigma_{eB}^{j'j}}{d\Omega} = \frac{k'}{k}\left|f_{eB}^{j'j}(\mathbf{k}',\mathbf{k})\right|^2 = \frac{4}{3}\frac{j_>}{2j+1}\frac{k'}{k}\left(\frac{\mathcal{D}}{ea_0}\right)^2\frac{1}{Q^2}, \tag{5.17}$$

where $j_> = \max(j,j')$ and $Q^2 = k^2 + k'^2 - 2kk'\cos\theta$. In the adiabatic approximation, when the energy of the incident electron is not very low $\epsilon \gg$

$|\Delta E_{j'j}| \sim j\mathcal{B}_0$ (where $\mathcal{B}_0$ is the rotational constant of molecule), one may neglect the difference between the k and k' values in (5.17). Then, the simple expressions of *Altshuler* [5.35] for the differential cross section, averaged over the orientations of the molecular axis, directly follow from (5.17)

$$\frac{d\sigma_{\mathrm{eB}}}{d\Omega} = \sum_{j'=j\pm 1} \frac{d\sigma_{\mathrm{eB}}^{j'j}}{d\Omega} = \left(\frac{\mathcal{D}}{ea_0}\right)^2 \frac{2}{3k^2\,(1-\cos\theta)}\ . \tag{5.18}$$

These expressions of the first-order Born approximation describe the contribution of the dipole interaction alone. Notice also that the divergence of the total cross section for the electron-point dipole scattering has the tendency of the momentum transfer to become infinite as $k \to 0$ is due to the long-range potential behavior $1/r^2$ but not to the Born-approximation [2.2]. This divergence is due to the use of the nonrotating polar molecule approximation, with the rotational constant taken to be $\mathcal{B}_0 \to 0$.

The first-order Born approximation with the point-dipole is certainly satisfactory for rotational excitation (de-excitation) $j \to j \pm 1$ processes at relatively small dipole moment $\mathcal{D} < (0.5-1)ea_0$ in the whole low-energy range of practical importance [5.4, 5]. However, as follows from the close coupling calculations, it yields reasonable cross sections even for $\mathcal{D} \sim ea_0$. On the other hand, at low energies the cross section for elastic electron-polar molecule scattering is primarily determined by the lowest partial S- and P-waves. Hence, it is very sensitive to the value of the short-range interaction and cannot be considered within the framework of perturbation theory. The differential cross section of inelastic $j \to j \pm 1$ transitions is also determined by the intermediate or the short-range parts of interaction at large scattering angles (i.e., at high momentum transfer Q), while at the small scattering angles it is due to the long-range dipole potential.

At large values of dipole moment $\mathcal{D} > (0.5-1)\,ea_0$ there are some interesting effects in the behavior of pure elastic scattering and rotational de-excitation of molecule at ultra-low electron energies. They are the results of the shape resonances on quasi-discrete levels or the presence of weakly bound or virtual dipole supported states of negative ions of polar molecules (see [5.8] and references therein). For example, the detailed studies of electron scattering by the strongly polar HF, CH_3Cl, and $C_6H_5NO_2$ molecules at sub-milli-electron-volt energies (down to a few microelectron volts) was recently performed in [5.36–38] using the experimental data with the Rydberg atoms at very high principal quantum numbers (up to 1100). It was suggested that at ultra-low energy region the electron scattering by these molecules are strongly influenced by the presence of dipole-supported virtual states.

In Fig. 5.3 (panel a) we present theoretical results [5.38] for the dependence of the rate constant $v\sigma_j(v)$ of rotational de-excitation of HF(j) with different values of $j = 0, ..., 7$ and its the incident electron velocity $v = \hbar k/m$. Theoretical analysis of rotationally inelastic $j \to j'$ transitions was based on the close-coupling calculations with the cut-off dipole, quadrupole and polar-

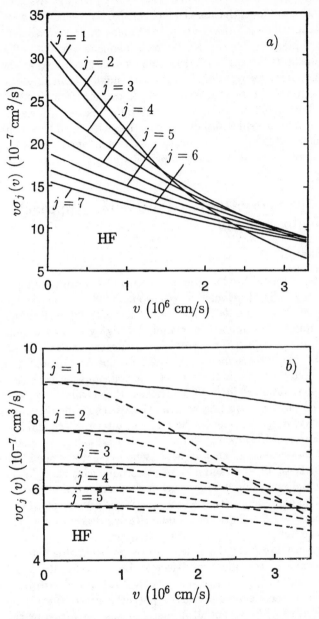

Fig. 5.3. The velocity dependence of the rate constants $v\sigma_j(v)$ for rotational de-excitation of the HF(j) molecule by electron impact. Full curves on panel a are the close-coupling calculations [5.38] assuming the presence of virtual state with energy 1.36 eV. Full curves on panel b are corresponding calculations [5.38] in the Born approximation

ization interaction and the presence of virtual state ($j = 0$) with an assumed energy of 1.36 meV. It was shown that for the incident partial s-electron wave the dipole-allowed transitions $j \rightarrow j - 1$ are predominant in de-excitation of molecules (the contribution of the $j \rightarrow j - 2$ transitions is about 4 %). The de-excitation rate constants increase with decreasing velocity, whereas at low v they turn out to be larger for rotational states with small j than those for high-j values. At higher v the reverse is true. As is evident from Fig. 5.3 (panel b) similar behavior follows from the Born approximation. However, the close coupling results are substansially higher than the Born rate constants and have a more pronounced electron velocity dependence that is the result of the presence of a virtual state with small energy.

5.3 Semiclassical Theory: Impact-Parameter Approach with Fermi Pseudopotential

5.3.1 Historical Sketch

Semiclassical impact-parameter method combined with the Fermi pseudopotential model was originally developed by *Gersten* [5.39], *Omont* [5.40], *Derouard* and *Lombardi* [5.41], and *de Prunelé* and *Pascale* [5.42] in the studies of quasi-elastic l-mixing and impact broadening of highly-excited atomic states by neutral atoms (see [5.2] for more details). Later *Kaulakys* [5.43] applied this approach to derive simple expressions for the broadening and shift of Rydberg isolated ns-states induced by elastic collisions with the rare gas atoms. The results obtained in [5.43] generalize the asymptotic (high n) expressions for the elastic scattering contribution to the shift [5.1] and broadening [5.40] of Rydberg ns-states in the range of intermediate and low enough values of n. Further development of the Fermi pseudopotential model was carried out by *Lebedev* and *Marchenko* [5.44], who presented a semiclassical theory for the inelastic $nl \rightarrow n'$ and $n \rightarrow n'$ transitions with changing of the orbital and principal quantum numbers within the framework of perturbation theory. Analytic expressions for the cross sections and rate constants obtained in this paper were applied to get a quantitative description of experimental data on the inelastic quenching of Rydberg atomic states by rare gas atoms and to demonstrate sharp dependence of the n, l-changing quenching processes on the value of the quantum defect δ_l. Later the results [5.44] for the inelastic $nl \rightarrow n'$ transitions were used [4.16, 2.14] in order to evaluate the contribution of the electron-perturber potential scattering to the broadening of high-Rydberg nl-levels of alkali-metal atoms perturbed by the ground-state parent atoms Li, Na, K, and Rb.

Using simple semiclassical expressions for the inelastic [5.44] and elastic [5.43] Rydberg atom-neutral collisions, *Sun* and *West* [5.45] calculated the impact broadening cross sections of the Rb(ns) by rare gas atoms. In another paper, *Sun* et al. [5.46] applied the multichannel-quantum defect method in order to generalize the above mentioned semiclassical results on the Rydberg atom having more than one valence electron, and to explain the irregular broadening rates of zhe $Sr(5sns^1S_0)$ series induced by collisions with Xe atoms. Simultaneously, *Sirko* and *Rosinski* [5.47] carried out numerical calculations of the J-mixing $(n^2D_{3/2} \rightarrow n^2D_{5/2})$ cross sections for the Rydberg states of Cs* by rare gas atoms on the basis of the semiclassical Fermi pseudopotential model in the weak coupling approximation $(n > 10$–$15)$. An analytic semiclassical description of the $nlJ \rightarrow nlJ'$ transitions between the fine-structure components of the Rydberg atom and of pure elastic $nlJ \rightarrow nlJ$ scattering has been performed by *Lebedev* [5.48, 49] who applied the normalized version of perturbation theory and the JWKB-approximation. The results obtained are valid for weak and close coupling, i.e., at high, intermediate and low enough n. Recently, an efficient unitarized semiclassical approach was developed [5.50–52], for simultaneous description of both the quasi-elastic l-mixing and the inelastic n, l-changing processes. It provides a successful quantitative explanation of major phenomena in these processes and their dependence on the energy ΔE transferred to the Rydberg atom, relative velocity V, and parameter of electron-perturber interaction in a wide range of the principal quantum number n.

5.3.2 Probabilities of the $nlJ \rightarrow n'l'J'$ and $nl \rightarrow n'l'$ Transitions

a) Basic Expressions of the First-Order Perturbation Theory. Below we present basic expressions of first-order perturbation theory (4.47) for the transition probabilities between the $nlJ \rightarrow n'l'J'$ and $nl \rightarrow n'l'$ states of the Rydberg atom with one valence electron over and above the closed shell of the parent core. Within the framework of the Fermi pseudopotential model for the short-range part of the electron-atom interaction

$$V_{\text{eB}}^{\text{sh.r.}}(\mathbf{r} - \mathbf{R}) = \frac{2\pi\hbar^2 L}{m} \delta(\mathbf{r} - \mathbf{R}),\tag{5.19}$$

the transition matrix elements over the wave functions of Rydberg atom A* are given by

$$\mathcal{V}_{fi}[\mathbf{R}(t)] = \langle\psi_f(\mathbf{r})|\,\mathcal{V}[\mathbf{r} - \mathbf{R}(t)]\,|\psi_i(\mathbf{r})\rangle = \frac{2\pi\hbar^2 L}{m}\psi_f^*[\mathbf{R}(t)]\,\psi_i[\mathbf{R}(t)].\tag{5.20}$$

The use of this expression and the impact-parameter approach (Sect. 4.4) allows us to derive simple analytic formulas for the cross sections of elastic, quasi-elastic and inelastic Rydberg atom-neutral collisions. For slightly polarizable perturbing atoms (for example, He) these formulas provide an appropriate qualitative description of such collisions and yield reliable quantitative results.

In the case of transitions between $nlJ \rightarrow n'l'J'$ -states with the given magnitudes of the principal and orbital quantum numbers and total angular momenta the matrix elements (5.20) should be calculated over the wave functions in the $nlsJM$-representation, i.e.,

$$|nlJM\rangle = \mathcal{R}_{n_*l}\left(r\right) Y_{JM}^{ls}\left(\theta_{\mathbf{r}}, \varphi_{\mathbf{r}}\right), \qquad Y_{JM}^{ls}\left(\mathbf{n}_r\right) = \sum_{m\sigma} C_{lms\sigma}^{JM} Y_{lm}\left(\theta_{\mathbf{r}}, \varphi_{\mathbf{r}}\right) \chi_{s\sigma}.$$

$$(5.21)$$

Here $\mathcal{R}_{n_*l}(r)$ is the radial wave function of the $|nlJM\rangle$-state with the effective quantum number $n_* = n - \delta_{lJ}$, δ_{lJ} is its quantum defect; $Y_{JM}^{ls}\left(\theta_{\mathbf{r}}, \varphi_{\mathbf{r}}\right)$ is the spherical spinor with the total angular momentum of J ($J = |l - 1/2|$ or $l + 1/2$) and its z-projection of M; $\chi_{s\sigma}$ is the spin function of the electron ($s = 1/2$) with the z-projection of $\sigma \equiv s_z = \pm 1/2$; and $C_{lms\sigma}^{JM}$ is the Clebsch-Gordan coefficient. Further, we use expansion of the zero-range pseudopotential (5.19) over the spherical harmonics (2.115) and the technique of nonreduced tensor operators [2.8] for calculation of the matrix elements $\langle n'l'J'M'| \mathcal{V}_{\text{eB}}^{\text{sh.r.}} |nlJM\rangle$. Then, the use of the basic formula (4.47) of first-order perturbation theory for the probability

$$W_{nlJ}^{n'l'J'} = \frac{1}{2J+1} \sum_{MM'} W_{nlJM}^{n'l'J'M'}$$

of the $nlJ \rightarrow n'l'J'$ transition and the summation of the matrix element squares over all possible M and M' magnitudes, leads to the following semiclassical result [1.3]:

$$W_{nlJ}^{n'l'J'}\left(\rho, V\right) = \frac{\hbar^2 L^2}{4m^2} \sum_{\varkappa=|l'-l|}^{l'+l} A_{l'J',lJ}^{(\varkappa)} \int_{-\infty}^{\infty} dt \int_{-\infty}^{\infty} dt' \, \exp\left[-i\omega_{fi}\left(t - t'\right)\right]$$

$$\times P_\lambda \left(\cos \Theta_{\mathbf{R}'\mathbf{R}}\right) \mathcal{R}_{n_*'l'}\left[R\left(t\right)\right] \mathcal{R}_{n_*l}\left[R\left(t\right)\right] \mathcal{R}_{n_*'l'}\left[R\left(t'\right)\right] \mathcal{R}_{n_*l}\left[R\left(t'\right)\right],$$

$$(5.22)$$

where $\omega_{fi} = |E_{n'l'J'} - E_{nlJ}|/\hbar$, and the angular $A_{l'J',lJ}^{(\varkappa)}$ coefficients are expressed in terms of $3j$- and $6j$-symbols

$$A_{l'J',lJ}^{(\varkappa)} = (2l+1)(2l'+1)(2J'+1)(2\varkappa+1) \left\{ \begin{array}{ccc} l' & J' & 1/2 \\ J & l & \varkappa \end{array} \right\}^2 \left(\begin{array}{ccc} l' & \varkappa & l \\ 0 & 0 & 0 \end{array} \right)^2.$$

$$(5.23)$$

Upon use of a straight-line trajectory, the relation between the internuclear $R \equiv R(t)$ [or $R' \equiv R(t')$] distances of heavy A^+ and B particles and the time moment t (or t') are reduced to

$$R(t) = \sqrt{\rho^2 + V^2 t^2}, \qquad dt = \frac{R dR}{V\sqrt{R^2 - \rho^2}}.$$

$$(5.24)$$

It is convenient to choose the coordinate system so that the ion core A^+ is placed at its origin O, the Z-axis is perpendicular to the collision plane of A^+ and B particles and the X and Y axes are directed parallel to the impact parameter vector ρ and relative velocity $\mathbf{V}$, respectively. Then, the Rydberg electron $\theta_{\mathbf{r}}$ and $\varphi_{\mathbf{r}}$ angles at the points; $\mathbf{r} = \mathbf{R}$ and $\mathbf{r}' = \mathbf{R}'$ and the included angle $\Theta_{\mathbf{R}'\mathbf{R}} = \varphi_{\mathbf{R}'} - \varphi_{\mathbf{R}}$ of the $\mathbf{R}$ and $\mathbf{R}'$ vectors in (5.22) are given by

$$\theta_{\mathbf{R}} = \pi/2, \quad \varphi_{\mathbf{R}} = \arctan\left(Vt/\rho\right), \qquad \cos\Theta_{\mathbf{R}'\mathbf{R}} = \left(\rho^2 + V^2 tt'\right)/RR'. \quad (5.25)$$

A similar semiclassical expression for the probability

$$W_{n'l',nl} = \frac{1}{2l+1} \sum_{mm'} W_{nlm}^{n'l'm'}$$

of the $nl \rightarrow n'l'$- transition directly follows from (5.22) if we neglect the energy $|\Delta E_{J'J}| = 2Ry\,|\delta_{J'} - \delta_J|\,/n_*^3$ of the fine-structure splitting as compared to the energy defect $|\Delta E_{n'l',nl}| = 2Ry\,|n'_* - n_*|\,/n_*^3$ of the initial nl and final $n'l'$-levels ($|\Delta\delta_{J'J}| \ll 1$). The final result is given by [1.3, 5.46]

$$W_{n'l',nl}\left(\rho,V\right) = \frac{(2l'+1)\hbar^2 L^2}{4m^2} \int\limits_{-\infty}^{\infty} dt \int\limits_{-\infty}^{\infty} dt' \ \exp\left[-\mathrm{i}\omega_{fi}\left(t-t'\right)\right] P_{l'}\left(\cos\Theta_{\mathbf{R}'\mathbf{R}}\right)$$

$$\times P_l\left(\cos\Theta_{\mathbf{R}'\mathbf{R}}\right) \mathcal{R}_{n'_*l'}[R(t)]\mathcal{R}_{n_*l}[R(t)]\mathcal{R}_{n'_*l'}[R(t')]\mathcal{R}_{n_*l}[R(t')],$$

$$(5.26)$$

where $n_* = n - \delta_l$, and $n'_* = n' - \delta_{l'}$.

b) Elastic Scattering and Quasi-elastic $nl \rightarrow nl'$ Transitions. The basic semiclassical formula (5.26) becomes particularly simple in the case of pure elastic scattering ($l' = l$, $n' = n$) of the Rydberg atom by a neutral particle and for quasi-elastic $nl \rightarrow nl'$ transitions with changing of the orbital angular momentum alone $l' \neq l$, $n' = n$

$$A^*(nl) + B \rightarrow A^*(nl') + B. \qquad (5.27)$$

In the simplest case of elastic scattering of the perturbing atom by the Rydberg atom in the ns-state, when $\omega = 0$ and $l' = l = 0$, it can be rewritten as

$$W_{ns}^{\mathrm{el}}(\rho,V) = \frac{\hbar^2 L^2}{4m^2} \left| \int\limits_{-\infty}^{\infty} \mathcal{R}_{ns}^2\left[R\left(t\right)\right] dt \right|^2 = \frac{\hbar^2 L^2}{m^2 V^2} \left| \int\limits_{\rho}^{\infty} \mathcal{R}_{ns}^2\left(R\right) \frac{R dR}{\left(R^2 - \rho^2\right)^{1/2}} \right|^2 .$$

$$(5.28)$$

Then, using the JWKB-approximation for the radial wave function $R_{ns}(r)$ of the Rydberg atom (2.65) and calculating the integral over the classically allowed range of r ($r = R$, $0 \leq R \leq 2n_*^2 a_0$), on the assumption that $\langle\cos^2\Phi_R\rangle = 1/2$, we obtain

$$W_{ns}^{\mathrm{el}}(\rho,V) = \frac{L^2}{\pi^2 n_*^6 \rho a_0} \left(\frac{v_0}{V^2}\right) K^2[\mathrm{k}(\rho)] \,, \qquad (5.29)$$

$$\mathrm{K}(k) = \int\limits_{0}^{\pi/2} \left(1 - k^2\sin^2\theta\right)^{-1/2} d\theta, \quad \mathrm{k}(\rho) = 2^{-1/2}\left(1 - \frac{\rho}{2n_*^2 a_0}\right)^{1/2} . \qquad (5.30)$$

Here $K(k)$ is the complete elliptic integral of the first kind [2.4], the value of which is slowly varying ($\pi/2 \leq K[k(\rho)] \leq 1.8$) in the range of $0 \leq \rho \leq 2n_*^2 a_0$. Due to the exponential decrease of the wave function $\mathcal{R}_{nl}(r)$ in the classically forbidden range, the probability strongly falls in the range of $\rho > 2n_*^2 a_0$. Therefore, its contribution to the cross section $\sigma_{ns}^{el}(V)$ can be neglected. Expression (5.29) is valid in the range of weak coupling, in which the probability obtained in first-order perturbation theory is small $W_{ns}^{el}(\rho, V) \ll 1$. Hence, it becomes inapplicable at small values of $\rho \leq \rho_0$ ($\approx a_0 \left(v_0 L/2a_0 V n_*^3\right)^2$), where ρ_0 is defined by the condition $W_{ns}^{el}(\rho_0, V) \approx 1$. At large values of $n_* \gg (v_0 |L|/4V a_0)^{1/4}$ the range of impact-parameters $0 \leq \rho \leq \rho_0$ corresponding to the close coupling is small ($\rho_0 \ll 2n_*^2 a_0$) and, hence, its contribution can also be neglected.

In a general case, the contribution of the perturber-quasi-free electron scattering mechanism to the cross section $\sigma_{nl',nl}(V)$ of quasi-elastic ($\omega_{nl',nl} = 0$) $nl \rightarrow nl'$ transition may be described at $l, l' \ll n$ by simple analytic formula [5.41, 3.20]

$$\sigma_{nl',nl} = \frac{1}{2l+1} \sum_{mm'} \sigma_{nlm}^{nl'm'} = \frac{2\pi C_{l'l} L^2}{(V/v_0)^2 n_*^4}, \qquad n_* \gg \left(\frac{v_0|L|}{4V a_0}\right)^{1/4}, \qquad (5.31)$$

which is applicable in the range of weak coupling. The magnitudes of the $C_{l'l}$ coefficients in the case for pure elastic scattering ($nl \rightarrow nl$ transition) are as follows $C_{ss} = 0.58$; $C_{pp} = 1.03$; $C_{dd} = 1.28$; and $C_{ff} = 1.46$.

5.3.3 Binary-Encounter Theory: $nl \rightarrow n'$ and $n \rightarrow n'$ Transitions

a) $nl \rightarrow n'$ Transitions. We shall now present the main results of *Lebedev* and *Marchenko* [5.44] semiclassical theory for the probabilities $W_{n',nl} = \sum_{l'} W_{n'l',nl}$, the cross sections $\sigma_{n',nl} = \sum_{l'} \sigma_{n'l',nl}$ and the rate constants $K_{n',nl} = \sum_{l'} K_{n'l',nl}$ of the inelastic $nl \rightarrow n'$ transitions from a given nl-state to all degenerate $n'l'$-sublevels of another hydrogen-like n'-level

$$A^*(nl) + B \rightarrow A^*(n') + B.$$

These quantities are needed, for one, to evaluate quenching and broadening cross sections in inelastic collisions between an atomic projectile B and a selectively excited Rydberg atom $A^*(nl)$ [or $A^*(nlJ)$]. An analytic semiclassical description of such $nl \rightarrow n'$ transitions can be made using the JWKB-approximation for the initial and final radial wave functions of Rydberg atom. Now we substitute expression (2.65) into (5.26) and use only the first term $\Phi_{R'} - \Phi_R = (d\Phi_R/dR)(R' - R)$ of series expansion for the semiclassical phase difference

$$\Phi_{R'} - \Phi_R = k_R(R' - R), \qquad k_R = \left[\varkappa_R^2 - (l+1/2)^2/R^2\right]^{1/2}, \qquad (5.32)$$

where $\varkappa_R = (2/Ra_0 - 1/n_*^2 a_0^2)^{1/2}$. Then, neglecting the rapidly-oscillating terms of $\cos(\Phi_{R'} + \Phi_R)$ and using (5.32), the probability of the $nl \to n'$ transition can be written as

$$W_{n',nl} = \frac{(v_0/a_0)^2 L^2}{2\pi^2 n_*^3 (n')^3} \int\limits_{-\infty}^{\infty} dt \int\limits_{-\infty}^{\infty} dt' \exp\left[i\omega_{n',nl}(t-t')\right] P_l(\cos\Theta_{R'R})$$

$$\times \frac{\cos\left[k_R(R'-R)\right]}{R^2 k_R} \sum_{l'} (l'+1/2) P_{l'}(\cos\Theta_{R'R}) \frac{\cos\left[k'_R(R'-R)\right]}{R^2 k'_R} .$$

(5.33)

The major contribution to the sum $W_{n',nl} = \sum_{l'} W_{n'l',nl}$ over all orbital momenta in the final state n' is determined by large values $l' \gg 1$. Thus, we may use an approximate expression for the Legendre polynomial

$$P_{l'}(\cos\Theta_{R'R}) \approx J_0\left[(l'+1/2)\Theta_{R'R}\right], \qquad \Theta_{R'R} \ll 1, \tag{5.34}$$

which directly proceeds from (2.76) at $m = 0$. This is equivalent to the application of the JWKB-approximation for the angular wave function $Y_{l'm'}$ in the final $n'l'm'$-state. Moreover, summation over all possible l' values may be replaced in (5.33) by integration over dl'. As a result, we have

$$\sum_{l'} (l'+1/2) P_{l'}(\cos\Theta_{R'R}) \frac{\cos[k'_R(R'-R)]}{R^2 k'_R}$$

$$\approx \int\limits_0^{\varkappa'_R R} \frac{(l'+1/2)\, dl'}{R^2} J_0\left[(l'+1/2)\Theta_{R'R}\right]$$

$$\times \frac{\cos\left[\left(\varkappa_R'^2 - (l'+1/2)^2/R^2\right)^{1/2}(R'-R)\right]}{\left[\varkappa_R'^2 - (l'+1/2)^2/R^2\right]^{1/2}}$$

$$= \frac{\sin\left(\varkappa'_R\left[(R'-R)^2 + R^2\Theta_{R'R}^2\right]^{1/2}\right)}{\left[(R'-R)^2 + R^2\Theta_{R'R}^2\right]^{1/2}} . \tag{5.35}$$

The use of the series expansion for the $R' - R$ and $\Theta_{R'R} = \varphi_{R'} - \varphi_R$ values (see (5.25)) in the first order of accuracy, yields

$$R' - R = \dot{V}\tau\sin\varphi_R, \qquad \sin\varphi_R = Vt/R,$$
$$R\Theta_{R'R} = V\tau\cos\varphi_R, \qquad \cos\varphi_R = \rho/R,$$

(5.36)

where $\tau = t' - t$ is the time interval.

Henceforth, we shall consider small orbital momenta $l \ll n$ of the initial Rydberg nl-states, because they are of greatest importance for experimental applications (i.e. ns, np, nd and nf-states). In this case the main contribution to the transition probability $W_{n',nl}$ is due to small enough time intervals $\tau \lesssim na_0/V$. Therefore, the characteristic values of the polar angles difference $\Theta_{R'R} = V\tau\rho/R^2 < 1/n \ll 1$, and we may assume $P_l(\cos\Theta_{R'R}) = 1$ in (5.34), if the initial orbital momentum $l \ll n$. Thus, with the aid of (5.25–36), the inelastic transition probability $W_{n',nl}$ in first-order perturbation theory can be described by the expression

$$W_{n',nl}(\rho, V) = \frac{2(v_0/a_0)^2 L^2}{\pi^2 n_*^3 (n')^3 V^2} \int_{R_{\min}}^{R_{\max}} \frac{dR}{Rk_R(R^2 - \rho^2)^{1/2}} \int_0^\infty \frac{d\tau}{\tau} \sin(\varkappa'_R V\tau)$$

$$\times \cos(\omega_{n',nl}\tau) \cos\left(\frac{V\tau k_R(R^2 - \rho^2)^{1/2}}{R}\right), \qquad l \ll n. \tag{5.37}$$

Here $R_{\max} = r_2$ and $R_{\min} = \max(\rho, r_1)$; while r_1 and r_2 are the classical turning points [see (2.66)], so that $r_2 \approx 2n_*^2 a_0$ and $r_1 \approx (l + 1/2)^2 a_0/2 \ll r_2$ at $l \ll n$ (i.e. we may assume $r_1 \approx 0$). Besides, it is evident that $k_R \approx \varkappa_R$ at $l \ll n$.

The cross section $\sigma_{n',nl}(V)$ of the $nl \to n'$ transition is determined from (5.37) by integrating the probability over the classically allowed range of the impact parameter. The result of the calculation for small orbital momentum $l \ll n$ of the initial Rydberg nl-state, is given by the simple analytic formula [5.44]

$$\sigma_{n',nl}(V) = \frac{2\pi L^2}{(V/v_0)^2 (n')^3} f_{n',nl}(\lambda), \qquad \lambda = n_* a_0 \omega_{n',nl}/V,$$

$$f_{n',nl}(\lambda) = \frac{2}{\pi}\left\{\arctan\left(\frac{2}{\lambda}\right) - \left(\frac{\lambda}{2}\right)\ln\left[1 + \left(\frac{2}{\lambda}\right)^2\right]\right\}. \tag{5.38}$$

Here $\omega_{n',nl} = |\Delta E_{n',nl}|/\hbar$ and $\Delta E_{n',nl} = E_{n'} - E_{nl}$ are the transition frequency and energy defect between the initial nl-state with quantum defect δ_l and final degenerate $n'l'$-states of hydrogen-like n'-level, respectively (i.e., $E_{nl} = -Ry/(n - \delta_l)^2$ and $E_{n'} = -Ry/n'^2$). The dimensionless parameter λ characterizes the inelasticity of the collisional $nl \to n'$ transition. This expression can also be derived [5.53] in the impulse approximation with constant electron-perturber scattering amplitude $f_{eB} = -L$ using the simple formula (3.54) of the binary-encounter theory [3.23] for the atomic form factor $F_{n',nl}(Q)$ of the $nl \to n'$ transition.

The asymptotic behavior of the $f_{n',nl}(\lambda)$ function at small and large inelasticity parameters, is given by

$$f_{n',nl}(\lambda) \to 1 \quad (\lambda \to 0), \qquad f_{n',nl}(\lambda) \approx (8/3\pi)\lambda^{-3} \quad (\lambda \gg 1). \tag{5.39}$$

Therefore, for high $n \gg (|\delta_l + \Delta n| v_0/V)^{1/2}$ (when $\lambda \ll 1$) the cross sections for inelastic $nl \to n'$ transitions ($\Delta n = n' - n$) tend to the quasi-elastic limit $\omega_{n',nl} \to 0$ of weakly coupled states

$$\sigma_{n',nl}(V) = \frac{2\pi L^2}{(V/v_0)^2(n')^3} \qquad (\lambda \to 0) \qquad (5.40)$$

because of the reduction in the energy defect $\Delta E_{n',nl} \propto 1/n^3$ with increasing n. Thus, analytic formula (5.38) contains the *Omont's* result [5.40] for the sum $\sigma_{n,nl} = \sum_{l'} \sigma_{nl',nl}$ of pure quasi-elastic $nl \to nl'$ transitions with change in the orbital angular momentum alone ($n' = n$, $\omega_{nl',nl} = 0$) as a special case. However, in the range of low $n \ll (|\delta_l + \Delta n| v_0/V)^{1/2}$, it leads to the rapid power fall ($\sigma_{n',nl} \propto n^3$) in the cross section for inelastic $nl \to n'$ transition

$$\sigma_{n',nl}(V) \approx \frac{16L^2Vn^3}{3v_0 |\delta_l + n' - n|^3} \qquad (\lambda \gg 1) \qquad (5.41)$$

with a decrease in the principal quantum number (i.e., with increasing in the energy defect $\Delta E_{n',nl} = 2Ry |\delta_l + \Delta n| /n^3$). For a given relative velocity V, the cross section $\sigma_{n',nl}(V)$ reaches its maximum

$$\sigma_{n',nl}^{\max}(V) \approx \frac{0.1\sigma_{eB}^{el}}{|\delta_l + n' - n|^{3/2} (V/v_0)^{1/2}} \qquad (\lambda \approx 1.2) \qquad (5.42)$$

at $n_{\max} \approx 0.9 (|\delta_l + \Delta n| v_0/V)^{1/2}$. Here $\sigma_{eB}^{el} = 4\pi L^2$ is the total cross section of elastic electron-atom scattering. One can see, that the particular magnitudes of $n_{\max}$ and $\sigma_{n',nl}^{\max}$ depend significantly on the value of the electron-perturber scattering length L, on relative velocity V of the colliding atoms, and on the quantum defect δ_l of the Rydberg state.

It is important to stress that for nonhydrogenic atom the cross section (5.38) depends on the orbital momentum l only through the frequency of the $nl \to n'$ transition. This dependence disappeares in the quasi-elastic limit $\omega_{n',nl} \to 0$ or at very high velocity in correspondence with (5.40).

The cross section $\langle \sigma_{n',nl} \rangle_T$ of inelastic transition averaged over the Maxwellian distribution of relative velocities of colliding A* and B particles is most interesting for the experimental applications at thermal energies. For de-excitation process $nl \to n'$ (when $E_{nl} > E_{n'}$), the result of calculation [5.44] can be written as

$$\langle \sigma_{n',nl} \rangle_T = \frac{\langle V\sigma_{n',nl}(V) \rangle_T}{\langle V \rangle_T} = \frac{2\pi L^2}{(V/v_0)^2(n')^3} \varphi_{n',nl}(\lambda_T),$$

$$\varphi_{n',nl}(\lambda_T) = \exp\left(\frac{\lambda_T^2}{4}\right) \text{erfc}\left(\frac{\lambda_T}{2}\right) - \frac{\lambda_T}{\pi} \int_0^\infty \ln\left(1 + \frac{4}{\lambda_T^2}u\right) \frac{e^{-u}du}{u^{1/2}} .$$

$$(5.43)$$

Here $V_T = (2kT/\mu)^{1/2}$ is the velocity corresponding to the kinetic energy $\mathcal{E} = \mu V^2/2 = kT$ of the colliding atoms relative motion for a given gas temperature T, $\langle V \rangle_T = (8kT/\pi\mu)^{1/2}$ is the mean thermal velocity, and erfc(z) is the additional probability integral. Figure 5.4 shows the plot of the $\varphi_{n',nl}(\lambda_T)$ function versus λ_T value. One can see, that the cross section reveals a strong drop with increase of the inelasticity parameter of transition.

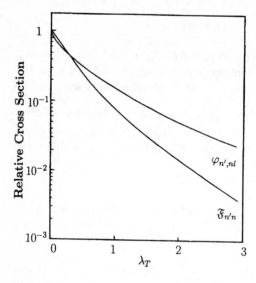

Fig. 5.4. The plots of the $\varphi_{n',nl}(\lambda_T)$ and $\mathfrak{F}_{n'n}(\lambda_T)$ functions that determine the dependencies of the cross sections of the $nl \to n'$ and $n \to n'$ transitions on the value of inelasticity parameter $\lambda_T = na_0\omega/V_T$. Theoretical curves represent the semiclassical calculations [5.44] of the cross sections averaged over the Maxwellian distribution of velocities

b) $n \to n'$ **Transitions.** Let us consider now inelastic transitions ($\omega_{n'n} \neq 0$) between the Rydberg hydrogen-like levels $n \to n'$

$$A^*(n) + B \to A^*(n') + B,$$

when different degenerate lm-sublevels corresponding to a given n being equally populated in accordance with their statistical weights $g_{nl} = 2l + 1$. This situation occurs even for relatively low buffer-gas densities because of high cross section values for the quasi-elastic l-mixing process ($\omega_{nl',nl} = 0$). The most interesting quantities are the total probability $W_{n'n}$, cross section $\sigma_{n'n}$ and rate constant $K_{n'n}$ of inelastic transition from level n to n' summed over all final l' and averaged over the initial l degenerate states of Rydberg atom. First numerical calculations of the rate constants for inelastic $n \to n'$ transitions with a change in the principal quantum number were carried out by *Bates* and *Khare* [5.54] and *Flannery* [5.55] within the framework of the semiclassical theory in the momentum representation which was described in review [4.18].

Simple analytic formulas for the probability $W_{n'n}$, the cross section $\sigma_{n'n}$, and the rate constant $K_{n'n}$ of the inelastic $n \to n'$ transition, which are valid in a wide range of n and energy defects $\Delta E_{n'n}$, were obtained [5.44] using the semiclassical impact-parameter method presented above. The final result for the cross section of the inelastic $n \to n'$ transition can be written as

$$\sigma_{n'n}(V) = \frac{2\pi L^2}{(V/v_0)^2(n')^3}F_{n'n}(\lambda), \quad \lambda = na_0\omega_{n'n}/V = |\Delta n|\, v_0/Vn^2,$$

$$\tag{5.44}$$

$$F_{n'n}(\lambda) = \frac{2}{\pi}\left[\arctan\left(\frac{2}{\lambda}\right) - \frac{2\lambda\left(3\lambda^2 + 20\right)}{3\left(4 + \lambda^2\right)^2}\right].$$

The same expression was simultaneously derived [5.56, 53] in the impulse approximation using the binary-encounter approximation (3.54) for the form factor of the $n \to n'$ transition and the scattering length approximation (when $f_{eB} = -L$ and $\sigma_{eB}^{el} = 4\pi L^2$).

Integration of (5.44) over the Maxwellian distribution of the colliding atoms relative velocity, leads to the following expressions for the rate constants of excitation $K_{nn'}$ ($n' \to n$ transition) and de-excitation $K_{n'n} = \langle V\sigma_{n'n}(V)\rangle_T$ ($n \to n'$ transition) of the Rydberg A* atom by neutral atom B ($n > n'$) [5.44]

$$K_{n'n}(T) = \frac{v_0 \sigma_{eB}^{el}}{\pi^{1/2}(V_T/v_0)n^3}\mathfrak{F}_{n'n}(\lambda_T) , \qquad \mathfrak{F}_{n'n}(\lambda_T) = \exp\left(\frac{\lambda_T^2}{8}\right)$$

$$\times \left[\exp\left(\frac{\lambda_T^2}{8}\right)\text{erfc}\left(\frac{\lambda_T}{2}\right) - \frac{\lambda_T^2}{(2\pi)^{1/2}}D_{-3}\left(\frac{\lambda_T}{2^{1/2}}\right) - \frac{5\lambda_T}{\pi^{1/2}}D_{-5}\left(\frac{\lambda_T}{2^{1/2}}\right)\right].$$

$$(5.45)$$

$$K_{nn'} = \frac{n^2}{(n')^2}K_{n'n}\exp\left(-\frac{|\Delta E_{n'n}|}{kT}\right) , \tag{5.46}$$

where $D_{-\nu}(z) \equiv U(\nu - 1/2, z)$ is the Weber parabolic cylinder function (see [2.4]), $\lambda_T = na_0\omega_{n'n}/V_T$ and $V_T = (2kT/\mu)^{1/2}$. Figure 5.4 shows a plot of the $\mathfrak{F}_{n'n}$ function versus the inelasticity parameter λ_T. Its asymptotic behavior is given by

$$\mathfrak{F}_{n'n}(\lambda_T) \approx 1 - 8\lambda_T/3\pi^{1/2} \quad (\lambda_T \ll 1), \quad \mathfrak{F}_{n'n}(\lambda_T) \approx 2^6/\pi^{1/2}\lambda_T^5 \quad (\lambda_T \gg 1).$$

$$(5.47)$$

Thus, in the asymptotic region of very high $n \gg (|\Delta n|\, v_0/V)^{1/2}$ the inelastic $n \to n'$ transition rate constant (5.45) reduces to the simple expression

$$K_{n'n}(T) = \left(\frac{\mu Ry}{\pi mkT}\right)^{1/2}\frac{v_0\sigma_{eB}^{el}}{n^3} \quad (\lambda_T \to 0), \tag{5.48}$$

which corresponds to the quasi-elastic limit ($\Delta E_{n'n} \to 0$) considered first by *Flannery* [5.55]. However, in the region of $n \ll (|\Delta n|\, v_0/V)^{1/2}$ (where $\lambda_T \gg 1$), general formula (5.45) yields [5.44]

$$K_{n'n}(T) = \frac{2^6 n^7}{\pi\,|\Delta n|^5}\left(\frac{2kT}{\mu v_0^2}\right)^2 v_0\sigma_{eB}^{el} \quad (\lambda_T \gg 1), \tag{5.49}$$

i.e., the rate constant falls rapidly $K_{n'n} \propto n^7$ with n decreasing. The $K_{n'n}$ value reaches its maximum for $\lambda_T \approx 1$.

It is interesting to note that formula (5.44) yields the well-known result of *Pitaevskii* [5.57] for the diffusion coefficient $D(|E_n|)$ in the energy space of the weakly bound electron (moving in the Coulomb field of the ion core with the energy $E_n < 0$) in the process of three-body electron-ion recombination in a buffer gas. In accordance with [5.58] it is determined by the averaged value of the energy transfer square ΔE^2 to the highly excited electron in its collisions with neutral atoms

$$D\left(|E_n|\right) = \frac{1}{2}\frac{\partial}{\partial t}\overline{\Delta E^2} = \frac{1}{2}\left\langle \sum_{n'} \Delta E_{n'n}^2 NV\sigma_{n'n}\left(V\right)\right\rangle_T. \tag{5.50}$$

Here the straight line denotes the averaging over a great number of collisions during the time t, which is equivalent to the averaging over the Maxwellian velocity distribution for a given gas temperature T. In the second equality we have rewritten (5.50) in terms of the transition cross section between the Rydberg levels $n \rightarrow n'$ and the gas density N. Let us substitute (5.44) into (5.50) and replace the summation over all final levels n' by integration over the energy transfer $\Delta E_{n'n} = 2Ry\Delta n/n^3$ (or over the inelasticity parameter λ). Then, the diffusion coefficient takes the form

$$D\left(|E_n|\right) = \left(\frac{m^2 e^6}{\hbar^3}\right)\frac{N\sigma_{eB}^{tr}}{n^3}\left\langle V^2 \int_0^\infty \lambda^2 F_{n'n}(\lambda)d\lambda\right\rangle_T, \tag{5.51}$$

where σ_{eB}^{tr} is the momentum transfer cross section in electron-atom collision, which is equal to $\sigma_{eB}^{tr} = \sigma_{eB}^{el} = 4\pi L^2$ in the ultra-low energy limit $\epsilon \rightarrow 0$. Calculating the integral and using the relations $\langle V^2 \rangle_T = 3kT/\mu$ and $|E_n| = me^4/2\hbar^2 n^2$, we finally obtain the result

$$D\left(|E_n|\right) = \frac{32\sqrt{2}NkT\sigma_{eB}^{tr}}{3\pi\mu}|E_n|^{3/2} \tag{5.52}$$

derived in [5.57] with the use of classical mechanics for calculation of the energy transfer and microcanonical distribution function for the electron energy in the Coulomb field. This expression was applied in [5.57] in order to evaluate the coefficient of the three-body recombination of an electron with an atomic ion in the three-body collision with a neutral atom.

5.4 Impulse Approximation for Rydberg Atom-Neutral Collisions

5.4.1 Introductory Remarks

For collisions of Rydberg atoms with neutral particles, the impulse approximation [4.14] was applied first by *Alekseev* and *Sobel'man* [5.59]. They obtained simple asymptotic formulas for the impact broadening and shift of highly excited levels in terms of the forward elastic scattering amplitude $f_{eB}(k, \theta = 0)$ of the quasi-free electron e by a perturbing atom B and the momentum distribution function in the Rydberg nl-state. Further applications of the quantal impulse approximation to collisions involving the Rydberg atoms and atomic or molecular targets were stimulated by series of papers by *Matsuzawa* [4.15, 3.17, 20, 5.60]. He has derived some expressions for the ionization and excitation cross sections in terms of the squared amplitude $|f_{eB}(Q)|^2$ of the electron-perturber scattering and the transition form factors $F_{fi}(Q)$ (see Sect. 3.4). They are valid given that the amplitude $f_{eB}^{\beta'\beta}$ of the elastic $(\beta' = \beta)$ or inelastic $(\beta' \neq \beta)$ electron-perturber scattering is either constant $(f_{eB} = const)$ or a function of the momentum transfer Q alone, i.e. $f_{eB} = f_{eB}(Q)$. A simple expression for the ionization cross section of the highly excited atom by polar or quadrupolar molecules was independently derived by *Fowler* and *Preist* [5.61, 62]. Another efficient semiclassical impulse approach for describing the energy transfer $n \rightarrow n'$ and ionization of Rydberg atoms by neutral atoms and molecules has been developed by *Pitaevskii* [5.57], *Bates* and *Khare* [5.54] and especially by *Flannery* [5.55, 4.19]. It is based on classical mechanics for calculation of the energy and momentum transfer in the binary encounter of the quasi-free electron with a neutral perturber and on the quantal (or semiclassical) description of the Rydberg atom. The total ionization cross section in this approach [5.55, 4.19] can be expressed through the differential cross section $d\sigma_{eB}/d\Omega = |f_{eB}(\mathbf{k}', \mathbf{k})|^2$ of the electron-perturber scattering and the momentum distribution function of the Rydberg electron in the initial state by means of the fourfold multiple integral. Significant simplifications in the basic expression appear, when the scattering amplitude f_{eB} is a function only of the momentum transfer $Q = |\mathbf{k}' - \mathbf{k}|$ or it is independent of the scattering angle θ (see [4.18]).

Hickman [5.63–65] and *Hugon* et al. [5.66–68] have made considerable use of the Born approximation combined with the Fermi pseudopotential for calculations of the l-mixing and n, l-changing processes with small energy defects $\Delta E_{n'l', nl}$ at thermal collisions of high-Rydberg atoms and the rare gas atoms. Such perturbative approach is equivalent to the quantal impulse approximation under the scattering length approximation $f_{eB} = -L = const$. The impulse-approximation calculations of the l-mixing and quasi-elastic state-changing cross sections in the Rydberg atom-rare gas atom collisions have been also carried out by Matsuzawa and coworkers [3.20, 5.69, 70]. All these results are applicable in the range of weak coupling (i.e. at high enough n for quasi-elastic collisions). At low and intermediate n, the simple power behavior of the l-mixing cross sections $(\sigma_{nl', nl} \propto n^{-4}$ and $\sigma_{nl}^{l-mix} = \sum_{l' \neq l} \sigma_{nl', nl} \propto n^{-3})$, predicted by the impulse approximation, is broken down. As was shown by *Hahn* [5.71], the high-order nonimpulsive correction terms to the impulse approximation [due to the interaction between the perturbing particle B and the parent core A^+ of the Rydberg atom A^* and by the effects of multiple scattering [4.1] become important at $n \leq n_{max}$, where $n_{max} \sim 10$–20 for the l-mixing thermal collisions with the rare gas perturbing atoms. Then, the l-mixing cross section behaves like $\sigma_{nl}^{l-mix} \propto n^4$ at $n \leq n_{max}$.

Gounand and *Petitjean* [3.23] have extended the binary-encounter approach [originally developed [3.17, 22] for the ionization and $n \rightarrow n'$ transitions with large values of Δn] to the inelastic $nl \rightarrow n'$ transitions from the initial nl-level to all degenerate sublevels of the final hydrogen-like state. They have applied the impulse approximation combined with the simple expression (3.54) for the sum of squared form factors $\sum_{l'} F_{n'l',nl}(Q)$ for theoretical analysis of quasi-elastic l-mixing, inelastic n-changing and ionizing processes involving the Rydberg atoms and neutral atoms or molecules [5.56, 44]. The use of this approach in the scattering length approximation $f_{eB} = -L$, leads to the analytic expressions for the cross sections of the $n \to n'$ [5.56, 53] and $nl \to n'$ [5.53] transitions, which have been simultaneously derived in [5.44] by the semiclassical impact-parameter method (see Sect. 5.3).

Further development of the collision theory of Rydberg atoms with neutral particles on the basis of the impulse approximation was made in a series of papers by *Lebedev* and *Marchenko* [4.16, 5.21] and *Lebedev* [4.17, 5.48, 49], who have formulated a general approach for describing various types of inelastic, quasi-elastic, elastic and ionizing collisions within the framework of the quasi-free electron model in the momentum representation. Quantal and semiclassical formulas for the cross sections, derived in these papers, are applicable to the electron-perturber scattering amplitude $f_{eB}^{\beta'\beta}(\mathbf{k}', \mathbf{k})$ of an arbitrary form (i.e., when the amplitude $f_{eB}^{\beta'\beta}$ depends simultaneously on both the electron momentum k and the scattering angle θ, in contrast to the Born approximation or the standard version of the impulse approximation developed in [4.15]). General semiclassical description of the bound-bound $nl \to n'$ and $nl \to n'l'$ transitions in the momentum representation was independently given by *Kaulakys* [5.72] on the basis of the free electron model, which is similar to the semiclassical theory developed in [4.18, 5.57, 54, 55]. Later Lebedev derived simple formulas for the cross sections and the rate constants of direct ionization [4.17] and developed a general analytic approach [5.48, 49] for the description of the inelastic transitions between the fine-structure $nlJ \to nlJ'$ components and for elastic Rydberg atom-neutral scattering.

5.4.2 Fast and Slow Collisions

Consider first the fast collisions ($V \gg v_0/n$) with the relative velocity V of neutral projectile B with respect to the center of mass of the Rydberg atom A being large as compared to the mean orbital electron velocity $v_n \sim v_0/n$. Then, the relative velocities and wave vectors of the Rydberg electron and the perturber B can be approximately represented as [4.15, 18])

$$\mathbf{v}_{eB} \approx -\mathbf{V}, \quad \mathbf{k} \approx -(m/\mu)\,\mathbf{q}, \quad \mathbf{Q} \equiv \mathbf{q} - \mathbf{q}' \approx \mathbf{k}' - \mathbf{k} \qquad (5.53)$$

since the first term on the right-hand side of (4.101) is small and can be neglected. This case corresponds to the weak binding condition in the impulse approximation (see [4.1]) owing to the kinetic energy of the heavy particle relative motion, $\mathcal{E} = \mu V^2/2$ is much greater than the Rydberg electron energy $|E_{nl}| = Ry/n_*^2$. Then, the asymptotic expression for the ionization cross section of the Rydberg atom by a neutral particle directly follows from the general equation (4.120) and can be written as

$$\sigma_{nl}^{\beta'\beta}(V) \underset{n\to\infty}{\longrightarrow} \int \left| f_{eB}^{\beta'\beta}(k,\theta) \right|^2 d\Omega = \sigma_{eB}^{\beta'\beta}(k \approx mV/\hbar). \qquad (5.54)$$

Summation of (5.54) over all possible quantum numbers β' of the perturbing particle yields the well-known result

$$\sigma_{nl,\beta}^{ion}(V) \underset{n\to\infty}{\longrightarrow} \sigma_{nl,\beta}^{tot}(V) = \sigma_{eB}^{tot}(k \approx mV/\hbar) \qquad (5.55)$$

obtained by *Butler* and *May* [5.73] and *B.Smirnov* [5.74]. Hence, at very high $n \gg v_0/V$, the cross section $\sigma_{nl,\beta}^{ion}(V)$ of ionization $A(nl) + B \to A^+ + B + e$ tends asymptotically ($n \to \infty$) to the total cross section $\sigma_{nl,\beta}^{tot}(V) = \sum\limits_{\alpha'\beta'} \sigma_{\alpha',nl}^{\beta'\beta}$ of Rydberg atom-neutral collision. In this mechanism, it is determined by the integral cross section $\sigma_{eB}(v_{eB} \approx V)$ of the electron-perturber scattering. The contribution of all inelastic bound-bound $nl \to n'$ transitions to the total cross section $\sigma_{nl,\beta}^{tot}$ is very small as compared to the ionization and can be neglected. However, at thermal energies, the limit of fast collisions (5.53) between the Rydberg atom and neutral particle is realized only for very high principal quantum numbers $n \gg v_0/V$, i.e., $n \gg 10^3$–10^4.

Since the main part of experiments with the Rydberg atoms and neutral species were performed at thermal energies $\mathcal{E} = \mu V^2/2$, the most important region of the principal quantum numbers $n \ll v_0/V$ corresponds to the case of slow collisions. Then, using the mass-disparity approximation ($M_B, M_A \gg m$), the general relations (4.101–107) for the relative velocities and wave vectors of electron e, ionic core A^+, and neutral perturber B can be rewritten as

$$\mathbf{v}_{eB} \approx \mathbf{v}_{eA^+}, \qquad \mathbf{v}_{A+B} \approx \mathbf{V}, \qquad \mathbf{k} \approx \boldsymbol{\kappa},$$
$$\mathbf{K} \equiv \boldsymbol{\kappa}' - \boldsymbol{\kappa} \approx \mathbf{k}' - \mathbf{k} \approx \mathbf{Q} \equiv \mathbf{q} - \mathbf{q}'. \qquad (5.56)$$

Hence, for slow collisions the basic expression (4.108) of the impulse approximation for the scattering amplitude is given by

$$f_{\alpha'\alpha}^{\beta'\beta}(\mathbf{q} - \mathbf{Q}, \mathbf{q}) \approx \frac{\mu}{m} \left\langle G_{\alpha'}(\mathbf{k} + \mathbf{Q}) \left| f_{eB}^{\beta'\beta}(\mathbf{k} + \mathbf{Q}, \mathbf{k}) \right| G_\alpha(\mathbf{k}) \right\rangle$$
$$= \frac{\mu}{m} \int G_{\alpha'}^*(\mathbf{k} + \mathbf{Q}) f_{eB}^{\beta'\beta}(\mathbf{k} + \mathbf{Q}, \mathbf{k}) G_\alpha(\mathbf{k}) \, d\mathbf{k}. \qquad (5.57)$$

This approximation gives the opportunity to obtain the general formulas of the impulse approximation for the cross sections of various types of the bound-bound and bound-free transitions, which can be used in practical calculations.

5.4.3 Cross Sections of Slow Collisions: General Expressions

a) Semiclassical Expressions: $nl \to n'$ Transitions and Ionization.
For slow collisions ($n \ll v_0/V$) the basic formula of the binary-encounter theory for the cross section of the inelastic (or quasi-elastic) $nl \to n'$ transitions [see (4.120, 122)] may be reduced to a rather simple form [4.16]

$$\sigma_{n',nl}^{\beta'\beta} = \frac{\pi v_0^2}{V^2(n')^3} \int\limits_{Q_{\min}}^{Q_{\max}} dQ \int\limits_{|k_0(Q)|}^{\infty} kdk \, |g_{nl}(k)|^2 \left|f_{eB}^{\beta'\beta}(k,Q)\right|^2 , \tag{5.58}$$

$$k_0(Q) = \frac{\left[m\left(E_{n'} - E_{nl}\right)/\hbar^2\right] - Q^2/2}{Q}, \quad Q^2 = q'^2 + q^2 - 2qq' \cos\theta_{q'q} ,$$

$$q = \mu V/\hbar, \qquad \left(q'\right)^2 = q^2 + \left(2\mu/\hbar^2\right)\left[\left(E_{n'} - E_{nl}\right) - \left(E_\beta - E_{\beta'}\right)\right].$$

Here $Q_{\min} = |q' - q|$ and $Q_{\max} = q' + q$ are the lower and upper limits of integration over the momentum $Q = |\mathbf{k}' - \mathbf{k}|$ transferred to the Rydberg atom in collision with neutral particle, and $g_{nl}(k)$ is the radial part of the momentum space wave function (see Sect. 2.4). This formula can be used in the general case, when the electron-perturber scattering amplitude $f_{eB}^{\beta'\beta}$ is a function not only of the momentum transfer Q, but also of the electron momentum k. Thus, formula (5.58) generalizes the result [3.17, 23] of the binary-encounter theory for the form factor $F_{n',nl}(Q) = \sum_{l'} F_{n'l',nl}(Q)$ and for the corresponding cross section $\sigma_{n',nl}$ of the $nl \to n'$ transition.

General expression (5.58) of the binary-encounter approximation becomes particularly simple in the case of slow collisions between the Rydberg atom and neutral atomic particle, when the internal energy of the perturber is not changed during the scattering, i.e., $\Delta E_{\beta'\beta} = 0$. In this case we can put $|k_0(Q)| \approx Q/2$ for all $Q \geq Q_{\min}$, since $|\Delta E_{n',nl}| \ll \hbar^2 Q_{\min}^2/2m$ for the most interesting range $n \ll (v_0/V)^{2/3}$ of the principal quantum numbers ($n \ll 100$–500 at thermal velocities). Then, using the substitution of the variables $k, Q \to k, \nu$ ($\nu = \cos\theta = 1 - Q^2/2k^2$), the resulting expression for the cross section of the inelastic $nl \to n'$ transition can be rewritten as [4.16]

$$\sigma_{n',nl} = \frac{\pi v_0^2}{2^{1/2} V^2 (n')^3} \int\limits_{k_{\min}}^{\infty} k^2 dk \, |g_{nl}(k)|^2 \int\limits_{-1}^{\nu_{\max}(k)} \frac{d\left(\cos\theta\right)}{\left(1 - \cos\theta\right)^{1/2}} |f_{eB}(k,\theta)|^2 , \tag{5.59}$$

$$\nu_{\max}(k) = 1 - 2k_{\min}^2/k^2, \qquad k_{\min} = Q_{\min}/2 \approx |\Delta E_{n',nl}|/2\hbar V.$$

It can be seen that the lower limit of integration over the quasi-free electron wave number dk (i.e. $|k_0(Q)| \approx Q/2$ or $k_{\min}$ in (5.58, 59), respectively) corresponds to the case of the electron-perturber backward scattering $\theta = \pi$ and $\cos\theta = -1$ ($\mathbf{k}' = -\mathbf{k}$ and $Q = |\mathbf{k}' - \mathbf{k}| = 2k$). An analogous semiclassical expression for the cross section of the $nl \to n'$ transition was also derived in [5.72] within the framework of the quasi-free electron model using the classical description of the energy and momentum transfer in the binary e–B encounter.

The general expression of the binary-encounter theory for the total ionization cross section of Rydberg atom in slow collision ($n \ll v_0/V$) with neutral atom proceeds from the basic equation (4.120) and relations (5.56). After integration over the ejected electron energy E ($E_{min} \leq E \leq E_{max}$, where $E_{min} = |E_{nl}| = Ry/n_*^2$ and $E_{max} = \mathcal{E} - |E_{nl}|$), the final result is given by [4.17]

$$\sigma_{nl}^{ion}(\mathcal{E}) = \frac{\pi a_0 v_0}{V} \int\limits_{Q_1}^{Q_2} dQ \left(Q - Q_0 - \hbar Q^2/2\mu V\right) \int\limits_{Q/2}^{\infty} k\,dk\, |g_{nl}(k)|^2 \, |f_{eB}(k, Q)|^2,$$

$$Q_{1,2} = \left(2\mu/\hbar^2\right)^{1/2} \left[\mathcal{E}^{1/2} \mp (\mathcal{E} - |E_{nl}|)^{1/2}\right], \qquad Q_0 = |E_{nl}|/\hbar V.$$

$$(5.60)$$

An important feature of the ionization process by an atomic projectile is the presence of the threshold of ionization ($\mathcal{E}_{min} = \hbar^2 q_{min}^2/2\mu = Ry/n_*^2$) for all principal quantum numbers n. In contrast to the molecular projectiles (see below Sect. 6.7), in this case the energy $\Delta E_{fi} = E + |E_{nl}|$ transferred to the Rydberg electron in the bound-free transition is determined by the change $\Delta \mathcal{E}_{fi} = \mathcal{E} - \mathcal{E}'$ of the kinetic energy of the colliding atoms relative motion (i.e., $\Delta E_{fi} = \Delta \mathcal{E}_{if}$), since the internal energy of atomic projectile remains unchanged $\Delta E_{\beta\beta'} = 0$.

b) Quantum Expressions: $nlJ \to n'l'J'$ and $nl \to n'l'$ Transitions. A solution of a number of problems in the theory of Rydberg-atom–neutral-particle collisions (for example, J-mixing processes and elastic scattering) need calculations of the transition cross sections between states with the given magnitudes of quantum numbers n, l, and total angular momentum J. In the range of $n \ll v_0/V$ corresponding to slow collisions, the general formula of the impulse approximation for the $nlJ \to n'l'J'$ transitions was derived in [5.49] using the basic equation (5.57) for scattering amplitude and expressions (2.83, 84) for the momentum space wave functions (with spin $s = 1/2$)

$$G_{nlJM}(\mathbf{k}) = \sum_{m\sigma} C_{lms\sigma}^{JM} g_{nl}(k) Y_{lm}(\theta_{\mathbf{k}}, \varphi_{\mathbf{k}}) \chi_{s\sigma} \qquad (5.61)$$

of the initial $|\alpha\rangle = |nlJM\rangle$ and final $|\alpha'\rangle = |n'l'J'M'\rangle$ states. The method of calculation is based on the expansion (see [2.8]) of the spherical $j_{l'}(k'r')Y_{l'm'}(\theta_{\mathbf{k'}}, \varphi_{\mathbf{k'}})$ wave of the $(l' + 1/2)$-order over the bipolar harmonics of the l'-rank . The final general expression for the cross section can be written as

$$\sigma_{nlJ,\beta}^{n'l'J',\beta'}(V) = 2\pi \left(\frac{\hbar}{mV}\right)^2 \sum_{\varkappa=|l'-l|}^{l'+l} A_{l'J',lJ}^{(\varkappa)} \int_{Q_{\min}}^{Q_{\max}} \left|\Phi_{n'l',nl}^{(\varkappa)}(Q)\right|^2 QdQ, \quad (5.62)$$

$$Q_{\min} = |q'-q|, \quad Q_{\max} = q'+q, \quad (q')^2 = q^2 + (2\mu/\hbar^2)[\Delta E_{fi} + \Delta E_{\beta'\beta}].$$

Here $\Delta E_{fi} = E_{n'l'J'} - E_{nlJ}$ and $\Delta E_{\beta'\beta} = E_{\beta'} - E_{\beta}$ are the energy changes of the Rydberg atom A^* and neutral perturber B, respectively. The angular $A_{l'J',lJ}^{(\varkappa)}$ coefficients are defined by (5.23), while the radial integral $\Phi_{n'l',nl}^{(\varkappa)}(Q)$ is given by [5.49]

$$\Phi_{n'l',nl}^{(\varkappa)}(Q) = i^{l-l'}\left(\frac{2}{\pi}\right)\int_0^\infty k^2 dk g_{nl}(k) \int_0^\infty (k')^2 dk' g_{n'l'}^*(k') f_{\mathrm{eB}}(k',k,Q)$$

$$\times \int_0^\infty r^2 dr j_l(kr) j_\varkappa(Qr) j_{l'}(k'r). \quad (5.63)$$

When the fine-structure splitting $\Delta E_{l'-1/2,l'+1/2}$ of the final $n'l'$-level with $J' = |l' \pm 1/2|$ can be neglected, we can perform summation of the cross section $\sigma_{nlJ,\beta}^{n'l'J',\beta'}$ over J' with the use of the known relation for the $6j$-symbols. Then, we obtain the following relation for the sum of squared matrix elements:

$$\frac{1}{2J+1}\sum_{MM'}\left|f_{nlJM,\beta}^{n'l'J'M',\beta'}(\mathbf{q}',\mathbf{q})\right|^2 = \frac{1}{2l+1}\sum_{mm'}\left|f_{nlm,\beta}^{n'l'm',\beta'}(\mathbf{q}',\mathbf{q})\right|^2 = \left(\frac{2\pi\mu}{m}\right)^2$$

$$\times \sum_{\varkappa=|l'-l|}^{l'+l} B_{l'l}^{(\varkappa)}\left|\Phi_{n'_*l',n_*l}^{(\varkappa)}(Q)\right|^2, \quad B_{l'l}^{(\varkappa)} = (2l'+1)(2\varkappa+1)\left\{\begin{array}{ccc}l' & \varkappa & l \\ 0 & 0 & 0\end{array}\right\}^2.$$

$$(5.64)$$

Thus, in this case the calculations of the cross sections $\sigma_{n'l',nl} = \sum_{J'}\sigma_{nlJ}^{n'l'J'}$ of the $nlJ \to n'l'J'$ transitions summed over $J' = |l' \pm 1/2|$ is reduced to the evaluation of the $\sigma_{n'l',nl}$ cross sections of the $nl \to n'l'$ transitions. As directly follows from (5.64), the general formula of the impulse approximation for the cross section of the $nl \to n'l'$ transition differs from (5.62) only by some other angular coefficient, i.e., one should replace $A_{l'J',lJ}^{(\varkappa)} \to B_{l'l}^{(\varkappa)} = \sum_{J'} A_{l'J',lJ}^{(\varkappa)}$.

A particularly simple formula may be obtained from the general expression (5.62) for pure elastic scattering of a neutral atom by the Rydberg atom in the ns-state. In this case the contribution to the sum over $\varkappa$ is determined only by one term with $\varkappa = 0$ (when $B_{ss}^{(0)} = 1$) and the integral over r' in (5.63) is taken analytically. As a result, for slow collisions we have [5.49]

$$\sigma_{ns}^{el}(V) = 2\pi \left(\frac{\hbar}{mV}\right)^2 \int_0^{2q} \left|\Phi_{n0,n0}^{(0)}(Q)\right|^2 QdQ, \qquad q = \mu V/\hbar,$$

(5.65)

$$\Phi_{n0,n0}^{(0)}(Q) = \frac{1}{2Q} \int_0^\infty kdk g_{n \cdot s}(k) \int_{|k-Q|}^{k+Q} k'dk' g_{ns}^*(k') f_{eB}(k,k',Q).$$

In the special case of the ns-state, a similar expression for the cross section of elastic scattering was obtained in [5.75].

5.4.4 Expressions Through the Form Factors and Scattering Length Approximation

If the electron-perturber scattering amplitude depends on the momentum transfer Q alone [i.e., $f_{eB} \equiv f_{eB}(Q)$] or it is a constant (for example, in the scattering length approximation $f_{eB} = -L$), the basic formula (5.57) for the transition amplitude $f_{\alpha'\alpha}^{\beta'\beta}(\mathbf{q} - \mathbf{Q}, \mathbf{q})$ in slow collisions ($n \ll v_0/V$) of the Rydberg atom with neutral particle can be rewritten as [4.15]

$$f_{fi}^{\beta'\beta}(\mathbf{q} - \mathbf{Q}, \mathbf{q}) \approx (\mu/m) f_{eB}^{\beta'\beta}(Q) \langle\psi_f(\mathbf{r})| \exp(i\mathbf{Q} \cdot \mathbf{r}) |\psi_i(\mathbf{r})\rangle. \quad (5.66)$$

Here $\psi_i(\mathbf{r})$ and $\psi_f(\mathbf{r})$ are the initial and final atomic wave functions in the coordinate representation. Then, the cross section (5.62) of the $|\alpha, \beta\rangle \to |\alpha', \beta'\rangle$ transition is reduced to a simple form

$$\sigma_{\alpha'\alpha}^{\beta'\beta}(V) = 2\pi \left(\frac{\hbar}{mV}\right)^2 \int_{Q_{min}}^{Q_{max}} \left|f_{eB}^{\beta'\beta}(Q)\right|^2 F_{\alpha'\alpha}(Q)QdQ$$

(5.67)

$$= 2\pi \left(\frac{\hbar}{mV}\right)^2 \sum_{\varkappa=|l'-l|}^{l'+l} A_{l'J',lJ}^{(\varkappa)} \int_{Q_{min}}^{Q_{max}} \left|f_{eB}^{\beta'\beta}(Q)\right|^2 |\langle n'l'|j_\varkappa(Qr)|nl\rangle|^2 QdQ.$$

Here $F_{\alpha'\alpha}(Q)$ is the atomic form factor (see Sect. 3.4), which is expressed through the radial matrix elements of the spherical Bessel function in accordance with (3.46). Comparison of expression (5.67) with the basic formula (4.32) of the Born approximation shows that they are in full agreement with each other. Hence, the impulse approximation in its general form contains the Born approximation as a special case.

In the simplest case of the scattering length approximation $f_{eB}(k,\theta) = -L = const$ for the electron-atom scattering amplitude formula (5.67) may be rewritten as

$$\sigma_{fi}(V) = 2\pi L^2 \left(\frac{\hbar}{mV}\right)^2 \int_{Q_{min}}^{Q_{max}} F_{fi}(Q) QdQ.$$

(5.68)

This formula gives the opportunity to obtain simple analytic expressions for the cross sections σ_{fi} of various types of the inelastic and quasi-elastic $|i\rangle \rightarrow |f\rangle$ transitions with the use of the appropriate approximations for the atomic form factors $F_{fi}(Q)$. Equivalent analytic expressions may be derived directly from the general formulas of Sect. 5.4.3 assuming $f_{eB} = -L$ and performing integration over the momentum transfer dQ [or over the scattering angle $d(\cos\theta)$] and over the electron momentum dk.

As was mentioned in Sect. 5.3, the impulse approximation in the scattering length approximation (5.68) allows us to derive analytic expressions (5.38) and (5.44) for the cross sections of the $nl \rightarrow n'$ and $n \rightarrow n'$ transitions summed over the orbital quantum numbers. For this it should be combined with the binary-encounter approximation (3.54) for the form factor and with simple semiclassical expressions (2.137) or (2.138) for the momentum distribution function. Thus, the binary-encounter approach in the impact-parameter representation [5.44] and in the momentum representation [5.56, 53] are equivalent to each other.

5.4.5 Total Scattering Cross Section

Now we turn to the behavior of the total cross section $\sigma_i^{\text{tot}} = \sum_f \sigma_{fi}$ of the Rydberg atom-neutral collision at high principal quantum numbers n. It is convenient to present it as the sum of the pure elastic scattering cross section σ_i^{el} and the total contribution σ_i^{in} of all inelastic bound-bound and bound-free $|i\rangle \rightarrow |f\rangle$ transitions ($f \neq i$)

$$\sigma_i^{\text{tot}} = \sum_f \sigma_{fi} = \sigma_i^{\text{el}} + \sigma_i^{\text{in}}, \qquad \sigma_i^{\text{in}} = \sum_{f\neq i} \sigma_{fi}, \qquad \sigma_i^{\text{el}} \equiv \sigma_{ii}, \qquad (5.69)$$

where the initial $i = \{\alpha, \beta\}$ and final $f = \{\alpha', \beta'\}$ states of the combined system $A^* + B$ include the sets of quantum numbers of the Rydberg atom (α) and projectile (β). Here we analyze the contribution of the electron-perturber scattering alone, while the role of elastic and inelastic perturber-core scattering mechanism will be discussed in Chap. 7. The asymptotic expression for the total cross section $\sigma_i^{\text{tot}} = \sum_f \sigma_{fi}$ of the Rydberg atom-neutral collision, which is applicable for an arbitrary relationship between the relative velocity V of colliding particles A^* and B and the mean orbital velocity of the valence electron $v_{nl} \sim v_0/n$, can be obtained from the basic equation (4.108). Assuming $\mathbf{q}' = \mathbf{q}$ for the imaginary part of the elastic scattering amplitude in the forward direction $\theta_{\mathbf{q}'\mathbf{q}} = 0$, i.e.,

$$\text{Im}\left\{f_{ii}\left(\mathbf{q}, \mathbf{q}\right)\right\} = (\mu/\mu_{eB}) \int \text{Im}\left\{f_{eB}^{\beta\beta}\left(\mathbf{k}, \mathbf{k}\right)\right\} |G_\alpha\left(\boldsymbol{\kappa}\right)|^2 d\boldsymbol{\kappa} \qquad (5.70)$$

and using the optical theorem for the total cross sections of the Rydberg atom-neutral collision $\sigma_i^{\text{tot}}(q)$ and for the electron-perturber scattering $\sigma_{eB}(k)$:

$$\mathbf{Im}\left\{\mathfrak{f}_{ii}\left(\mathbf{q},\mathbf{q}\right)\right\} = \frac{q}{4\pi}\sigma_i^{\text{tot}}\left(q\right), \qquad \mathbf{Im}\left\{f_{\text{eB}}^{\beta\beta}\left(\mathbf{k},\mathbf{k}\right)\right\} = \frac{k}{4\pi}\sigma_{\text{eB}}^{\beta\beta}\left(k\right), \qquad (5.71)$$

we finally obtain the well-known asymptotic expression [4.15, 18]

$$\sigma_i^{\text{tot}}(V) = \frac{1}{V}\int v_{\text{eB}}\sigma_{\text{eB}}^{\text{tot}}\left(k\right)\left|G_i\left(\boldsymbol{\kappa}\right)\right|^2 d\boldsymbol{\kappa}, \qquad n \gg \left(v_0/V\right)^{1/2}. \ (5.72)$$

For fast collisions $V \gg v_0/n$, expression (5.72) yields the result (5.55) of *Butler* and *May* [5.73] discussed in Sect. 5.4.2. In the opposite case of slow collisions $V \ll v_0/n$ (when the electron velocity $v = v_{\text{eA+}} \approx v_{\text{eB}} = \hbar k/m$ and momentum $\boldsymbol{\kappa} \approx \mathbf{k}$), expression (5.72) directly leads to the following formula:

$$\sigma_{nl}^{\text{tot}}(V) \approx \int\limits_0^\infty \left(\hbar k/mV\right)\sigma_{\text{eB}}^{\text{el}}(k)\left|g_{nl}\left(k\right)\right|^2 k^2 dk = \frac{\left\langle v\sigma_{\text{eB}}^{\text{el}}\left(v\right)\right\rangle_{nl}}{V}, \quad (5.73)$$

derived by *Alekseev* and *Sobel'man* [5.59] for collisions of the Rydberg atom with an atomic particle. Here

$$\sigma_{\text{eB}}^{\text{el}}(k) = 2\pi\int\limits_0^\pi \left|f_{\text{eB}}(k,\theta)\right|^2 \sin\theta d\theta = \frac{4\pi}{k}\,\mathbf{Im}\left\{f_{\text{eB}}\left(k,\theta=0\right)\right\} \quad (5.74)$$

is the integral cross section for elastic electron-atom scattering. Asymptotic formula (5.73) is applicable in the region of high enough principal quantum numbers $(v_0/V)^{1/2} \ll n \ll (v_0/V)$, i.e., $40-60 \ll n \ll 10^3$–10^4 at thermal energies. In this range of n the cross section σ_{nl}^{el} for elastic Rydberg atom-neutral collision becomes very small [since $\sigma_{nl}^{\text{el}} \propto n^{-4}V^{-2}$, see (5.31) at $l' = l$] as compared to the total contribution σ_{nl}^{in} of all inelastic $nl \to n'$ transitions. Hence, its contribution to the total cross section σ_{nl}^{tot} can be neglected, so that $\sigma_{nl}^{\text{tot}} \approx \sigma_{nl}^{\text{in}}$. Furthermore, in this range of n the inelastic cross section σ_{nl}^{in} is determined by the contribution of a great number of final n' states with $\Delta n = 0, \pm 1, \pm 2, \dots$. As was shown in [5.21], formula (5.73) proceeds directly from the general equation of the impulse approximation (5.59), if we approximately replace the summation over all n' values by integration over the transition energy defects $d(\Delta E_{n',nl})$.

In the scattering length approximation $f_{\text{eB}} = -L$ formula (5.73) reduces to simple analytic expressions for the total cross sections σ_{nl}^{tot} $(l \ll n)$ and $\sigma_n^{\text{tot}} = \sum\limits_{l=0}^{n-1}\left[(2l+1)/n^2\right]\sigma_{nl}^{\text{tot}}$ of the Rydberg atom-neutral atom collision

$$\sigma_{nl}^{\text{tot}}(V) \approx \sigma_{nl}^{\text{in}}(V) \approx \frac{8L^2}{n(V/v_0)}, \qquad (l \ll n), \quad (5.75)$$

$$\sigma_n^{\text{tot}}(V) \approx \frac{32L^2}{3n(V/v_0)}. \quad (5.76)$$

These expressions were first derived [5.40] by the semiclassical impact-parameter method using the Fermi pseudopotential model and the JWKB-approximation for the atomic wave functions.

It should be worthwhile to point out that the general formula of the impulse approximation (5.59) is applicable within the framework of the quasi-free electron model not only in the asymptotic region of high n but also at much lower magnitudes of the principal quantum number. In contrast to the asymptotic expressions (5.73, 75) it provides a reliable quantitative result for the total contribution $\sigma_{nl}^{\text{in}} = \sum_{n'} \sigma_{n',nl}$ of all inelastic $nl \rightarrow n'$ transitions with the use of appropriate values of the energy defects $\Delta E_{n',nl}$. Therefore, formula (5.59) can also be used for the calculations of the quenching $\sigma_{nl}^{\text{Q}} \equiv \sigma_{nl}^{\text{in}}$ cross sections, when the main contribution is determined by an $nl \rightarrow n'$ transition to a single (nearest to the initial nl state) or several n' levels.

5.4.6 Resonance on Quasi-discrete Level

In the presence of low-energy resonance on the quasi-discrete level of the perturbing atom B the standard scattering length approximation is certainly inapplicable for the description of collisions involving Rydberg atoms. Due to strong energy dependence of the scattering amplitude $f_{\text{eB}}(\epsilon, \theta)$ the cross sections of transitions between highly excited states cannot be expressed in terms of the form factors. Hence an appropriate description of the Rydberg atom-neutral collisions at high principal quantum numbers needs application of the general impulse-approximation approach [4.16, 5.21, 49] presented in Sect. 5.3. The mechanism of the resonance scattering in such collisions have been the subject of intensive theoretical and experimental research (see review [1.3]). The first theoretical works were devoted to the study of the possible influence of extremely narrow ($\Gamma_r \ll E_r$) resonances in electron-molecule [5.76] and electron-atom [5.77] scattering upon the behavior of the pressure width and shift of high Rydberg levels. They were based on the asymptotic theory [5.59] of spectral-line broadening. The next theoretical works on the quenching [4.16, 5.75] and broadening [5.21, 19] of Rydberg states in alkali vapors have demonstrated the major contribution of the ^3P-resonance scattering by alkali-metal atoms (for which $\Gamma_r \sim E_r = 10\text{--}100$ meV) in these processes. This resonance leads to sharp energy and angular ($\propto \cos\theta$) dependence of the scattering amplitude $f_{\text{eB}}(\epsilon, \theta)$ in the range above 10 meV (see Sect. 5.2.1). As a result, the quenching and broadening cross sections become particularly large. Furthermore, the resonance scattering leads to the appearance of some new phenomena in dependencies of the broadening and shift of spectral lines on the principal quantum number (such as their oscillatory behavior at intermediate $n \sim 15\text{--}25$ for K, Rb, and Cs). This phenomena will be discussed in Chap. 9 within the framework of the adiabatic quasi-molecular approach [5.78, 26, 79]. Below we present simple results of calculations [4.16, 5.21] for the contributions of the resonance $\sigma_{\text{r}}^{\text{in}}(nl)$ and potential $\sigma_{\text{p}}^{\text{in}}(nl)$ electron-atom scattering into the total cross section $\sigma_{nl}^{\text{in}} = \sum_{n'} \sigma_{n',nl}$ of all inelastic $nl \rightarrow n'$ transitions between highly excited states induced by collisions with the ground-state alkali-metal atoms.

We shall proceed from the general equation (5.59) for slow collisions. In contrast to the asymptotic expression (5.73), this equation gives the opportunity to explain quantitatively the essential influence of the quantum defect value δ_l on the magnitude of the cross section σ_{nl}^{in}, which was observed in the experiments on the quenching [5.80, 29] and broadening [5.27, 28] of high-Rydberg atomic levels by the ground-state alkali-metal atoms. It is convenient to present the sum σ_{nl}^{in} of all inelastic $nl \to n'$ transitions as

$$\sigma_{nl}^{in} = \sigma_r^{in} + \sigma_p^{in} + \text{interference term} \tag{5.77}$$

in accordance with expression (5.6). Each term in this expression is a result of substitution of the corresponding term in (5.6) into the general formula (5.59).

Let us discuss the contribution of the ^3P-resonance scattering to the total inelastic cross section. With the use of simple expression (5.8) for the resonance part of the differential cross section, the final result can be written as [4.16]

$$\sigma_r^{in}(nl) = \frac{63\pi v_0^2}{10V^2} \sum_{n'} \frac{1}{(n')^3} \int_{\epsilon_{min}}^{\infty} |f_+^r(\epsilon)|^2 \mathfrak{D}(\epsilon) W_{nl}(\epsilon) d\epsilon,$$

$$\mathfrak{D}(\epsilon) = 1 - \frac{15}{7}\left(\frac{\epsilon_{min}}{\epsilon}\right)^{1/2} + \frac{20}{7}\left(\frac{\epsilon_{min}}{\epsilon}\right)^{3/2} - \frac{12}{7}\left(\frac{\epsilon_{min}}{\epsilon}\right)^{5/2}, \tag{5.78}$$

where $\epsilon_{min} = \hbar^2 k_{min}^2/2m = |\Delta E_{n',nl}|^2/8mV^2$. The distribution function of the Rydberg electron kinetic energy $\epsilon = \hbar^2 k^2/2m$ in (5.78) is expressed through the momentum distribution function, considered in Chap. 2, by the relation

$$W_{nl}(\epsilon) = \frac{\sqrt{2m^3\epsilon}}{\hbar^3} |g_{nl}[k(\epsilon)]|^2, \qquad \int_0^{\infty} W_{nl}(\epsilon)d\epsilon = 1. \tag{5.79}$$

Formula (5.78) follows directly from (5.59) by performing a trivial integration of the differential cross section $d\sigma_{el}^r/d\Omega \propto \cos^2\theta$ over the scattering angle $\cos\theta$. Thus, the very appearance of the $\mathfrak{D}(\epsilon)$ factor in this formula is due to the p-type of the resonance partial scattering wave. A simple asymptotic estimate for the resonance part of the total inelastic cross section can be obtained from (5.78) as a special case. The final asymptotic expression is given by [5.21]

$$\sigma_r^{in}(nl) \approx \frac{9\pi\gamma^2\hbar^2}{2^{1/2}m^{3/2}V} \int_0^{\infty} \frac{W_{nl}(\epsilon)\epsilon^{5/2}d\epsilon}{[(\epsilon - \epsilon_0)^2 + \gamma^2\epsilon^3]}. \tag{5.80}$$

As was mentioned in Sect. 5.2.1, for the alkali-metal atoms the energies and widths of the ^3P-resonances prove to be of the same order of magnitude ($\Gamma_r \sim E_r \sim 10^{-1}$–$10^{-2}$ eV). Therefore, at sufficiently high n the period of oscillations of the energy distribution function $W_{nl}(\epsilon)$ [described by (2.136)] is much smaller than the width of such resonances. Therefore, these oscillations practically fully cancel each other in (5.78, 80). Hence in the region of applicability of the quasi-free electron model ($n > 25$–30 for thermal collisions with K, Rb, and Cs, see below Sect. 5.4.7) the oscillations in the dependencies of the broadening cross sections upon n do not arise. Thus, the simple expression (2.137) for the average energy distribution function can be certainly used in calculations of the broadening and quenching cross sections of Rydberg atoms perturbed by alkali-metal atoms.

At sufficiently high n the resonance $\sigma_r^{in}(nl)$ contribution to the broadening or quenching cross sections falls with increase of the principal quantum number. It is due to the decrease of the energy distribution function $W_{nl}(\epsilon)$ for the resonance energies $\epsilon \sim E_r$ at $n \gg (Ry/E_r)^{1/2}$. As a result, in the asymptotic region of very high n the main contribution to the total cross section σ_{nl}^{in} of the Rydberg nl level by alkali atoms is due to the electron-perturber potential scattering. Thus, a reliable quantitative description of the quenching and impact broadening processes at high principal quantum numbers needs an account of both the ^3P-resonance and potential triplet and singlet scattering. Since the partial phase shifts $\eta_\ell^{(+)}(k)$ and $\eta_\ell^{(-)}(k)$ strongly fall near the zero energy ($k \ll a_0^{-1}$) with ℓ increasing ($\ell > 1$), it is enough to take into account the contribution of only the first few partial waves to the scattering amplitude (5.1) and corresponding differential cross section (5.2). We will show in Chap. 6 that the general impulse-approximation approach [4.16, 5.21] combined with the theory of resonance and potential electron-alkali atom scattering provides a quantitative description of these processes at high principal quantum numbers [$n > 15$–20 for Li($2s$), Na($3s$), and $n > 25$–30 for K($4s$), Rb($5s$), and $n > 30$ for Cs]. In this range of n the quasi-free electron model is valid and practically no oscillations occur. Therefore, the theory presented above allows us to obtain the values for the ^3P-resonance energies E_r and widths Γ_r, which are in reasonable agreement with the results [5.18, 19] of the modified effective range theory (see Table 5.2). These semiempirical data have been obtained for K and Rb [4.16, 5.21] from the comparison of theoretical results for the monotone component of the broadening cross sections of Rydberg levels in alkali vapors with experimental data measured in [5.27, 28].

5.4.7 Validity Criteria of Quasi-free Electron Model and Impulse Approximation

The applicability conditions of the impulse approximation and the quasi-free electron model for quasi-elastic, inelastic and ionizing collisions between the Rydberg atom A* and neutral perturbing particle were discussed in reviews [4.15, 18, 1.3]. Here we discuss only the physical meaning of the basic approximations. In the impulse approximation for the independent treatment of the two possible mechanisms of the Rydberg atom-neutral collisions caused by the scattering of the perturber B on the outer electron and on the ionic core A$^+$, the following conditions should be satisfied:

$$r_{eA^+} \gg A = \max \{|f_{eB}|, \; |f_{A^+B}|, \; \lambda_{eB}\}, \tag{5.81}$$

$$T_n \gg \tau_{eB}, \; \tau_{A^+B}, \tag{5.82}$$

where f_{eB} and f_{A^+B} are the amplitudes for the scattering of the perturbing particle B on the outer electron and on the parent core A^+, respectively; while λ_{eB} is the de Broglie wavelength for the electron-perturber relative motion. Due to the large characteristic dimension r_{eA^+} of the interaction region between the ionic core A^+ and Rydberg electron (which is responsible for the main contribution to the collisional process) the interference effects in the e-B and A^+−B scattering as well as the effects of multiple scattering can be neglected at high enough n. The second condition (5.82) means that in order for the "impulse" treatment to be applicable, both the electron-perturber τ_{eB} and the core-perturber τ_{A^+B} collision times should be small compared to the period $T_n \sim 2\pi (a_0/v_0) n^3$ of the electron orbital motion. As is apparent from (5.82), the applicability of the impulse approximation in calculations of various elementary processes involving Rydberg atoms and neutral particles is determined not only by the nature of interparticle e − B and A^+−B interactions but also the relative velocity V of the colliding partners A^* and B, as well as, the orbital electron velocity v_0/n. It is important to emphasized that the impulse approximation may be used both for fast $V \gg v_0/n$ and for slow $V \ll v_0/n$ collisions provided the principal quantum number n is large enough.

One more important condition of the applicability of the impulse approximation, which clarifies the physical meaning of the quasi-free electron model, can be written as

$$\hbar Q_{eB} \gg |F_{eA^+}| \tau_{eB} = \left(e^2/r_{eA^+}^2\right) \tau_{eB} . \tag{5.83}$$

Here $\hbar Q_{eB}$ is the characteristic momentum transferred to the Rydberg electron in its collision with the perturbing particle B; $(|F_{eA^+}| \tau_{eB})$ represents the impulse of the Coulomb force acting on this electron from the parent core A^+ during the collision time τ_{eB} of e and B particles. The collision time of the quasi-free electron with the neutral projectile B is determined by the relation $\tau_{eB} \sim r_{eB}/v_{eB}$, in which the characteristic dimension of the interparticle interaction region may be estimated as $r_{eB} \sim \max(\lambda_{eB}, |f_{eB}|)$. Condition (5.83) can be approximately rewritten as (5.82) or $\omega_{n,n\pm1} \tau_{eB} \ll 1$, where $\omega_{n,n\pm1} = 2Ry/\hbar n^3$ is the characteristic transition frequency. Thus, it is clear that the validity conditions (5.82, 83) of the impulse approximation (or the "quasi-free" electron scattering) corresponds to the applicability criterion of the sudden perturbation theory.

Note also that for the impulse approximation to be applicable it is necessary that the cross sections of the Rydberg atom-neutral collisions do not exceed the geometrical cross section of the highly excited atom

$$\sigma_{fi} \ll \sigma_{geom} \sim \pi a_0^2 n_*^4. \tag{5.84}$$

This condition is particularly important in the case of elastic and quasi-elastic collisions, since it is automatically satisfied for the inelastic collisions accompanied by large enough energy transfer.

The value of the electron-perturber scattering amplitude in (5.81) may be estimated as $|f_{eB}| \sim (\langle \sigma_{eB} \rangle /4\pi)^{1/2}$. It is usually reduced to the scattering length $|f_{eB}| \sim |L|$ for the potential electron-atom scattering. The perturber-core scattering amplitude $|f_{A+B}|$ is determined by the characteristic dimension of the polarization interaction $V^{l.r.}_{A^+B} = -\alpha_B e^2/2R^4$ of perturbing atom B and ionic core A^+. Its value may be estimated by the Weisskopf radius $\rho_W = (\pi \alpha_B v_0/4V)^{1/3}$ or by the impact parameter $\rho_{cap} = (4e^2\alpha_B/\mu V^2)^{1/4}$ corresponding to the capture of atom B by the core A^+ of the Rydberg atom (i.e., $|f_{A+B}| \sim \rho_W$ or $|f_{A+B}| \sim \rho_{cap}$ depending on the inelasticity of one or another process). At thermal energies the values of ρ_W and ρ_{cap} are usually greater than the corresponding magnitudes of the electron-atom scattering lengths L, so that $A \sim \rho_W$ or $A \sim \rho_{cap}$ in (5.81). The characteristic radius r_{eA^+} and the de Broglie wavelength λ_{eB} depend on the principal quantum number n, relative velocity V of the colliding particles, and the energy $\Delta E_{fi} = \hbar \omega_{fi}$ transferred to the Rydberg electron. As was shown in [5.44], their values can be estimated as

$$r_{eA^+} \sim \frac{2n_*^2 a_0}{1 + (n_* a_0 \omega_{fi}/V)^2}, \qquad \lambda_{eB} \sim (a_0 r_{eA^+}/2)^{1/2} \qquad (5.85)$$

for the $nl \to n'$ and $n \to n'$ transitions. Thus, for quasi-elastic collisions (when the inelasticity parameter is small $\lambda = n_* a_0 \omega_{fi}/V \ll 1$) the r_{eA^+} value is of the order of the atomic orbital radius $2n_*^2 a_0$ and the de Broglie wavelength of the electron-perturber relative motion is $\lambda_{eB} \sim n_* a_0$. However, for essentially inelastic collisions (with the parameter $\lambda \gg 1$) these values are determined by the relations $r_{eA^+} \sim 2n_*^2 a_0/\lambda^2$ and $\lambda_{eB} \sim n_* a_0/\lambda$, so that $r_{eA^+} \ll 2n_*^2 a_0$ and $\lambda_{eB} \ll n_* a_0$.

Specific estimates show that at thermal collisions of the Rydberg atoms with the ground-state rare gas atoms the impulse approximation is valid for $n \gg 10$–20 (and $n \gg 20$–30 for ionization) depending on the kind of colliding particles and on the energy defect $|\Delta E_{fi}|$ of the transition under consideration. The validity conditions of the impulse approximation in the presence of the ultra-low energy resonance on the quasi-discrete level of the perturbing atom was discussed in [4.16, 5.21]. It was shown that in order for it to be applicable, it is also necessary that the resonance width Γ_r be significantly greater than the energy level spacing $\Delta E_{n,n\pm1}$, so that $\Gamma_r \gg 2Ry/n^3$. This condition is equivalent to (5.82), since the collision time in this case is determined by the lifetime $\hbar/\Gamma_r$ of the quasi-discrete level. The detailed analysis of conditions (5.81, 82) shows that at thermal collisions of the Rydberg atoms with the ground-state alkali-metal atoms the impulse approximation is applicable for $n > n_{low}$, where $n_{low} \sim 15$–20 for Li, Na, $n_{low} \sim 25$–30 for K, Rb, and $n_{low} \sim 30$–35 for Cs. For collisions with strongly polar molecules (such as

NH_3, HF, HCl etc.), the typical magnitudes of the scattering amplitudes $|f_{eB}|$ and $|f_{A+B}|$ are much greater than for atomic perturbers since they are determined by the long-range dipole potential (5.16). As a result, the permissible values of n turn out to be shifted toward high principal quantum numbers and the quasi-free electron model is justified only for $n > 50$–60. Nevertheless, reasonable agreement with the experimental data of the impulse-approximation calculations may be sometimes achieved in the range of n, in which it is formally inapplicable (see below Sect. 6.5).

6. Elementary Processes Involving Rydberg Atoms and Neutral Particles: Effects of Electron-Projectile Interaction

In this chapter our attention will be focused on further development of the theory of collisions between the Rydberg atoms and neutral particles (atoms and molecules) and its application to studies of various types of bound-bound and bound-free transitions. We shall consider different processes of excitation, de-excitation, and ionization of the highly-excited atomic states in both the weak- and strong- coupling regions. However, as in the preceding chapter, we consider here situations in which these processes are due to the mechanism associated with the scattering of a Rydberg electron on the target atom or molecule. The main emphasis will be put on the detailed description of the total and orbital angular momentum transfer processes, the inelastic transitions with a change in the principal quantum number since they were the subject of intensive experimental and theoretical research. We shall present a careful analysis of the dependence of cross section behavior on the principal quantum number, relative velocity of colliding particles, and the inelasticity parameter of one or another process. The final formulas will be given in the form convenient for specific calculations. A considerable part of this chapter is devoted to comparison of the theoretical results with available experimental data on quenching and ionization of Rydberg atomic states by the ground-state rare gas and alkali-metal atoms as well as by the polar, nonpolar, and electron-attaching molecules.

6.1 Classification of Processes and Theoretical Treatments

The elementary processes induced by collisions between the Rydberg atoms and neutral targets (atoms or molecules) can be divided into two main groups. The first one involves the so-called state-changing processes (i.e., various types of quasi-elastic ánd inelastic bound-bound transitions between highly excited states) and elastic scattering. The second group involves different ionization processes (i.e., bound-free transitions of the Rydberg electron). The summary table containing the most important collisional processes, which will be the subject of this and the next chapters, is presented below.

Some of these processes (e.g. elastic scattering, quasi-elastic l-mixing collisions, and inelastic n, l-changing transitions) have been already considered in

Table 6.1. Collisional processes of Rydberg atoms with neutral atoms and molecules

---

State-changing collisions and elastic scattering

---

1. Elastic scattering processes
   $$A(nl) + B \rightarrow A(nl) + B, \quad A(nlJ) + B \rightarrow A(nlJ) + B$$

2. Transitions between the fine-structure components
   $(J - \text{mixing process } J = |l \pm 1/2| \rightarrow |l \mp 1/2|)$
   $$A(nlJ) + B \rightarrow A(nlJ') + B, \quad (J' \neq J)$$

3. Transitions with change in the orbital angular momentum (l-mixing process)
   $$A(nl) + B \rightarrow A(nl') + B, \quad (l' \neq l)$$

4. Transitions with change in the principal quantum number (n-changing process)
   (a) Collision with atom
       $$A(\alpha) + B \rightarrow A(\alpha') + B, \quad (n' \neq n)$$
   (b) Rotationally inelastic collision with molecule
       $$A(\alpha) + B(j) \rightarrow A(\alpha') + B(j'), \quad (n' \neq n, \ j' \neq j)$$

Ionizing collisions

---

1. Direct and associative ionization
   $$A(\alpha) + B(\beta) \rightarrow \begin{cases} A^+ + B(\beta') + e \\ BA^+ + e \end{cases}$$

2. Ionization by electron attaching atom or molecule
   $$A(\alpha) + B(\beta) \rightarrow A^+ + B^-(\gamma)$$

3. Ionization accompanied by dissociative electron attachment
   $$A(\alpha) + CD \rightarrow A^+ + C^- + D$$

---

the preceding chapter. However, all theoretical results on collisions between the Rydberg atoms and neutral particles presented above were obtained in the weak-coupling approximation like the Born approximation or quantal impulse approximation and its semiclassical version (binary-encounter theory in the impact-parameter or momentum representations). Such approaches are valid if the transition probability between highly excited states and the corresponding cross section are sufficiently small ($W_{fi} \ll 1$ and $\sigma_{fi} \ll \sigma_{\text{geom}} \sim \pi a_0^2 n_*^4$). At the same time, the behavior of transition probabilities and cross sections of Rydberg-atom–neutral collisions depends drastically on both the principal quantum number n and the energy defect ΔE_{fi} of the process as well as on a particular type of colliding particles and their relative velocity V. In particular, the cross sections and rate constants turn out to be quite different for transitions with small and large energy transferred to the highly excited electron from the relative motion of the neutral projectile and ionic core of the Rydberg atom.

Analysis of the inelastic n, l-changing transitions with sufficiently large energy transfer ΔE_{fi} and especially ionizing collisions involving the Rydberg and the ground-state atoms shows that these processes may be reasonably described within the framework of first-order of perturbation theory or in the impulse approximation in a wide enough range of the principal quantum number. This means that in this case the range of weak coupling involves simultaneously high, intermediate and sufficiently low principal quantum number (provided the value of n is not too small so that the quasi-free electron model is broken down). Close coupling arises here only at very low magnitudes of the impact parameters $0 \leq \rho \leq \rho_0$ (corresponding to large values of the momentum transfer Q), which do not exert any appreciable influence on the final result for the cross section in the integration over all possible values of ρ ($0 \leq \rho < \infty$).

However, the approaches based on perturbation theory can be applied to calculations of the cross sections of collisions with small energy transfer (e.g., quasi-elastic l-mixing and J-mixing processes) only at high principal quantum numbers $n \gg n_{\max}$ which correspond to the weak-coupling range. At low $n \ll n_{\max}$ and intermediate $n \sim n_{\max}$ the probabilities W_{fi} of these processes become of the order of unity and the cross sections are about the geometrical area of the Rydberg atom $\sigma_{\text{geom}} \sim \pi a_0^2 n_*^4$. Use of first-order perturbation theory in the strong-coupling range of $n \lesssim n_{\max}$ leads to appreciable overestimation of the transition probabilities and to incorrect qualitative behavior of the corresponding cross sections. A consistent quantitative description of such collisions at low and intermediate n requires inclusion of all couplings between near-degenerate states, but the application of the close-coupling method to collisions involving Rydberg atoms becomes very difficult at high principal quantum numbers due to the presence of a great number of closely-spaced levels.

A simple version of this method at low $n(< 10)$ was used for the quasi-elastic l-mixing process in thermal Rydberg atom–rare gas atom collisions [5.64]. Semiclassical calculations [6.1] of the n-changing process in thermal Na(nS)+He collisions were performed by the close-coupling method at $n = 6$ and 9. The same approach was used [6.2] for description of l-mixing in rotationally elastic collisions of Rydberg atoms with strongly polar molecules HF and HCl in a wide range of n. However, reliable semiclassical calculations based on numerical integration of the close-coupling equations for the transition amplitudes were carried out only for few specific processes involving Rydberg atoms. Furthermore, numerical close-coupling calculations have not given a general picture for different processes with regard to their dependence on the quantum numbers of the Rydberg atom, the relative velocity V of the colliding particles, and the transition energy defect $|\Delta E_{fi}|$. On the other hand, the perturbative approaches violate the conservation of transition probability at low and intermediate n if the energy transfer is not sufficiently large.

In a general case, for arbitrary magnitudes of the principal quantum number n, and energy defect ΔE_{fi}, an efficient method for describing Rydberg atom–neutral collisions is to use the so-called normalized perturbation theory. The simplest way to restore the conservation of probability is to use the unitarized version of first-order of perturbation theory in the impact-parameter representation (see Sect. 4.3.2). It is based on separation of the whole range of ρ into two regions ($\rho \leq \rho_0$) and ($\rho_0 < \rho$) with qualitatively different behavior of the transition probability (4.49) and its integration in accordance with (4.50).

Such a semiclassical approach combined with the JWKB-approximation for the Rydberg atom wave functions was developed for description of the J-mixing collisions [5.49] as well as the n, l-changing and l-mixing processes [5.50–52] in a wide range of the principal and orbital quantum numbers and transition energy defects. The identified approach allows us to perform most of the calculations of probabilities, cross sections and rate constants in the analytical form. Hence, it turns out to be especially effective for understanding their dependencies on the principal quantum number n, energy defect of process ΔE_{fi}, relative velocity V, and parameter of electron-perturber interaction.

6.2 Transitions Between the Fine-Structure Components and Elastic Scattering

The inelastic transitions between the fine-structure $nlJ \rightarrow nlJ'$ components of Rydberg atom induced by collisions with neutral projectiles

$$A^*(nlJ) + B \rightarrow A^*(nlJ') + B \qquad (6.1)$$

are accompanied by small energy $\Delta E_{J'J}$ transferred to the highly excited electron from the translational motion of heavy particles. Therefore, they are characterized by large values of the cross sections, which can be of the same order of magnitude as in the process of elastic scattering. Calculations of the J-mixing cross sections in the range of weak coupling (high principal quantum numbers n) has been performed by *Sirko* and *Rosinski* [5.47], *Lebedev* [5.48], and *Liu* and *Li* [6.3] for the $n^2D_{3/2} \rightarrow n^2D_{5/2}$ transitions in the Rydberg Cs and Rb induced by thermal collisions with the rare gas atoms. Analytic description of the inelastic $nlJ \rightarrow nlJ'$ transitions between the fine-structure components in the range of weak and close coupling has been carried out in [5.49] on the basis of the impulse approximation and the semiclassical impact-parameter method.

6.2.1 Weak Coupling Limit

In the range of weak coupling the semiclassical formula (5.22) for the probability of the $nlJ \rightarrow nlJ'$ transition in the impact-parameter representation can be reduced to a rather simple form

$$W_{nlJ}^{nlJ'}(\rho, V) = \frac{L^2 v_0^2}{\pi^2 n_*^6 V^2} \sum_{s=0}^{l} A_{lJ',lJ}^{(2s)} \int_{r_{min}}^{r_{max}} \frac{\cos(\Delta\Phi_R) dR}{R k_R (R^2 - \rho^2)^{1/2}}$$

$$\times \int_{r_{min}}^{r_{max}} \frac{\cos(\Delta\Phi_{R'}) dR'}{R' k_{R'} (R'^2 - \rho^2)^{1/2}} \cos\left[\omega_{J'J}\left(t(R) - t(R')\right)\right] P_{2s}\left(\cos\Theta_{\mathbf{R'R}}\right).$$

(6.2)

if we use the JWKB-approximation for the radial wave functions (2.65) and the Fermi pseudopotential model for the electron-perturber interaction. Here $r_{max} = r_2$, $r_{min} = \max(\rho, r_1)$, and $r_{1,2} = n_*^2 a_0 (1 \mp \epsilon)$ are the classical turning points, i.e., $r_2 \approx 2n_*^2 a_0$ and $r_1 \approx (l + 1/2)^2 a_0/2 \ll 2n_*^2 a_0$ at $l \ll n$. The transition frequency $\omega_{J'J} = |\Delta E_{J'J}|/\hbar$ between the fine-structure components $J = |l - 1/2|$ and $J' = l + 1/2$ (and corresponding quantum defect difference $\Delta\delta_{J'J}$) of the Rydberg atom with one valence electron may be estimated from the relation [3.2]

$$|\Delta E_{J'J}| = \frac{2Ry|\delta_{lJ} - \delta_{lJ'}|}{n_*^3}, \qquad |\delta_{lJ} - \delta_{lJ'}| = \frac{\alpha^2 Z_i^2 H_{rel}(l, Z)}{2l(l + 1)}.$$

(6.3)

Here $\alpha = e^2/\hbar c \approx 1/137$; $H_{rel}(l, Z)$ is the relativistic correction which is relevant only for heavy Rydberg atoms with the spectroscopic symbols $Z \gtrsim 50$ (and $H_{rel} \approx 1$ for $Z \lesssim 50$); $Z_i e^2$ is the effective charge of the ionic core A^+ for the valence electron being at short distance from the nucleus ($Z_i \equiv Z_i(l, Z)$). As a rule: $Z - Z_i(l = 1) \approx 4-6$ and $Z - Z_i(l = 2) \approx 10-16$ for the p- and d-states of alkali-metal atoms.

One can see that the quantum defect difference of the fine-structure components is very small $\Delta\delta_{J'J} \ll 1$. Hence, the phase difference $\Delta\Phi_R$ of the semiclassical wave functions (2.68) in formula (6.2) may be neglected, and we may assume $\cos(\Delta\Phi_R) \approx 1$ and $\cos(\Delta\Phi_{R'}) \approx 1$. The magnitudes of the angular $A_{lJ',lJ}^{(2s)}$ coefficients are presented in [5.49] for transitions between the fine-structure components of the np-, and nd-levels. In the special case, when the energy splitting of the fine-structure components is very small and can be neglected, the transition probability of the $nlJ \to nlJ'$ can be approximately presented as

$$W_{nlJ}^{nlJ'}(\rho, V) \approx \frac{C_{J'J}^{(l)}}{2n_*^6}\left(\frac{v_0}{V}\right)^2\left(\frac{L^2}{\rho a_0}\right), \qquad \Delta E_{J'J} \to 0.$$

(6.4)

This formula differs from the case of pure elastic scattering of the perturbing atom by the Rydberg atom in the ns-state (see Sect. 5.3.2) only by a specific value $C_{J'J}^{(l)}$ of constant coefficient.

Integration of (6.2) over the $2\pi\rho d\rho$ yields the required result for the cross sections of the J-mixing process $\sigma_{nlJ}^{J-\text{mix}} \equiv \sigma_{nlJ}^{nlJ'}$ (transition between the fine-structure $nlJ \to nlJ'$-components with $J' \neq J$) and of pure elastic scattering $\sigma_{nlJ}^{el} \equiv \sigma_{nlJ}^{nlJ}$ ($nlJ \to nlJ$ transition) in the range of weak coupling. However, to derive a simple result for the cross section and its dependence on the energy of the fine-structure splitting it is more convenient to use directly expression

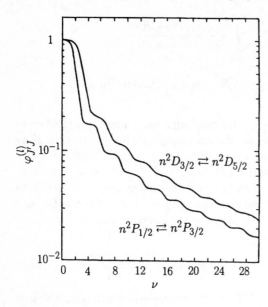

Fig. 6.1. The plots of the $\varphi^{(l)}_{J'J}(\nu)$ functions that determine the dependencies of the J-mixing cross sections for the $n^2P_{1/2} \rightarrow n^2P_{3/2}$ and $n^2D_{3/2} \rightarrow n^2_{5/2}$ transitions on the value of the inelasticity parameter $\nu = n_*^2 a_0 \omega_{J'J}/V$. Theoretical curves represent the impulse-approximation results [5.49]

(5.67) of the impulse approximation. As has been shown in [5.48], at high enough n the final analytic formulas are given by

$$\sigma^{J-\mathrm{mix}}_{nlJ} = \frac{2\pi C^{(l)}_{J'J} L^2 v_0^2}{V^2 n_*^4} \varphi^{(l)}_{J'J}(\nu), \quad \sigma^{\mathrm{el}}_{nlJ} = \frac{2\pi C^{(l)}_{JJ} L^2 v_0^2}{V^2 n_*^4} \quad (n_* \gg n^{(0)}_*), \quad (6.5)$$

where $\nu = n_*^2 a_0 \mid \Delta E_{J'J} \mid /\hbar V$ is the dimensionless inelasticity parameter of the J-mixing process. The $C^{(l)}_{J'J}$ coefficients and the $\varphi^{(l)}_{J'J}(\nu)$ function in (6.5) may be expressed in terms of spherical Bessel functions

$$C^{(l)}_{J'J} = \xi^{(l)}_{J'J}(0), \quad \varphi^{(l)}_{J'J}(\nu) = \xi^{(l)}_{J'J}(\nu)/\xi^{(l)}_{J'J}(0),$$

$$\xi^{(l)}_{J'J}(\nu) = \sum_{s=0}^{l} A^{(2s)}_{lJ',lJ} \int_{\nu}^{\infty} j_s^2(z) J_s^2(z) z\, dz, \quad \xi^{(l)}_{JJ'}(\nu) = \frac{2J+1}{2J'+1} \xi^{(l)}_{J'J}(\nu). \quad (6.6)$$

The plot of the $\varphi^{(l)}_{J'J}(\nu)$ function versus the value of the inelasticity parameter $\nu_{J'J} = \mid \delta_{lJ'} - \delta_{lJ} \mid v_0/V n_*$ is shown in Fig. 6.1 for the most interesting cases of $n^2P_{1/2} \rightleftarrows n^2P_{3/2}$ and $n^2_{3/2} \rightleftarrows n^2D_{5/2}$ transitions. The magnitudes of coefficients for such transitions are as follows: $C^{(p)}_{3/2,1/2} = 0.451$ and $C^{(d)}_{5/2,3/2} = 0.474$ for the $1/2 \rightarrow 3/2$ and $3/2 \rightarrow 5/2$ transitions, while $C^{(p)}_{1/2,3/2} = 0.225$ and $C^{(d)}_{3/2,5/2} = 0.316$ for the inverse transitions, respectively.

As is evident from (6.5, 6), at small magnitudes of the inelasticity parameter $\nu_{J'J} \ll 1$ the behavior of the J-mixing cross section $\sigma^{J-\mathrm{mix}}_{nlJ}$ is similar to the case of pure elastic scattering ($nlJ \rightarrow nlJ$) or quasi-elastic $nl \rightarrow nl'$ transitions ($\omega_{nl',nl} = 0$) with change in the orbital angular momentum alone (5.31). However, the magnitudes of the constant coefficients differ essentially from each other (e.g. $C^{(p)}_{1/2,3/2}/C_{pp} \approx 0.218$ and $C^{(d)}_{3/2,5/2}/C_{dd} \approx 0.246$). Hence, at $\nu_{J'J} \ll 1$ formulas (6.5,6) of the weak-coupling approximation leads to the following expression for the sum of the J-mixing and elastic scattering cross sections:

$$\sigma^{el}_{nlJ} + \sigma^{J-mix}_{nlJ} \equiv \sum_{J'} \sigma^{nlJ'}_{nlJ} \xrightarrow[\nu \to 0]{} \sigma_{nl,nl} = \frac{2\pi C_{ll} L^2}{n^4_*(V/v_0)^2} , \qquad n_* \gg n^{(0)}_* , \qquad (6.7)$$

$$C_{ll} = \sum_{J'} C^{(l)}_{J'J} = \sum_{s=0}^{l} B^{(2s)}_{ll} \int_0^\infty j^2_s(z) J^2_s(z) z dz , \qquad (6.8)$$

where the angular coefficients $B^{(2s)}_{ll}$ are defined by (5.64).

6.2.2 Extension to Strong- Coupling Region

If the energy splitting of the fine-structure components of Rydberg atom is not too large than the value of the inelasticity parameter $\nu_{J'J} = | \delta_{lJ'} - \delta_{lJ} |$ v_0/Vn_* of the J-mixing process turns out to be small enough $\nu_{J'J} \lesssim 1$ not only at high n, but also at the intermediate and low n. As is evident from (6.3), this situation occurs usually for Rydberg atoms with small spectroscopic symbols Z in their thermal collisions with both the light and the heavy perturbing atoms. However, even for the $n^2 D_{3/2} \to n^2 D_{5/2}$ transitions in the Rydberg Rb- and Cs-atoms (for which the experimental data are available) the influence of the value $\nu_{J'J}$ on the behavior of the J-mixing collisions with He is not practically important in the whole studied range of n.

Here we show that simple analytic description [5.49] of such quasi-elastic J-mixing transitions with $\nu_{J'J} \lesssim 1$ and $\varphi^{(l)}_{J'J}(\nu) \approx 1$ may be simultaneously given in the range of weak and close coupling, i.e., at high, intermediate, and low enough principal quantum numbers. This approach is based on the results of first-order perturbation theory for the transition probability $W^B = W^{nlJ'}_{nlJ}$ (6.4) and their normalization at low and intermediate n in accordance with basic semiclassical formulas of the unitarized perturbation theory (4.47–50).

According to [5.49], the probability of the $nlJ \to nlJ'$ transition between the fine-structure components with $l \ll n$ may be described by the following expression:

$$W^{nlJ'}_{nlJ}(\rho, V) = \begin{cases} (g_{J'}/g_s g_{nl})c , & 0 \le \rho \le \rho_0, \\ \dfrac{C^{(l)}_{J'J} L^2}{2n^6_* \rho a_0} \left(\dfrac{v_0}{V}\right)^2 , & \rho_0 \le \rho \le 2n^2_* a_0 \\ 0 & 2n^2_* a_0 < \rho \end{cases} \qquad (6.9)$$

if we neglect its fast exponentially decaying tail in the classically forbidden range of the impact parameters ρ out of the right turning point $r_2 = 2n^2_* a_0$. Here $c < 1$ is the normalization constant whose value may be chosen from comparison with the available experimental data or with the results of the close-coupling calculations at low enough n. It should be emphasized that c is the only empirical parameter of the theory and its value might slightly vary when we switch from one collision system to another. For example, $c = 5/8$ if the cross section $\sum_{J'} \sigma_{nlJ}^{nlJ'}$ is normalized to the geometrical area of the Rydberg atom defined by the relation $\sigma_{\text{geom}} = \pi \langle r^2 \rangle_{nl} \approx \left(5\pi a_0^2/2\right) n_*^4$ at $l \ll n$. The value of $\rho_0 \equiv \rho_0(V, n_*)$ is determined from the condition $W^B(\rho_0, V) = (g_{J'}/g_s g_{nl})c$, where $g_{J'} = 2J' + 1$ and $g_{nl}g_s = \sum_J (2J + 1) = 2(2l + 1)$ are the statistical weights of the final nlJ'-sublevel and the total nl-level (involving spin statistical weight $g_s = 2$), respectively. As is apparent from (6.9), ρ_0 is given by

$$\rho_0(V, n_*) = \begin{cases} 2n_*^2 a_0, & n_* \leq n_0 \\ \dfrac{(2l + 1)C_{J'J}^{(l)}L^2}{(2J' + 1)cn_*^6 a_0}\left(\dfrac{v_0}{V}\right)^2, & n_* \leq n_0 \end{cases} \tag{6.10}$$

Further, following the paper [5.49], we introduce the magnitude of the principal quantum number $n_0(V)$, so that the value of the impact parameter ρ_0 becomes equal to the radius $r_2 \approx 2n_*^2 a_0$ of the Rydberg atom at $n_* = n_0(V)$. As a result, we obtain the relation

$$n_0^8 = \frac{(2l + 1)C_{J'J}^{(l)}L^2}{2(2J' + 1)c}\left(\frac{v_0 L}{V a_0}\right)^2 \tag{6.11}$$

for the $n_0 = n_0(V)$ value, which separates the regions of weak $n_* \gg n_0$ from that of close $n_* \lesssim n_0$ coupling.

For low principal quantum numbers $n_* \leq n_0(V)$, the whole classically allowed range of integration ($0 \leq \rho \leq 2n_*^2 a_0$) over the $2\pi\rho d\rho$ in (4.50) corresponds to the case of close coupling, i.e. $W_{nlJ}^{nlJ'} = const$. Hence, the value of $\sigma_{nlJ}^{nlJ'}(V)$ is determined by the geometrical cross section σ_{geom} of the Rydberg atom. On the contrary, with $n_* > n_0(V)$ [when $\rho_0(V, n_*) < 2n_*^2 a_0$] it is necessary to take into account the reduction of the transition probability $W^B(\rho, V)$ with increasing ρ in the range of $\rho_0 \leq \rho \leq 2n_*^2 a_0$. As a result, performing the integration over the $2\pi\rho d\rho$ in the classically allowed range of ρ, we derive a general semiclassical expression for the cross section of the $nlJ \to nlJ'$ transition [5.49]

$$\sigma_{nlJ}^{nlJ'}(V) = \begin{cases} \dfrac{2J' + 1}{2(2l + 1)}4c\pi a_0^2 n_*^4, & n_* \leq n_0(V), \\ \dfrac{2\pi C_{J'J}^{(l)}L^2 v_0^2}{n_*^4 V^2}\left[1 - \dfrac{n_0^8}{2n_*^8}\right], & n_* \geq n_0(V), \end{cases} \tag{6.12}$$

which is applicable simultaneously at high, intermediate, and low enough n_*. As is evident from (6.12), this result leads to the simple expression (6.5) of the impulse approximation at high n.

One can also see, that the cross section $\sigma_{nlJ}^{nlJ'}$ reaches its maximum at $n_*^{\max}(V) = (3/2)^{1/8} n_0(V)$, if the inelasticity parameter of the J-mixing process is very small $\nu \ll 1$ (or equals to zero, when $J' = J$), so that we may assume $\varphi_{J'J}^{(l)}(\nu) \approx 1$. Then, the cross section $\sigma_{nlJ}^{nlJ'}$ of the quasi-elastic J-mixing collisions behaves like n_*^4 and n_*^{-4} at low and high principal quantum numbers, respectively. However, the situation becomes quite different at large values of the fine-structure splitting $\Delta E_{J'J}$ and fairly low relative velocities V of the colliding A* and B atoms. In this case, the inelasticity parameter for the $nlJ \to nlJ'$ transition ($J' \neq J$) is large, which leads to an appreciable reduction of the J-mixing cross section as compared to the quasi-elastic case ($\nu \ll 1$), since $\varphi_{J'J}^{(l)}(\nu) \ll 1$ at $\nu \gg 1$ (see Fig. 6.1).

The averaged cross sections of excitation ($nlJ \to nlJ'$, $E_{nlJ} < E_{nlJ'}$) and de-excitation $nlJ' \to nlJ$ are related through the detailed balance ($g_J = 2J + 1$ and $g_{J'} = 2J' + 1$):

$$\left\langle \sigma_{nlJ}^{nlJ'} \right\rangle_T = \frac{g_{J'}}{g_J} \left\langle \sigma_{nlJ'}^{nlJ} \right\rangle_T \exp\left(-\frac{\Delta E_{J'J}}{kT} \right) , \tag{6.13}$$

where $\left\langle \sigma_{nlJ}^{nlJ'} \right\rangle_T = \left\langle V \sigma_{nlJ}^{nlJ'} \right\rangle_T / \left\langle V \right\rangle_T$. In the special case of pure elastic scattering ($J' = J$ and $\nu = 0$) or the J-mixing of slightly splitted fine-structure components ($\nu \ll 1$), the averaging expression (6.12) may be performed analytically [5.49]

$$\left\langle \sigma_{nlJ}^{nlJ'} \right\rangle_T = \left(\frac{4 c g_{J'} C_{J'J}^{(l)}}{g_s g_l} \right)^{1/2} \pi a_0^2 \left(\frac{v_0 |L|}{V_T a_0} \right) \mathcal{F}(\eta) , \quad \eta = \frac{g_s g_l C_{J'J}^{(l)}}{4 c g_{J'} n_*^8} \left(\frac{v_0 L}{V_T a_0} \right)^2 , \tag{6.14}$$

where $g_{J'} = 2J' + 1$, $g_l = 2l + 1$, and $g_s = 2$ are the statistical weights, and the $\mathcal{F}(\eta)$ function is given by

$$\mathcal{F}(\eta) = \eta^{1/2} \left\{ E_2(\eta) + \eta^{-1} \left[1 - \exp(-\eta) \right] \right\} , \quad E_2(\eta) = \int\limits_1^\infty \frac{\exp(-\eta t)}{t^2} dt . \tag{6.15}$$

Here $E_2(\eta)$ is the integral exponential function of the 2nd order; $\eta = n_0^8(V_T)/n_*^8$ is the parameter characterizing the strength of the collision; $V_T = (2kT/\mu)^{1/2}$.

The use of the normalized perturbation theory and impact-parameter approach for the calculation of the the elastic scattering cross section for a given nl-state leads to a similar expression. As it was shown in [5.49], the final result $\left\langle \sigma_{nl}^{el} \right\rangle$ directly follows from (6.14) if we put the statistical weights ratio equal to unity ($g_{J'}/g_s g_l \to 1$). The form of $\mathcal{F}(\eta)$-function is similar to the result obtained [5.43] in the intermediate range of n_* for the elastic scattering cross section in the Rydberg ns-state. In particular, $\mathcal{F}(\eta)$ behaves like $\eta^{-1/2}$ and $\eta^{1/2}$ at large $\eta \gg 1$ (strong collision) and small $\eta \ll 1$ (weak collision) values of η, and reaches its maximum $\mathcal{F}_{\max} \approx 0.8$ at $\eta_{\max} \approx 0.68$.

6.3 Orbital Angular Momentum and Energy Transfer: $l-$Mixing and $n, l-$Changing Processes

In the theory of the J-mixing process of the fine-structure components, presented above, we deal with transition between only two Rydberg states with the given quantum numbers nlJ and nlJ' (or with one state in the case of elastic scattering). Here we shall continue theoretical analysis of transitions

$$A^*(nl) + B \rightarrow A^*(n') + B$$

from state nl to a degenerate manifold n', considered previously in Sects. 5.3.3, 4.3 for weak collisions. As was mentioned above, this transition is characterized by the energy defect $\Delta E_{n',nl} \approx 2Ry|\delta_l + \Delta n|/n^3$ (where $\Delta n = n' - n$). If the noninteger part of quantum defect of the nl-state is substantial, the transition energy defect is relatively large even between the nearest nl and n' levels and the process is inelastic. If the quantum defect δ_l is close to zero and the principal quantum number is not changed $n' = n$, we have a quasi-elastic (l-mixing) process.

The focus of this section is to present recent results of the semiclassical impact-parameter approach developed in [5.50–52] for the simultaneous description of inelastic n, l-changing, and quasi-elastic l-mixing processes in collisions of Rydberg atoms with neutral atomic targets induced by the electron–perturber interaction. As in the preceding Sect. 6.2.2, it is based on the normalized perturbation theory and, hence, may be used at high, intermediate, and low enough n.

6.3.1 Semiclassical Unitarized Approach to Inelastic $nl \rightarrow n'$ Transitions

We assume now that the interaction between the perturbing atom B and the ion core can be ignored, and the short-range interaction between the Rydberg electron and B can be described by the zero-range Fermi pseudopotential $V_{eB}(\mathbf{r} - \mathbf{R}) = (2\pi\hbar^2/m) L_{eff}\delta(\mathbf{r} - \mathbf{R})$, where L_{eff} is the effective scattering length for electron-perturber scattering. In contrast to the standard scattering length L, L_{eff} depends on collision parameters and n. This dependence will be discussed in the next subsection.

Further, starting from the basic formula of first-order perturbation theory (5.37) in the impact-parameter representation we apply the standard normalization procedure (4.49) for the probability of the $nl \rightarrow n'$ transition, which was described in Sect. 4.3. As a result, the final equation for the transition probability can be expressed in terms of the incomplete elliptic integrals $F(\varphi, k)$ of the first kind [5.50]

$$W_{n',nl}(\rho) = \begin{cases} c, & 0 \le \rho \le \rho_0 \\[2mm] \dfrac{L_{\text{eff}}^2 v_0^2}{2\pi a_0^2 n_*^6 V^2}\sqrt{\dfrac{a_0}{\rho}}\,[F(\varphi_2,\mathrm{k}) - F(\varphi_1,\mathrm{k}) & \rho_0 \le \rho \le \rho_{\max} \\ \quad + 2\Theta(\tilde{\rho}-\rho)F(\varphi_1,\mathrm{k})], & \\[2mm] 0, & \rho_{\max} < \rho. \end{cases} \qquad (6.16)$$

Here $\Theta(z) = 1$ for $z \ge 0$ and $\Theta(z) = 0$ for $z < 0$; and

$$F(\varphi,\mathrm{k}) = \int_0^{\varphi}(1 - \mathrm{k}^2\sin^2\theta)^{-1/2}d\theta, \qquad \mathrm{k} = [(1 - \rho/2n_*^2 a_0)/2]^{1/2}, \qquad (6.17)$$

(see, for example, [2.4]), while the arguments φ_1 and φ_2 in (6.16) are

$$\varphi_s = \arcsin\left[\left(\frac{R_s(\rho) - \rho}{(1 - \rho/2n_*^2 a_0)R_s(\rho)}\right)^{1/2}\right], \qquad s = 1,2. \qquad (6.18)$$

Parameters $R_1^{(\lambda)}(\rho) = 2n_*^2 a_0 x_1^{(\lambda)}(y)$ and $R_2^{(\lambda)}(\rho) = 2n_*^2 a_0 x_2^{(\lambda)}(y)$ $(R_1 \le R_2)$ are determined from the equation

$$y = \phi_\lambda(x), \qquad \phi_\lambda(x) = (2\lambda)^{1/2}\frac{x^{5/4}}{(1-x)^{1/4}}\left[1 - \frac{\lambda}{2}\frac{x^{1/2}}{(1-x)^{1/2}}\right]^{1/2} \qquad (6.19)$$

for a fixed value of the scaled impact parameter $y = \rho/2n_*^2 a_0$. Here $x = R/2n_*^2 a_0$ is the scaled internuclear separation, and $\lambda = n_* a_0|\Delta E_{n',nl}|/\hbar V$ is the inelasticity parameter for the $nl \to n'$ transition, introduced in Sect. 5.3. The impact parameter $\rho_{\max}$ in (6.16) is the maximum possible value of ρ in the classically allowed region determined by the inelasticity parameter λ of the $nl \to n'$ transition and by the principal quantum number n_*. Within the framework of the semiclassical approach, the transition probability $W_{n',nl}(\rho)$ becomes zero for $\rho > \rho_{\max}(\lambda)$. It corresponds to the maximum value of $\phi_\lambda(x)$. The impact parameter $\tilde{\rho} = 2n_*^2 a_0 \tilde{y}$ in (6.16) is determined by the relation $\tilde{y} = \phi_\lambda(\tilde{x}) = \tilde{x}$, where $\tilde{x} = \bar{R}/2n_*^2 a_0$. From (6.19), we obtain

$$\begin{aligned} \tilde{\rho}(\lambda) &= 2n_*^2 a_0\tilde{y}(\lambda), & \tilde{y}(\lambda) &= 1/(1 + \lambda^2), \\ \rho_{\max}(\lambda) &= 2n_*^2 a_0\phi_\lambda^{\max}, & y_{\max}(\lambda) &= \phi_\lambda^{\max}. \end{aligned} \qquad (6.20)$$

Thus, $\tilde{\rho}$ and $\rho_{\max}$ are determined only by the inelasticity parameter λ and by the principal quantum number n, where $\tilde{\rho} < \rho_{\max}$ for given n_* and λ.

Parameter ρ_0 can be calculated from the equation $W_{fi}^{B}(\rho_0) = c$ using expression (6.16) for the transition probability. It is determined by the principal quantum number n_* and the energy defect $|\Delta E_{fi}|$, as well as by the particular type of colliding partners and their relative velocity V. It should be noted that the roots $x_1^{(\lambda)}(y)$ and $x_2^{(\lambda)}(y)$ of (6.19) depend significantly on λ. Integration of the general expression (6.16) over all impact parameters leads to the following result for the cross section of the inelastic $nl \to n'$ transition [5.50]:

$$\sigma_{n',nl} = \begin{cases} c\pi\rho_{max}^2(\lambda), & \rho_0 = \rho_{max} \\[2mm] c\pi\rho_0^2 + \dfrac{2\pi L_{eff}^2 v_0^2}{V^2 n_*^3}\mathfrak{F}_\lambda(\rho_0/2n_*^2 a_0), & \rho_0 \leq \rho_{max} \end{cases} \tag{6.21}$$

where the function $\mathfrak{F}_\lambda(y_0)$ can be written as

$$\mathfrak{F}_\lambda(y_0) = \frac{1}{\pi}\left\{2\Theta(\tilde{y} - y_0)\int_{y_0}^{\xi_1}\left(\frac{x^2 - y_0^2}{x - x^2}\right)^{1/2}dx\right.$$

$$\left. + \int_{\xi_1}^{\xi_2}\left(\frac{x^2 - y_0^2}{x - x^2}\right)^{1/2}dx + \mathfrak{Q}(\xi_2) - \mathfrak{Q}(\xi_1)\right\},$$

$$\mathfrak{Q}(z) = \arctan\left[\left(\frac{z}{1-z}\right)^{1/2}\right] - [z(1-z)]^{1/2}$$

$$+ \lambda\left[z - \ln\left(\frac{1}{1-z}\right)\right]. \tag{6.22}$$

Here $y_0 = \rho_0/2n_*^2 a_0$ is determined from the normalization condition for the transition probability $W_{n',nl}^B(y_0) = c$ calculated by first-order perturbation theory. Parameters $\xi_{1,2} = x_{1,2}^{(\lambda)}(y_0)$ are the roots of (6.20) for a given value of the scaled impact parameter $y_0 = \rho_0/2n_*^2 a_0$, and $\tilde{y} = \tilde{y}(\lambda)$ is given by expression (6.20).

The expression for the total contribution of all inelastic transitions (quenching cross section σ_{nl}^Q) for a given nl-level (with $l \ll n$) can be written as

$$\sigma_{nl}^Q \equiv \sigma_{nl}^{in} = \sum_{n'}\zeta_{l_0,n'}\sigma_{n',nl}, \tag{6.23}$$

where $\zeta_{l_0,n'}$ is a correction factor for a nonhydrogenic Rydberg atom which takes into account the fact that not all final states are hydrogen-like. There are several prescriptions for incorporating this factor into calculations discussed in [5.39, 42, 45, 1.3]. A natural choice of the correction factor is based on a simple physical picture according to which it is determined by the statistical weights ratio $\zeta_{l_0,n'} = (g_{n'} - g_0)/g_{n'}$, where $g_{n'} = (n')^2$ is the total statistical weight of the final n'-level ($n' = n$ for quasi-elastic l-mixing process), and $g_0 = \sum_{l'=0}^{l_0-1}(2l' + 1) = l_0^2$ is the statistical weight of the first few nonhydrogen-like Rydberg $n'l'$-states (with $l' < l_0$). We assume that all other Rydberg $n'l'$-states with $l' \geq l_0$ with very small values of the quantum defects $\delta_{l'}$ are practically degenerated [so that their energy is equal to $E_{n'} = Ry/(n')^2$]. According to this assumption the correction factor can be written as [1.3, 5.45]

$$\zeta_{l_0,n'} = 1 - l_0^2/(n')^2. \tag{6.24}$$

As is apparent from (6.24), $\zeta_{l_0,n'} \approx 1$ in the range of sufficiently large principal quantum number since all Rydberg $n'l'$-states with $l' \gtrsim 2$–3 are practically hydrogen-like, i.e., the l_0 value usually does not exceed 2–3 (see review [1.3] for more details).

6.3.2 Quasi-elastic Limit: l–Mixing Process

In the particular case of a pure quasi-elastic transition ($|\Delta E_{n',nl}| = 0$ and hence $\lambda = 0$), when $\tilde{y} = 1$ and $y_{max} = 1$, we have $x_{1,2}^{(\lambda)}(y) = 1$. The basic expression for the transition probability can be written as

$$W_{n,nl}(\rho, V) = \begin{cases} c, & 0 \leq \rho \leq \rho_0 \\ \dfrac{L_{\mathrm{eff}}^2 v_0^2}{\pi a_0^2 n_*^6 V^2} \sqrt{\dfrac{a_0}{\rho}} K[k(\rho)], & \rho_0 \leq \rho \leq 2n_*^2 \\ 0, & 2n_*^2 < \rho < \infty, \end{cases} \quad (6.25)$$

where $K(k)$ is the complete elliptic integral of the first kind and $k(\rho)$ is determined by (5.30). The basic semiclassical formula of normalized perturbation theory for the cross section (6.21) is also significantly simplified and takes the form

$$\sigma_{n,nl}(V) = \begin{cases} c 4\pi n_*^4 a_0^2, & n_* < n_0(V) \\ c\pi \rho_0^2 + \dfrac{2\pi L_{\mathrm{eff}}^2 v_0^2}{V^2 n_*^3} \mathfrak{F}_{\lambda=0}(\rho_0/2n_*^2 a_0), & n_* \geq n_0(V). \end{cases} \quad (6.26)$$

The function $\mathfrak{F}_\lambda(y_0)$ of the scaled impact parameter $y_0 = \rho_0/2n_*^2 a_0$ at $\lambda = 0$ is reduced to the simple expression

$$\mathfrak{F}_{\lambda=0}(y_0) = \frac{2}{\pi} \int\limits_{y_0}^{1} \left(\frac{x^2 - y_0^2}{x - x^2} \right)^{1/2} dx = \frac{16}{\pi} \int\limits_0^{k_0} K(k) \left(1/2 - k^2 \right)^{1/2} k\, dk, \quad (6.27)$$

where $k_0 = [(1 - y_0)/2]^{1/2}$. Using the approximate relation $K(k) \approx K(0) = \pi/2$ and performing integration over k this formula can be rewritten in the analytical form

$$\sigma_{n,nl}(V) \approx \begin{cases} c \left(\dfrac{3\sqrt{2}}{4} \right) 4\pi n_*^4 a_0^2, & n_* \leq n_0(V) \\ \pi c \left(\dfrac{3\sqrt{2}}{4} \right) \rho_0^2 + \dfrac{2\pi L_{\mathrm{eff}}^2 v_0^2}{V^2 n_*^3} \left[1 - \left(\dfrac{\rho_0}{2n_*^2 a_0} \right)^{3/2} \right], & n_* \geq n_0(V) \end{cases}$$

$$(6.28)$$

Here ρ_0 is determined from the normalization condition for the Born's probability of the quasi-elastic $nl \to n$ transition (6.25). Its value is given by

$$\rho_0(V, n_*) \approx \begin{cases} 2n_*^2 a_0 & n_* \leq n_0 \\ a_0 \left(L_{\mathrm{eff}}^2 v_0^2 / 2c a_0^2 V^2 n_*^6 \right)^2 & n_* \geq n_0 \end{cases}. \quad (6.29)$$

The upper relation reflects the fact that ρ_0 can not exceed the maximal possible value of the impact parameter $\rho_{max} (\lambda = 0) = 2n_*^2 a_0$, for which the quasi-elastic transition occurs in the classically allowed range. Thus, relation $\rho_0 (V, n_*) = 2n_0^2 a_0$ yields the value of the principal quantum number n_0 for which the value of ρ_0 becomes equal to the radius of the Rydberg atom, which is determined by the right turning point $r_2 = n_*^2 a_0 (1 + \epsilon) \approx 2n_*^2 a_0$ at $l \ll n$. As a result, we have

$$n_0 \approx \left(\frac{L_{eff}^2 v_0^2}{2^{3/2} c a_0^2 V^2} \right)^{1/7}.$$ (6.30)

The maximum of the quasi-elastic cross section $\sigma_{n,nl} = \sum_{l'} \sigma_{nl',nl}$ appears at $(n_{max}/n_0)^{21} = 2$ and its value is given by

$$\sigma_{n,nl}^{max} \approx \pi a_0^2 \frac{c^{3/7} 7 \sqrt{2}}{4} \left(\frac{|L_{eff}| v_0}{a_0 V} \right)^{8/7}.$$ (6.31)

The cross section of quasi-elastic transition $\langle \sigma_{n,nl} \rangle_T = \langle V \sigma_{n,nl} (V) \rangle_T / \langle V \rangle_T$ averaged over the Maxwellian velocity distribution can be expressed in terms of the collision strength parameter ς

$$\langle \sigma_{n,nl} \rangle_T = \pi a_0^2 \frac{3 c^{3/7}}{2^{5/14}} \left(\frac{|L_{eff}| v_0}{a_0 V_T} \right)^{8/7} \mathcal{P}(\varsigma), \quad \varsigma = \frac{1}{c} \left[\frac{1}{2^{2/3} n_*^7} \left(\frac{v_0 L_{eff}}{V_T a_0} \right)^7 \right],$$

$$\mathcal{P}(\eta) = \varsigma^{-4/7} \left\{ 1 + \left(\frac{\varsigma}{3} - 1 \right) \exp(-\varsigma) - \frac{\varsigma^2}{3} E_3(\varsigma) \right\}, \quad E_3(z) = \int\limits_1^\infty \frac{e^{-zt}}{t^3} dt.$$
(6.32)

Here $E_3(z)$ is the integral exponential function of the 3rd order. The plot of the $\mathcal{P}(\varsigma)$ function versus the collision strength parameter ς is presented in Fig. 6.2. The asymptotic behavior of this function is given by

$$\mathcal{P}(\varsigma) \xrightarrow[\varsigma \to 0]{} \mathcal{P}_1(\varsigma) = \frac{4}{3} \varsigma^{3/7}, \qquad \mathcal{P}(\varsigma) \xrightarrow[\varsigma \to \infty]{} \mathcal{P}_2(\varsigma) = \varsigma^{-4/7}.$$ (6.33)

The maximum $\mathcal{P}_{max} \approx 0.73$ appears at $\varsigma_{max} \approx 0.77$. As it follows from (6.32), the range of weak coupling $n \gg n_{max}$ and close coupling $n \ll n_{max}$ correspond to small ($\varsigma \ll 1$) and large ($\varsigma \gg 1$) values of the collision strength parameter, respectively. With the use of (6.33) the averaged cross section can be rewritten as

$$\langle \sigma_{n,nl} \rangle_T = \begin{cases} 3\sqrt{2} c \pi a_0^2 n_*^4 & n \ll n_{max} \\ 5.37 a_0^2 c^{3/7} \left(|L_{eff}| v_0/a_0 V_T \right)^{8/7} & n = n_{max} \\ 2\pi L^2 v_0^2 / n_*^3 V_T^2 & n \gg n_{max} \end{cases} .(6.34)$$

As is evident from (6.34) the cross section of the l-mixing process is proportional to the geometrical area of the Rydberg atom $\pi a_0^2 n_*^4$ at low enough principal quantum number and reaches its maximum in the range

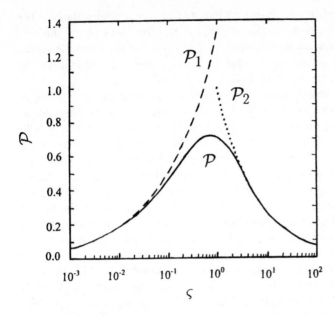

Fig. 6.2. The plot of the $P(\varsigma)$ function that determine the dependence of quasi-elastic $(\Delta E_{n,nl} = 0)$ cross section of the l-mixing process on the collision strength parameter $\varsigma = n_0^7 (V_T) / n_*^7$. Full curve is the semiclassical calculation of the normalized cross section averaged over the Maxwellian distribution of velocities. Dashed and dotted curves represent its asymptotic behavior $P_1(\varsigma)$ and $P_2(\varsigma)$ at small and large values of ς

of $n \sim (v_0 |L_{\mathrm{eff}}| / a_0 V_T)^{2/7}$. Then, at high $n \gg n_{\mathrm{max}}$, where first-order perturbation theory becomes valid, the total l-mixing cross section behaves like $\sigma_{nl}^{l-\mathrm{mix}} \propto L^2 / V^2 n^3$ in accordance with the result (5.40).

6.3.3 Effective Scattering Length

If the scattering length approximation $f_{\mathrm{eB}} = -L$ does not hold, it is necessary to take into account the actual momentum and angular dependencies of the amplitude $f_{\mathrm{eB}}(k, \theta)$ for scattering of a free electron by the ground state target atom. This may be achieved by combining the semiclassical impact parameter approach based on the normalized perturbation theory with the results of the impulse approximation at high n. The first semiempirical attempts to include this dependence in the theory of Rydberg atom–ground-state atom collisions [5.64, 65, 42] incorporate the dependence of the cross section for elastic electron–perturber scattering on n. To improve the results of calculations for quasi-elastic l-mixing processes induced by collisions with the rare-gas atoms, they include in their models the n-dependent parameter $L_{\mathrm{eff}}(n_*) = \left[\sigma_{\mathrm{eB}}^{\mathrm{el}}(\epsilon_n)/4\pi\right]^{1/2}$ instead of standard scattering length L. According to [5.65, 42] the effective scattering length $L_{eff}(n_*)$ is determined by the total elastic scattering cross section $\sigma^{el}_{eB}(\epsilon_n)$ of the free electron by the perturbing atom for the mean kinetic energy $\epsilon_n = Ry/n_*^2$ of the orbital electron motion [5.63] (or by its averaged value $\langle \sigma^{el}_{eB}(k) \rangle_{nl}$ over the momentum distribution function in the nl-state [5.42]).

A method which also includes the dependence on the inelasticity parameter of the process was suggested in [5.51]. Note first that in the strong-coupling region the transition probabilities do not depend significantly on the specific form of the electron-perturber interaction. It can be seen directly from (6.21) as $\rho_0 \rightarrow \rho_{max}$ [see also (6.28)] and reflects a general feature of any collisional processes involving a large number of closely-spaced levels. Indeed for strong collisions the quenching cross section for a given energy level is practically the same as the total scattering cross section, which is determined by the unitarity condition. Thus, the required modification of the scattering amplitude can be accomplished in the weak-coupling limit. In accordance with the results of [5.51], we will present it here within the framework of the quasi-free electron model. We proceed from the general semiclassical formula (5.59) for the cross section of the $nl \rightarrow n'$ transition in the momentum representation [4.16]. In the scattering length approximation ($f_{eB} = -L = const$) it directly yields the analytical result (5.38).

Comparison of (5.38) with the general formula (5.59) allows us to introduce a parameter $L^2_{eff}(n_*, \lambda)$ which can be incorporated into all formulas (6.21–34) in order to take into account the actual behavior of the electron-perturber scattering amplitude $f_{eB}(k, \theta)$. According to [5.51], the final result for the effective scattering length square $L^2_{eff}(n_*, \lambda)$ can be written as

$$L^2_{eff} = \frac{1}{2^{3/2} f_{n',nl}(\lambda)} \int_{k_{min}}^{\infty} k^2 dk \, |g_{nl}(k)|^2 \int_{-1}^{\nu_{max}(k)} \frac{d(\cos\theta)}{(1 - \cos\theta)^{1/2}} |f_{eB}(k, \theta)|^2, \quad (6.35)$$

where $\nu_{max}(k)$ and k_{min} are given by (5.59), and $f_{n',nl}(\lambda)$ is the function (5.38) of the inelasticity parameter $\lambda = n_* a_0 |\Delta E_{n',nl}| / \hbar V$, introduced in Sect. 5.3.3, whereas $f_{n',nl}(\lambda = 0) = 1$ is for the quasi-elastic transitions without energy transfer. As is evident from (6.35), the procedure for calculation of the effective scattering length $L_{eff}(n_*, \lambda)$ for a given $nl \rightarrow n'$ transition includes the average of the differential scattering cross section $d\sigma_{eB}/d\Omega = |f_{eB}(k, \theta)|^2$ over the momentum distribution function $W_{nl}(k) = k^2 |g_{nl}(k)|^2$ taking into account its actual momentum and angular dependencies.

The ratio L^2_{eff}/L^2 can also be written as $\sigma_{n',nl}/\sigma^L_{n',nl}$ where $\sigma_{n',nl}$ is calculated in the impulse approximation (5.59) with the actual scattering amplitude $f_{eB}(k, \theta)$, while $\sigma^L_{n',nl}$ is the corresponding value obtained for $f_{eB} = -L = const$ [see (5.38)]. In the scattering length approximation $f_{eB} = -L$, formula (6.35) yields $L_{eff}(n_*, \lambda) = L = const$. In a general case it incorporates both the short and long range parts of the electron-perturber interaction. The parameter $L_{eff}(n_*, \lambda)$ characterizes the actual electron-atom

$$W_{\lambda\varsigma}(y) =$$

$$\begin{cases} c, & 0 \le y \le y_0 \\ \frac{\varsigma}{\pi\sqrt{2}} \left\{ 2\Theta(\tilde{y} - y)\Im\left[y, x_1^{(\lambda)}(y)\right] + \Im\left[x_1^{(\lambda)}(y), x_2^{(\lambda)}(y)\right] \right\}, & y_0 \le y \le y_{\max} \\ 0, & y_{\max} \le y \end{cases}$$ (6.39)

Here $\varsigma = L_{\text{eff}}^2 v_0^2 / 2^{3/2} a_0^2 V^2 n_*^7$ is a scaled parameter characterizing the collision strength (which differs from the previously introduced in Sect. 6.3.2 dimensionless parameter ς only by the value of normalization constant c, i.e. $\varsigma = \zeta/c$); and

$$\Im(u, v) = \int\limits_u^v \frac{dx}{\sqrt{(x - x^2)(x^2 - y^2)}}.$$

Analysis of the transition probability behavior as a function of the scaled impact parameter $y = \rho/2n_*^2 a_0$ and dependence on the inelasticity parameter $\lambda = n_* a_0 |\Delta E_{n',nl}| / \hbar V$ and the collision strength ζ was given in [5.51]. In the region $0 \le y \le y_0$ the first-order probability $W_{\lambda\varsigma}^B(\rho)$ becomes large and should be normalized to a constant $c = 0.25$ according to the normalized perturbation theory. The probability $W_{\lambda\varsigma}^B(\rho)$ becomes zero at $y > y_{\max} = \rho_{\max}/2n^2 a_0$. This happens because the semiclassical approach neglects the exponentially decaying tail of the electron wave function.

The maximum impact parameter $\rho_{\max}$ depends substantially on the inelasticity parameter λ. To show this we shall proceed from the conservation of energy law for collisional $nl \to n'$ transition according to which the minimum possible value of the momentum transfer $Q = |\mathbf{q}' - \mathbf{q}|$ is determined by the relation $Q_{\min} \approx |\Delta E_{n',nl}| / \hbar V$ if the kinetic energy $\mathcal{E} = \hbar^2 q^2 / 2\mu \gg |\Delta E_{n',nl}|$. The minimum value of the Rydberg electron momentum k for the $nl \to n'$ transition with the energy transfer $|\Delta E_{n',nl}|$ is $k_{\min} \approx |\Delta E_{n',nl}| / 2\hbar V$. It corresponds to the backward scattering ($\mathbf{k}' = -\mathbf{k}$) of the Rydberg quasi-free electron by the perturber B. Substitution of $k_{\min} = (\lambda/2n_*) a_0^{-1}$ into the classical expression

$$\hbar^2 k^2 / 2m - e^2/r = -Ry/n_*^2$$ (6.40)

for the energy of the Rydberg electron in the Coulomb field of the ion core A^+, yields

$$r_{\max}(\lambda) = \frac{2n_*^2 a_0}{1 + (n_* k_{\min} a_0)^2} = \frac{2n_*^2 a_0}{1 + (\lambda/2)^2}.$$ (6.41)

Within the framework of the Fermi pseudopotential model for the electron-perturber interaction, the collisional transition of the Rydberg electron occurs when its radius r relative to the ion core A^+ is equal to separation R between A^+ and B. Thus, expression (6.41) for the radius $r_{\max}(\lambda)$ corresponds to the maximum possible value of the internuclear separation, and therefore can be considered as an upper bound for ρ_{max}. In accordance with (6.20), its value may be estimated as

$$\left[1 + \lambda^2\right]^{-1} < \rho_{max}\left(\lambda\right)/2n_*^2 a_0 < \left[1 + (\lambda/2)^2\right]^{-1}.$$

The results of calculation [5.51] of the dependence $\rho_{max}/2n_*^2 a_0$ on λ is shown in Fig. 6.4 by the dashed curve. The dependence of the impact parameter $\rho_0(\lambda)$ (which separates the regions of weak and strong coupling) is shown by full curves for different values of the collision strength ζ. As is evident from Fig. 6.4, at large $\zeta > 1$ practically the whole range of classically allowed impact parameters $0 < \rho < \rho_{max}(\lambda)$ corresponds to the strong coupling of Rydberg states. Therefore, the first-order perturbation theory becomes inapplicable for almost all possible values of $y = \rho/2n_*^2 a_0$.

Basic formula (6.21) for the cross section of the inelastic $nl \rightarrow n'$ transition can be rewritten in terms of the scaled parameters λ and ζ [5.51]

$$\sigma\left(\lambda, \zeta\right) = \pi n_*^4 a_0^2 \cdot \begin{cases} 4cy_{max}^2(\lambda), & y_0 = y_{max} \\ 4cy_0^2(\lambda, \zeta) + 2^{5/2}\zeta\mathfrak{F}_\lambda(y_0), & y_0 \leq y_{max}. \end{cases} \tag{6.42}$$

Here $y_{max} = \rho_{max}(\lambda)/2n_*^2 a_0$ and $y_0 = \rho_0(\lambda, \zeta)/2n_*^2 a_0$ should be calculated from equation $W_{\lambda\zeta}(y_0) = c$ in which the first-order transition probability is given (6.39); and the $\mathfrak{F}_\lambda$-function is determined by (6.22). Note that $\mathfrak{F}_\lambda(y_0) \rightarrow 0$ when $y_0(\lambda, \zeta) \rightarrow y_{max}(\lambda)$, i.e., for large collision strength ζ. The ratio of the cross section $\sigma_{n',nl}$ to the geometric area of the Rydberg atom depends only on λ and ζ. Thus, it is interesting to analyze the scaled cross section $\sigma_{n',nl}/\pi n_*^4 a_0^2$ as a function of the inelasticity parameter λ and the collision strength ζ.

The scaling formula (6.42) has an analytical form in two limiting cases. The first corresponds to quasi-elastic transitions without energy transfer ($\lambda = 0$) to the Rydberg atom. In this case $y_{max} = 1$ in (6.42), the $\mathfrak{F}_{\lambda=0}(y_0)$-function is given by the simple formula (6.27) in which the $y_0(\zeta)$ is determined from the equation $(2\zeta/\pi y_0^{1/2})K[k(y_0)] = c$. Hence, it is approximately equal to $(\zeta/c)^2$ since $K(k) \approx \pi/2$ for $0 \leq k \leq 2^{-1/2}$, when $0 \leq y \leq 1$. For $\lambda = 0$ there is a certain boundary value of the collision strength $\zeta_0 = c$ (and hence the principal quantum number $n_* = n_0$) for which the scaled parameter y_0 reaches y_{max}. This results from the nonzero value of the transition probability at $y_{max}(\lambda = 0) = 1$, in contrast to the general case of $\lambda \neq 0$ when $W[(y_{max}(\lambda)] = 0$. The strong coupling region corresponds to $\zeta \geq c$. In the weak coupling region $\zeta \ll c$ ($n_* \gg n_0$) the total cross section (6.42) approaches the asymptotic expression $\sigma_{n,nl} = 2^{5/2}\zeta\pi n_*^4$. In the scattering length approximation $L_{eff} = L$ this limiting expression is in full agreement with the Omont's result (5.40). In this case the magnitude of the quasi-elastic cross section is much lower than the geometric area of the Rydberg atom.

The second case corresponds to inelastic $nl \rightarrow n'$ transitions in the range of weak coupling $\zeta \ll 1$. The contribution of the strong-coupling region $0 \leq$

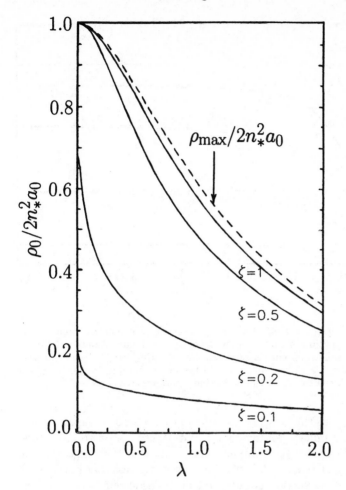

Fig. 6.4. Semiclassical calculations [5.51] of scaled impact parameter $y_0 = \rho_0/2n_*^2 a_0$, separating the close- and weak-coupling regions, as a function of the inelasticity parameter λ for different collision strengths $\zeta = 0.1$, 0.2, 0.5, and 1 (full curves). All data were obtained for the value of normalized constant $c = 1/4$. Dashed curve represents the maximum possible scaled impact parameter $\rho_{max}(\lambda)/2n_*^2 a_0$

$\rho \leq \rho_0$ can be neglected. Thus, assuming $y_0 = \rho_0/2n_*^2 a_0 \to 0$ in (6.22, 42), we have $\mathfrak{F}_\lambda(y_0 = 0) \to \mathfrak{f}_{n',nl}(\lambda)$, where f is defined by (5.38). Hence, for the scaled cross section we obtain the result

$$\sigma(\lambda, \zeta) = \left(\pi n_*^4 a_0^2\right) 2^{5/2} \zeta \mathfrak{f}_{n',nl}(\lambda), \qquad (6.43)$$

which is in full agreement with the general semiclassical formula (5.59) of the weak coupling limit [4.16] in the momentum representation. It is reduced to the simple analytical formula (5.38) of first-order perturbation theory [5.44, 53] if the scattering length approximation is applicable. For pure

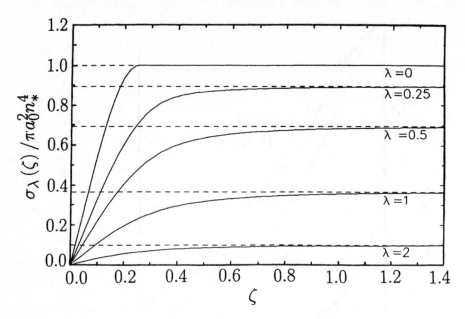

Fig. 6.5. The ratio $\sigma_\lambda(\zeta)/\pi n_*^4 a_0^2$ of the cross section to geometric area of the Rydberg atom as a function of the collision strength ζ. Full curves were calculated in [5.51] at $c = 0.25$. Dashed curves represent the limiting value of this ratio at $\zeta \to \infty$ for $\lambda \neq 0$ and $\zeta \geq \zeta_0 \approx c$ for $\lambda = 0$ in the close-coupling range. Numbers near the curves mark the values of the inelasticity parameter λ

quasi-elastic transitions the $f_{n',nl}(\lambda)$-function in (6.43) becomes equal to one, while the collision strength parameter $\zeta_0 \approx c$. Hence, the aforementioned Omont's formula [5.40] for the l-mixing cross section in the range of weak coupling may also be derived (6.43). It is important to stress that in the weak coupling limit the cross section is independent of the specific choice of the normalized constant c.

The general case, given by the scaling formula (6.42), is presented in Figs. 6.5, 6. Figure 6.5 demonstrates the scaled cross section $\sigma_\lambda(\zeta)/\pi n_*^4 a_0^2$ as a function of ζ for different values of λ. Note that the inelastic cross section strongly falls with increasing inelasticity parameter λ both in the range of weak and strong coupling.

To demonstrate the failure of the perturbative approach in the strong-coupling region, we make a comparison between the semiclassical calculations [5.51] and first-order perturbation theory in Fig. 6.6. The results are presented for $\zeta = 0.1$, 0.25, and 1. For each couple of curves, the lower one has been calculated using the scaling formula (6.42), while the upper curves have been obtained in the first-order perturbation theory. The first-order perturbation theory gives satisfactory results at $\zeta < 0.1$. In this case the cross sections are very close to those calculated by the normalized perturbation theory. The

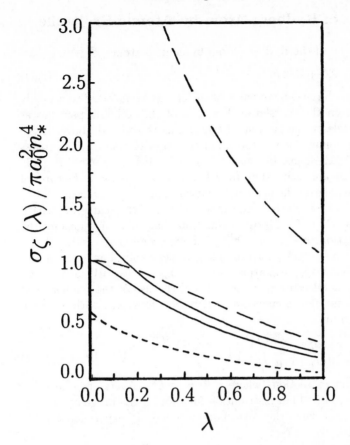

Fig. 6.6. The scaled cross section $\sigma_\zeta(\lambda)/\pi n_*^4 a_0^2$ as a function of the inelasticity parameter λ. Long-dashed, full, and short-dashed couples of curves are the semi-classical results [5.51] for three values of the collision strength parameter $\zeta = 0.1$, 0.25, and 1, respectively. The lower curves for each couple were calculated using the scaling formula, while the upper curves correspond to the first-order perturbation theory

difference between these two methods becomes particularly small for large λ. However, at intermediate ζ (full curves for $\zeta = 0.25$) both methods give close results only for inelastic transitions with large λ. The difference becomes very large at small λ and large $\zeta > 1$. In this strong-coupling case first-order perturbation theory (or impulse approximation in the momentum representation) leads to significantly overestimated magnitudes and to qualitatively incorrect behavior of the cross section (see long-dashed curves for $\zeta = 2$).

6.4 Ionization of Rydberg Atom by Atomic Projectile

The direct ionization of the Rydberg atom by neutral atom

$$A^* + B \rightarrow A^+ + B + e \qquad (6.44)$$

is accompanied by a large enough value of the energy transferred to a highly excited electron from the translational motion of the colliding particles as compared to all inelastic bound-bound transitions considered above. If the principal quantum number is not too small the cross section of this process can be evaluated in the quasi-free electron model using the general formula (5.60) of the impulse approximation. In order to clarify the major features of this process dependence on the principal quantum number and relative velocity we restrict here our analysis to the simplest case of the scattering length approximation $f_{eB} = -L$. In this approximation analytic expressions for the ionization cross sections $\sigma_{nl}^{ion}(\mathcal{E})$ and $\sigma_n^{ion}(\mathcal{E})$ have been derived in [4.17].

In the range of $n \ll v_0/V$, corresponding to slow collisions, one can use in (5.60) the following approximation $na_0k \geq na_0k_1 \approx v_0/4nV \gg 1$ for all possible values of the electron momenta. Then, the use of the semiclassical expression (2.137) for the momentum distribution function (with $Z = 1$), yields a simple formula

$$\sigma_{nl}^{ion}(\mathcal{E}) = \frac{8n_* \sigma_{el}^{eB}}{3\pi v_0} \left(\frac{2\mathcal{E}}{\mu} \right)^{1/2} \left(1 - \frac{|E_{nl}|}{\mathcal{E}} \right)^{3/2} \qquad (l \ll n) \qquad (6.45)$$

for the ionization cross section (where $\sigma_{el}^{eB} = 4\pi L^2$). Averaging the value $V\sigma_{nl}^{ion}(V)$ over the Maxwellian distribution function in the range of $|E_{nl}| \leq \mathcal{E} = \mu V^2/2 < \infty$, we have the corresponding analytic expression for the ionization rate constant [4.17]

$$K_{nl}^{ion}(T) = \langle V\sigma_{nl}^{ion}(V)\rangle_T = \frac{8kTn_*}{\pi\mu v_0} \sigma_{el}^{eB} \exp\left(-\frac{Ry}{kTn_*^2} \right) \qquad (l \ll n). \qquad (6.46)$$

Here T is the gas temperature. Formulas (6.45, 46) describe the case of selectively excited nl levels with given values of the principal n and orbital l ($l \ll n$) quantum numbers. If equally populated nlm sublevels of the Rydberg $A(n)$ atom are provided at a definite value of n, similar calculations lead to the following expressions for the ionization cross section and the rate constant [4.17]

$$\sigma_n^{ion}(\mathcal{E}) = \frac{2^{10}n^3\sigma_{el}^{eB}}{15\pi v_0^3} \left(\frac{2\mathcal{E}}{\mu} \right)^{3/2} \left(1 - \frac{|E_{nl}|}{\mathcal{E}} \right)^{1/2} \left[1 - \frac{7}{6}\frac{|E_n|}{\mathcal{E}} + \frac{1}{6}\left(\frac{|E_n|}{\mathcal{E}} \right)^2 \right],$$

$$\qquad (6.47)$$

$$K_n^{ion}(T) = \frac{2^{10}n^3(kT)^2\sigma_{el}^{eB}}{\pi\mu^2 v_0^3} \left(1 + \frac{Ry}{3kTn^2} \right) \exp\left(-\frac{Ry}{kTn^2} \right). \qquad (6.48)$$

It can be seen from (6.46, 48), the ionization rate constants calculated in the quasi-free electron model are changed exponentially K_{nl}^{ion}, $K_n^{ion} \propto$

$nT \exp(-Ry/n^2kT)$ for low $n \ll (Ry/kT)^{1/2}$ (when $\mid E_{nl} \mid \gg kT$). In the range of $(Ry/kT)^{1/2} \ll n \ll v_0/V_T$ the K_{nl}^{ion} and K_n^{ion} values are proportional to the first and third powers of the principal quantum number (i.e., $K_{nl}^{\text{ion}} \propto n$ ($l \ll n$) and $K_n^{\text{ion}} \propto n^3$).

Analysis of the validity conditions of the impulse approximation (see Sect. 5.4.7) for the direct ionization process shows that formulas presented above are applicable for a wide enough range of the principal quantum numbers

$$(A/32a_0)^{1/4}(v_0/V)^{1/2} = n_{\text{low}} \ll n \ll v_0/V, \tag{6.49}$$

where $A = \max(\mid L \mid, \rho_W)$ and $\rho_W = (\pi \alpha v_0/4V)^{1/3}$ is the Weisskopf radius, and α is the polarizability of the perturbing atom.

For very high principal quantum numbers of $n > n_{\text{up}} \sim v_0/V$ (where $n_{\text{up}} \sim 5 \cdot 10^2$–$5 \cdot 10^4$ for thermal collisions) expressions (6.45–48) are inapplicable due to condition (6.49). For such values of n, the increase of the ionization σ_{nl}^{ion} and σ_n^{ion} cross sections (or the rate constants) with growing n would become slower compared with their behavior for $n \ll v_0/V$. According to (5.55), the limiting values of σ_{nl}^{ion} and σ_n^{ion} are determined by the total cross section $\sigma_{\text{el}}^{\text{eB}} = 4\pi L^2$ for elastic scattering of the ultra-low energy electron e on the perturbing atom B:

$$\sigma_{nl}^{\text{ion}} \xrightarrow[n \to \infty]{} \sigma_{\text{el}}^{\text{eB}} = 4\pi L^2, \qquad K_{nl}^{\text{ion}} \xrightarrow[n \to \infty]{} \langle V \rangle_T \sigma_{\text{el}}^{\text{eB}}. \tag{6.50}$$

Here $\langle V \rangle_T = (8kT/\pi\mu)^{1/2}$ is the average thermal velocity of the colliding atoms.

Thus, at thermal velocities of the Rydberg atoms with neutral atomic particles, the traditional mechanism of the perturber-quasi-free electron scattering can be effective for the direct ionization (6.44) only for high enough n. As a rule, in the range of $n < 20$–40 other physical mechanisms of ionization induced by the perturber-core scattering prove to be predominant at thermal collisions of Rydberg atoms with neutral atomic particles (see Chap. 7).

6.5 Quenching of Rydberg States: Thermal Collisions with Atoms

The cross section σ_i^Q of collisional quenching (depopulation) of a given Rydberg state $|i\rangle$ is determined by the total contribution of all inelastic bound-bound and bound-free $|i\rangle \to |f\rangle$ transitions, i.e. $\sigma_i^Q \equiv \sigma_{nl}^{\text{in}} = \sum_f {}' \sigma_{fi}$. Here the prime at the summing symbol cancels the contribution of pure elastic scattering (when $|f\rangle = |i\rangle$) from the summation over all final states f. Thus, for the Rydberg atom with one valence electron the quenching cross section σ_{nl}^Q of the nl-state can be represented as the sum of the total cross section $\sigma_{nl}^{n,l-\text{ch}} = \sum_{n'l'} {}' \sigma_{n'l',nl}$ of the bound-bound $nl \to n'l''$ transitions with change both in the principal and orbital quantum numbers (including the contribution of the l-mixing $nl \rightarrow nl'$ transitions with $n' = n$) and the ionization cross section, i.e.,

$$\sigma_{nl}^{Q} \equiv \sigma_{nl}^{\text{in}} = \sigma_{nl}^{n,l-\text{ch}} + \sigma_{nl}^{\text{ion}} \ . \tag{6.51}$$

The quenching cross section σ_{nl}^{Q} of selectively excited Rydberg nlJ-state with given magnitudes of n, l, and the total angular momentum J $(=|\,l - 1/2\,|$ or $l + 1/2)$ is determined by the relation

$$\sigma_{nlJ}^{Q} \equiv \sigma_{nlJ}^{\text{in}} = \sigma_{nlJ}^{J-\text{mix}} + \sigma_{nlJ}^{n,l-\text{ch}} + \sigma_{nlJ}^{\text{ion}} \ , \tag{6.52}$$

i.e., it additionally includes the J-mixing cross section of the fine-structure components. At thermal collisions between the Rydberg atoms and neutral atoms the influence of the direct ionization process on the value of the total depopulation cross section is usually very small and can be neglected in calculations of the quenching cross sections.

Below we present a number of theoretical results for the different types of inelastic and quasi-elastic quenching processes of highly excited states in thermal collisions between the Rydberg atom and ground-state rare gas and alkali-metal atoms. Our major goal is to demonstrate the characteristic features of one or another process and to make a detailed comparison of theory with experiment. It may be worthwhile to remember that in the most part of available experiments on the quenching and ionization of Rydberg states by neutral atoms or molecules, the measured value is the corresponding cross section $\langle \sigma \rangle_T = \langle V\sigma \rangle_T / \langle V \rangle_T$ averaged over the Maxwellian velocity distribution for a given gas temperature T, where $\langle V \rangle_T = (8kT/\pi\mu)^{1/2}$ is the mean relative velocity of the colliding particles.

6.5.1 Collisions with Rare Gas Atoms

a) Depopulation of nl-States: Quasi-elastic l-Mixing and Inelastic n, l-Changing Processes. We start the analysis of the quenching processes for the Rydberg nl-states of the alkali-metal atoms in collisions with the rare gas atoms. The cross sections behavior and their magnitudes depend essentially on the value of the quantum defect δ_l. The quenching of Rydberg nl-states with very small quantum defects $\delta_l \approx 0$ (for example, n^1P-states of helium; nd and nf states of sodium; nf states of rubidium and xenon) is due, mainly, to quasi-elastic $(\Delta E_{nl',nl} \approx 0)$ transitions with a change in the orbital angular momentum $nl \rightarrow nl'$, but not in the principal quantum number. Therefore, the calculation of the total quenching cross section $\sigma_{nl}^{Q} = \sum_{n'l'}{}' \sigma_{n'l',nl}$ is usually reduced to the evaluation of the contribution of all quasi-elastic l-mixing collisions $\sigma_{nl}^{l-\text{mix}} = \sum_{l'}{}' \sigma_{nl',nl}$ if the principal quantum number is not too large. The analysis of the available theoretical and

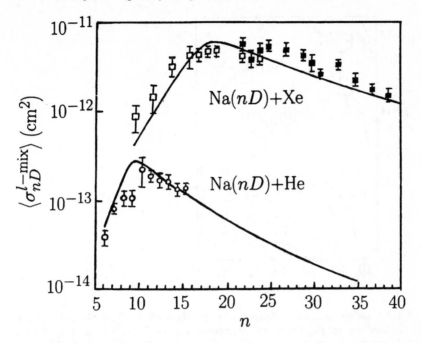

Fig. 6.7. The quenching cross sections $\langle \sigma_{nD}^{Q} \rangle$ of the Rydberg Na(nD) states in thermal collisions with He (lower curve) and Xe (upper curve) plotted against n. Full curves are the semiclassical results [5.50, 52] calculated by formula (6.21) of the normalized perturbation theory at $T = 430$ K (for He) and $T = 423$ K (for Xe). Empty circles are the experimental data [6.4] for He. Empty and full squares are the experimental data for Xe obtained in [6.6, 5], respectively

experimental results on quasi-elastic quenching was given in [5.2], [1.3]. Here we discuss some recent data illustrating the major features of this process.

Figure 6.7 shows the experimental data [6.4–6] and the results of calculations [5.50] for the quenching cross sections of Rydberg Na(nD)-states ($\delta_d^{Na} = 0.015$) for thermal collisions with He and Xe (see also the data for Rb(nF) + He, Xe in Figs. 6.8, 9). One should note that the qualitative behavior of the l-mixing cross sections is similar for all perturbers. At low n the cross sections first increase with n growing roughly as n^4 and then decrease slowly at high principal quantum numbers after reaching a maximum. However, the position of maximum $n_{\max}$ and the magnitudes of $\sigma_{\max}^{Q}$ differ significantly from each other due to a considerable difference in the values of the electron scattering cross sections $\sigma_{el}(\epsilon)$ at small energies (see Sect. 5.2.1 and Fig. 5.1). In particular, as is evident from Figs. 6.7–9 the l-mixing cross sections $\sigma_{\max}^{Q}$ for He and Xe perturbers differ by nearly two orders of magnitude ($\sigma_{\max}^{Q} \sim 10^{-13}$ cm^2 for He and $\sigma_{\max}^{Q} \sim 10^{-11}$ cm^2 for Xe), while $n_{\max}^{He} \approx 10$ and $n_{\max}^{Xe} \approx 20$. The results of available theoretical calculations of the l-mixing cross sections are, on the whole, in quite reasonable quantita-

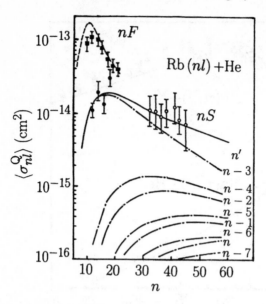

Fig. 6.8. Cross sections for collisional depopulation of the Rydberg nS and nF states of rubidium by He. Full curve represents the semiclassical results [5.49] for the total quenching cross section $\langle \sigma_{ns}^Q \rangle = \sum_{n'} \langle \sigma_{n',ns} \rangle$ of the inelastic Rb(nS)+He collisions at $T = 520$ K averaged over the Maxwellian velocity distribution. Dashed-dotted curves indicate the contribution $\langle \sigma_{n',ns} \rangle$ of individual $nS \to n'$ transitions. Empty and full circles are the experimental data [5.66] ($T = 520$ K) and [5.67,68] ($T = 296$ K), respectively. Dashed curve is the quenching cross section $\langle \sigma_{nf}^Q \rangle_T$ for quasi-elastic l-mixing Rb(nf)+He collisions calculated by scaling formula [5.65]. Full squares are the corresponding experimental data [5.66]

tive agreement with experimental data for the total quenching cross section at low and intermediate n. As it was recently shown in [5.51, 52], an appropriate theoretical description of the quenching processes for Rydberg states with small quantum defects at high enough n should additionally include the contribution of the inelastic n, l-changing transitions ($n > 50$ for collisions of Na(nD) and Rb(nF) with Ar, Kr, and Xe).

There is also a series of experimental works (see [1.2, 3] and references therein) on the quenching of Rydberg nl levels of alkali-metal atoms with low values of the orbital angular momentum l (for example, ns-levels of sodium, np-levels of potassium, and ns-, np-, and nd-levels of rubidium). Due to the large quantum defects δ_l the quenching process of these levels is accompanied by a substantial energy transfer from the Rydberg atom to the kinetic energy of the colliding particles relative motion. Therefore, quenching occurs preferentially with $nl \to n'l'$ transitions to the $n'l'$ states that is the closest in energy to the initial nl level (i.e., with minimum energy defects $\Delta E_{n'l',nl}$). As a result, it gives rise to a change both in the orbital angular momentum

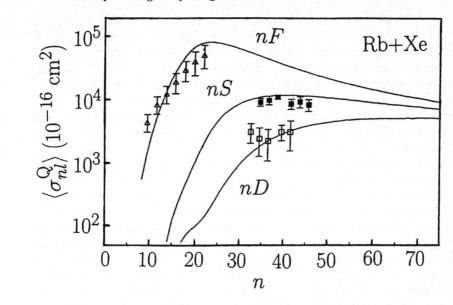

Fig. 6.9. Cross sections $\langle \sigma_{nl}^Q \rangle$ for quenching of the nS-, nD-, and nF-states of the Rydberg Rb atom by Xe averaged over the Maxwellian velocity distribution. Full curves are the semiclassical results [5.51] of the normalized perturbation theory ($T = 296$ K). Dashed curve is the same calculation for the nF level at $T = 520$ K. Full squares, empty squares and triangles are the experimental data for the nS-, and nD-levels ([5.68], $T = 296$ K), and nF-levels ([5.66], $T = 520$ K), respectively

and principal quantum number not only at high enough n, but also at low and intermediate n. The presence of the appreciable energy defects means that the quenching process can no longer be regarded as quasi-elastic.

Figure 6.8 shows the experimental data [5.66–68] and the results of theoretical calculations [5.49] for the total quenching cross section $\langle \sigma_{ns}^Q \rangle = \sum_{n'} \langle \sigma_{n',ns} \rangle$ (full curve) of the Rydberg Rb(nS) states by helium. One can see that the theoretical results are in good agreement with the measured cross sections in the entire investigated range of n. The contribution of individual inelastic $ns \rightarrow n'$ transitions (dashed-dotted curves) for a number of final n'-levels was calculated by a simple semiclassical formula of perturbation theory (5.43). It is clear that the quenching cross section at low $n \ll n_{\max}$ and intermediate $n \sim n_{\max}$ principal quantum numbers, is mainly determined by the Rb(ns) + He $\rightarrow$ Rb($n-3, l' > 2$) + He transitions with $n' = n - 3$, for which the energy defect $\Delta E_{n',ns} = 2Ry(\delta_s + \Delta n)/n^3$ turns out to be minimum among all possible transitions ($\delta_s^{Rb} \approx 3.13$). The maximum value of the inelastic quenching cross section $2 \cdot 10^{-14}$ cm^2, which appears at $n_{\max} \approx 18$, turns out to be appreciably smaller than for the quasi-elastic quenching (dashed curve) of Rb(nF) levels by He. This is due to the significant energy defect of the $ns \rightarrow n - 3$ transition, which is substantial even for $n \sim$ 30–40. As is apparent from Fig. 6.8 that for a reliable description of the quenching cross section at high n it is necessary to take into account additionally the contribution of other $ns \to n'$ transitions with $n' \neq n - 3$.

To demonstrate the significant difference between inelastic n, l-changing and quasi-elastic l-mixing processes we present in Fig. 6.9 recent results of semiclassical calculations [5.51] for quenching $Rb(nl) + Xe$ collisions. They are based on the semiclassical approach and normalized perturbation theory, presented in Sect. 6.3. In this case the experimental data are available for both the nF-level [5.66] (having a small quantum defect, $\delta_f = 0.02$), and for isolated nS-, nD-levels [5.68] (for which $\delta_s = 3.13$, and $\delta_d = 1.34$). One can see that the quenching cross sections reveal strong dependence on the quantum defect δ_l of the initial Rydberg nl-state in the whole experimentally studied range of n. This behavior is reasonably reproduced by semiclassical calculations for quasi-elastic l-mixing quenching of nF-levels as well as for inelastic n, l-changing quenching of nS and nD-levels. Note that the impulse approximation does not describe the observed cross sections for the quenching of nF-states in the range $n < 25$–30, where it is mainly determined by the quasi-elastic l-mixing collisions. On the other hand, for $n > 40$–50 the cross sections, corresponding to different values of the initial orbital angular momentum, start to merge. This occurs due to the contribution of a large number of different $nl \to n'$ transitions ($\Delta n = 0, \pm 1, \pm 2, ...$) which makes the total quenching cross section independent of quantum defect in accordance with the well-known results of the asymptotic theory [5.40].

It is important to stress that actual energy and angular dependencies of the amplitude $f_{eB}(\epsilon, \theta)$ for electron scattering by the heavy rare gas atoms were incorporated into the calculations [5.51] using expression (6.35) and the modified effective range theory (5.3). The results obtained clearly demonstrate that the Ramsauer-Townsend effect affects significantly the values of the cross sections, especially, for the inelastic transitions with large energy transfer. This means that the scattering length approximation (which provides quite good results for the He atom as a perturber) does not allow us to get a reliable quantitative description of collisions between the Rydberg atom and the ground-state heavy rare gas atoms.

Although the ground-state neon atom does not exhibit the Ramsauer-Townsend effect in free electron scattering, the case of the Rydberg atom-neon collisions is special due to a very small value of the scattering length ($L = 0.21$ a.u.). The electron–Ne scattering cross section grows sharply at low energies (see Fig. 5.1). This increase makes the scattering-length approximation completely unreliable, even at relatively large values of n. To demonstrate this fact we present in Fig. 6.10 the results of semiclassical calculations [5.52] of the quenching cross sections for $Rb(ns)$–Ne collisions and their comparison with available experimental data [5.68] at $n =$32, 36, and 38. One can see that the model incorporating the exact energy and angular dependence of the e-Ne scattering amplitude $f_{eB}(k, \theta)$ (upper curve) works much better than the scattering length approximation $f_{eB} = -L = const$

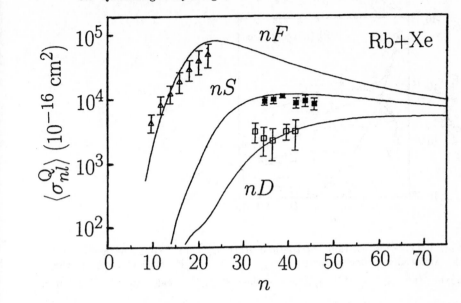

Fig. 6.10. Quenching cross sections of Rb(ns) states by Ne. Full curves are the results of semiclassical theory [5.52] for $V = 3.41 \cdot 10^{-4}$ a.u. Lower curve is the scattering length approximation; upper curve represents calculations incorporating the energy and angular dependence of the electron-neon scattering amplitude. Full circles, experimental data [5.68]

(lower curve). In particular, in the range of $n = 15 - 20$ near the maximum of the inelastic quenching process the calculated cross sections exceed those obtained in the scattering length approximation for more than one order of magnitude.

Thus, the semiclassical unitarized approach [5.50–52], presented in Sect. 6.3, provides quite reasonable quantitative description of major phenomena in inelastic and quasi-elastic collisions of Rydberg atoms with both the light and heavy rare gas atoms.

b) Depopulation of nlJ States: J-Mixing and n, l-Changing Processes. The quenching of Rydberg nlJ-states with the given principal n and orbital l quantum numbers and the total angular momentum $J = |\, l-1/2\,|$ (or $J = l + 1/2$) may be accounted for not only by the n, l-changing processes, but also by the $nlJ \rightarrow nlJ'$ transitions between the fine-structure components. As follows from the results of experimental [5.66, 6.7, 8] and theoretical [5.47, 48, 6.3, 5.49] studies, the depopulation of the nlJ-states with relatively small n [$n < 15$–20 at thermal collisions of Rb($n^2 D_{3/2}$) and Cs($n^2 D_{3/2}$) with the rare gas atoms] is primarily the result of the total angular momentum transfer ($J' \neq J$) without change in the orbital l and principal n quantum numbers. However, at high enough n the n, l-changing processes predominate. Therefore, the total cross section for quenching of selectively excited nlJ-level

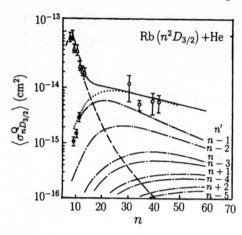

Fig. 6.11. Cross sections of thermal $Rb(n^2D_{3/2})$+He collisions averaged over the Maxwellian velocity distribution at $T = 520$ K. Dashed curve is the J-mixing $(n^2D_{3/2} \rightarrow n^2D_{5/2})$ cross sections $\left\langle \sigma_{nD_{3/2}}^{J-\text{mix}} \right\rangle$ calculated [5.45, 49] using the normalized perturbation theory. Dotted curve is the cross section $\left\langle \sigma_{nD_{3/2}}^{n,l-\text{ch}} \right\rangle = \sum_{n'} \left\langle \sigma_{n',nD_{3/2}} \right\rangle$ of all inelastic transitions that determines the depopulation of the nD-level out of the doublet. Empty circles ($T = 520$ K) and full circles ($T = 296$ K) are the corresponding experimental data [5.66–68]. Dashed-dotted curves indicate the contribution $\left\langle \sigma_{n',nD_{3/2}} \right\rangle$ of individual $n^2D_{3/2} \rightarrow n'$ transitions. Full curve represents the total quenching cross section $\left\langle \sigma_{nD_{3/2}}^Q \right\rangle = \left\langle \sigma_{nD_{3/2}}^{J-\text{mix}} \right\rangle + \left\langle \sigma_{nD_{3/2}}^{n,l-\text{ch}} \right\rangle$ of the selectively excited $n^2D_{3/2}$ level. Empty ($T = 520$ K) and full ($T = 380$ K) triangles are corresponding experimental data [5.66, 6.7], respectively

has qualitatively different dependence on the principal quantum number n than for the nl-states. In particular, two clearly pronounced maxima can occur here at low and high values of n.

The available experimental data [5.66, 6.7] and theoretical results [5.48, 49] for the quenching of $Rb(n^2D_{3/2})$ states by helium ($T = 520$ K) are shown in Fig. 6.11. The dashed curve shows the n-dependence of the Maxwell-averaged J-mixing cross section $\left\langle \sigma_{nD_{3/2}}^{J-\text{mix}} \right\rangle$. It was calculated by the simple formula (6.14) corresponding to the quasi-elastic limit of the J-mixing process. It is justified because of the inelasticity parameter $\nu = |\delta_{D_{3/2}} - \delta_{D_{5/2}}|v_0/V_T n_*$ for the $n^2D_{3/2} \rightarrow n^2D_{5/2}$ transitions of rubidium is small ($\nu_{3/2,5/2} \ll 1$) in the entire considered range of n (the quantum defects difference is $\delta_{D_{3/2}} - \delta_{D_{5/2}} = 0.002$). As is evident from Fig. 6.11, the J-mixing cross section reaches its maximum $\left\langle \sigma_{nD_{3/2}}^{J-\text{mix}} \right\rangle_{\text{max}} \approx 1 \cdot 10^{-13}$ cm^2 at $n_{\text{max}}^{J-\text{mix}} = 8$. With increase of $n > n_{\text{max}}^{J-\text{mix}}$ the cross section of the $n^2D_{3/2} \rightarrow n^2D_{5/2}$ transition tends rapidly to the asymptotic limit of weakly coupled states $\left\langle \sigma_{nD_{3/2}}^{J-\mathrm{mix}} \right\rangle \propto n_*^{-4}$. However, at low principal quantum numbers the J-mixing cross section behaves like $\propto n_*^4$ in accordance with (6.12).

The averaged cross sections $\left\langle \sigma_{n',nD_{3/2}} \right\rangle$ of the inelastic $n^2 D_{3/2} \to n'$ transitions with change in the principal and orbital quantum numbers ($2 < l' \leq n' - 1$, $J' = 3/2, 5/2$) and their total contribution $\left\langle \sigma_{nD}^{n,l-\mathrm{ch}} \right\rangle$ calculated in [5.49] are shown in Fig. 6.11 by dashed-dotted and dashed curves, respectively. One can see that the maximum $\left\langle \sigma_{nD_{3/2}}^{n,l-\mathrm{ch}} \right\rangle^{\max} \approx 8.8 \cdot 10^{-15}$ cm^2 of the n, l-changing cross section is reached at $n_{\max} \approx 22$. The main contribution is determined by $n^2 D_{3/2} \to n - 1$ and $n^2 D_{3/2} \to n - 2$ transitions, which have the smallest energy defects $\Delta E_{n-1,nD}$ and $\Delta E_{n-2,nD}$ ($\delta_D = 1.34$). At large $n > 40 - 50$, the inelastic $n^2 D_{3/2} \to n'$ transitions to other degenerate hydrogen-like sublevels $n'l'J'$ ($l' > 2$) with $n' \neq n - 1, n - 2$ become important.

The total quenching cross section $\left\langle \sigma_{nD_{3/2}}^{Q} \right\rangle = \left\langle \sigma_{nD_{3/2}}^{J-\mathrm{mix}} \right\rangle + \left\langle \sigma_{nD_{3/2}}^{n,l-\mathrm{ch}} \right\rangle$, which is determined by the sum of the J-mixing and n, l-changing contributions, is shown by a full curve in Fig. 6.11. Thus, for high $n > 20$ the quenching of selectively excited $n^2 D_{3/2}$ states is primarily due to inelastic $n^2 D_{3/2} \to n'$ transitions with change of the principal and orbital quantum numbers. On the other hand, at low $n < 15$ it is mainly determined by the $n^2 D_{3/2} \to n^2 D_{5/2}$ transitions with change of only the total angular momentum. These processes make comparable contributions in the region $n \sim 15$–20.

To demonstrate an appreciable dependence of the J-mixing cross sections on the energy splitting of the fine-structure components we shall compare in Fig. 6.12 theoretical results [5.49] for the $n^2 D_{3/2} \to n^2 D_{5/2}$ (panel: a) and $n^2 P_{1/2} \to n^2 P_{3/2}$ (panel: b) transitions (the dashed curves) induced by thermal collisions ($T = 353$ K) of Rydberg Cs* and the ground-state He atoms. It is seen that calculations (6.12) are in good agreement with available experimental data [6.8] for $n^2 D_{3/2} \to n^2 D_{5/2}$ transitions in both the range of weak coupling ($n > 10$–12) and strong coupling ($n \sim 8$–11), i.e., the close vicinity of the maximum $\left\langle \sigma_{nD}^{J-\mathrm{mix}} \right\rangle \approx 1.03 \cdot 10^{-13}$ cm^2 at $n_{\max} = 9$. There are no experimental data so far for the J-mixing of the fine-structure $n^2 P_{1/2}$ and $n^2 P_{3/2}$ components of Rydberg Cs atom in helium, so that of particular interest are the corresponding calculations and their comparative analysis with the case of the $n^2 D_{3/2} \to n^2 D_{5/2}$ transition.

The quantum defects differences of these states are as follows: $\delta_{P_{1/2}} - \delta_{P_{3/2}} = 0.033$ and $\delta_{D_{3/2}} - \delta_{D_{5/2}} = 0.009$, while $\delta_P = 3.58$ and $\delta_D = 2.47$ for Cs, see Table 2.1). Thus, the substantial differences in the behavior of the J-mixing processes are primarily due to the qualitatively different influence of the inelasticity parameter $\nu_{J'J} = |\delta_{J'} - \delta_J| v_0 / V n_*$ in these two cases. For Cs($n^2 D_{3/2}$)+He collisions the value of $\nu_{3/2,5/2}$ is small enough already at $\dot{n} \geq 12$, where $\nu_{3/2,5/2}^{(d)} \leq 1.7$ and, hence, we can put $\varphi_{3/2,5/2}^{(d)} \approx 1$ (see

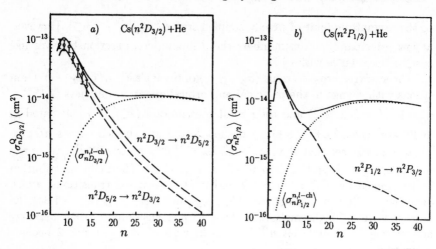

Fig. 6.12. Semiclassical calculations [5.49] of the depopulation cross sections of the selectively excited nlJ-states of Rydberg Cs atom in thermal collisions with He averaged over the Maxwellian velocity distribution at $T = 353$ K. Panel a shows the results for the $n^2D_{3/2}$ states. Dashed curves are the J-mixing cross sections $\left\langle \sigma_{nD_J}^{J-\text{mix}} \right\rangle$ for direct $n^2D_{3/2} \to n^2D_{5/2}$ and inverse $n^2D_{5/2} \to n^2D_{3/2}$ transitions. Dotted curve is the sum $\left\langle \sigma_{n3/2}^{n,l-\text{ch}} \right\rangle = \sum_{n'} \left\langle \sigma_{n',nD_{3/2}} \right\rangle$ of all inelastic $n^2D_{3/2} \to n'$ transitions with a change in the principal and orbital quantum numbers. Full curves represent the total quenching cross sections $\left\langle \sigma_{nD_{3/2}}^{Q} \right\rangle = \left\langle \sigma_{nD_{3/2}}^{J-\text{mix}} \right\rangle + \left\langle \sigma_{nD_{3/2}}^{n,l-\text{ch}} \right\rangle$. Empty circles are the corresponding experimental data measured in [6.8]. Panel b presents the same calculations for the $n^2P_{1/2}$ states

Fig. 6.1). Small deviations in the behavior of the J-mixing process from the pure quasi-elastic case ($\nu = 0$) arises here only at low $n \sim 8$-11. However, due to an appreciable increase of the energy splitting between the fine-structure components of the Rydberg n^2P-levels compared with that for the n^2D levels ($\Delta E_{1/2,3/2}^{(p)}/\Delta E_{3/2.5/2}^{(d)} \approx 3.67$), the inelasticity parameter turns out to be large ($\nu_{1/2,3/2} \gg 1$) practically in the whole range of n under consideration. This leads to substantial reduction of the J-mixing cross sections for the $n^2P_{1/2} \to n^2P_{3/2}$ transition compared with the cross sections for the $n^2D_{3/2} \to n^2D_{5/2}$ transition in spite of insignificant difference between the values of $C_{3/2,1/2}^{(p)}$ and $C_{5/2,3/2}^{(d)}$ coefficients and corresponding statistical weights.

The cross sections $\left\langle \sigma_{nlJ}^{n,l-\text{ch}} \right\rangle$ of the n, l-changing processes for $Cs(n^2D_{3/2})$ + He and $Cs(n^2P_{1/2})$+He collisions taking into account the contribution of many inelastic $nlJ \to n'$ transitions, are shown in Fig. 6.12 by dotted curves. It is seen that because of the close values of the energy defects ($\Delta E_{n-2,nd}$, $\Delta E_{n-3,nd}$ and $\Delta E_{n-4,np}$, $\Delta E_{n-3,np}$) of the most important transitions, the cross sections $\left\langle \sigma_{nD_{3/2}}^{n,l-ch} \right\rangle$ and $\left\langle \sigma_{nP_{1/2}}^{n,l-ch} \right\rangle$ turn out to be close to each other. Therefore, the total quenching cross section $\left\langle \sigma_{nlJ}^{Q} \right\rangle$ of the selectively excited $n^2 D_{3/2}$ and $n^2 P_{1/2}$ states of Cs (full curves in Fig. 6.12) differ considerably only in the region of $n < 20$, where the major role is played by the J-mixing transitions without a change of both the principal and orbital quantum numbers.

As follows from the experimental studies [6.8], the behavior of the J-mixing cross sections for the $Cs(n^2 D_{3/2}) \rightarrow Cs(n^2 D_{5/2})$ transitions induced by thermal collisions with the heavy rare gas atoms remains qualitatively similar to the case of He atom as a perturber. However, their magnitudes increase appreciably in the range of $n > 11$–12 (see also review [1.3] and [5.47], [6.3]).

6.5.2 Collisions with Alkali-Metal Atoms

The main qualitative features in the experimentally observed behavior of the cross sections and the rate constants of thermal collisions between the Rydberg atoms and the ground-state alkali atoms remain the same as those considered above. However, the maxima of the quenching cross sections become much greater in magnitudes and are shifted toward higher n as compared to the collisions with the ground-state rare gas atoms, particularly so, for the heavy elements K, Rb, and Cs as the perturbers. This is the result of extremely large value of the cross section for elastic scattering of the ultra-low-energy electron by the alkali-metal atom and, in particular, of the ^3P-resonance on the quasi-discrete level of the corresponding negative ion (see Fig. 5.2 and Table 5.2). Below we shall illustrate some features in the quenching of high-Rydberg states by the ground-state sodium and rubidium atoms.

Let us first consider thermal collisions ($\mathcal{E} = \mu V^2/2 = 0.037$ eV) of Rydberg sodium in the nS and nD states with the ground-state parent atoms Na(3S). The impulse-approximation results [4.16] for the total quenching cross section $\sigma_{nl}^{Q} = \sum_{n'}' \sigma_{n',nl}$, obtained using the general formula (5.59), are shown in Fig. 6.13 by full curves. These calculations were made in the range of applicability of the quasi-free electron model on the basis of the theory of resonance and potential scattering, presented in Sects. 5.2.1, 4.6. The parameters of elastic scattering of a ultra-slow electron by Na(3S) were taken from [5.20] (see Table 5.2). It can be seen that the relative role of the potential and resonance electron-perturber scattering in the quenching of Rydberg levels depends essentially on both the principal quantum number n and the value of the transition energy defect $\Delta E_{n',nl}$. Because of the small quantum defect $\delta_d^{Na} = 0.0145$, the quenching of the Na(nD) levels is mainly determined by the quasi-elastic $nD \rightarrow nl'$ ($l' > 2$) transitions with a change in the orbital angular momentum if the principal quantum number is not too large. In this

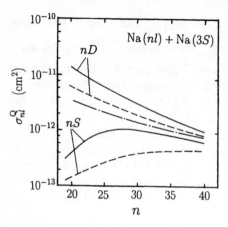

Fig. 6.13. The impulse-approximation results [4.16] for the quenching cross sections $\sigma_{nl}^{Q} = \sum_{n'} \sigma_{n',nl}$ of the Rydberg nD and nS levels of sodium by the ground-state Na(3S) atoms ($\mathcal{E} = 0.037$ eV). Dashed curves indicate the contribution σ_{nl}^{P} of the potential electron-perturber scattering alone. Full curves represent the total quenching cross section $\sigma_{nl}^{Q} = \sigma_{nl}^{P} + \sigma_{nl}^{r}$ including the contribution of the ^{3}P-resonance scattering. Dashed-dotted curve is the corresponding result for the cross section σ_{nl}^{Q} obtained using simple asymptotic formula [5.59]

case the contributions of the resonance and potential scattering are of the same order of magnitude in a wide range of n.

The quenching of the Rydberg Na(nS) states ($\delta_{s}^{Na} = 1.35$) is due to the inelastic $nS \rightarrow n'l'$ transitions with a change in the orbital and principal quantum number (primarily, with $n' = n - 1$, $n - 2$). In this case the resonance scattering plays a much greater relative role near the maximum of the quenching cross sections σ_{nS}^{Q}. As is apparent from the comparison of full and dashed curves in Fig. 6.13, the contribution of resonance scattering to the quenching of the nS levels at $20 < n < 35$ is considerably greater than that of the potential scattering. As the principal quantum number increases further $n > 35$–40, the role of the ^{3}P-resonance begins to decrease. This is a result of a substantial drop of the value of the energy distribution function $W_{nl}(\epsilon)$ for the resonance energies $\epsilon \sim E_{r}$ at high principal quantum numbers $n \gg (Ry/E_{r})^{1/2}$ (see Sect. 5.4.6 and [4.16], [5.21] for more details). Note that at high enough n the total quenching cross section $\sigma_{nl}^{Q} = \sum_{n'} \sigma_{n',nl}$ is due not only to transitions to the nearest energy levels, but also to a large group of final levels n'.

As is evident from Fig. 6.13, the quenching cross sections σ_{nS}^{Q} and σ_{nD}^{Q} calculated in [4.16] for nS and nD levels (full curves) reveal a strong dependence on the quantum defect values in the region of $n \sim 20$–30 (the impulse approximation is valid for Na*–Na collisions at $n > 20$). This dependence cannot be described on the basis of the asymptotic theory [5.59] supplemented by the ^3P-resonance contribution of the electron-perturber scattering, which yields practically the same values $\sigma_{nS}^Q = \sigma_{nD}^Q$ (dashed-dotted curve) for the quenching cross sections. The most important difference between the values of σ_{nS}^Q and σ_{nD}^Q arises, as expected, in the range $n \sim 20$–30 in which the main contribution to the quenching processes is determined by the inelastic $nS \to n'$ or quasi-elastic $nD \to n$ transitions to several nearest energy levels ($n' = n - 1$, $n - 2$ for nS and $n' = n$ for nD levels). On the other hand, the calculations within the framework of the asymptotic theory agree reasonably with the results of the quenching theory developed in [4.16] with n growing ($n > 40$). This is due to the increasing role of the inelastic $nl \to n'$ transitions to a large group of final n' levels.

Since the energy E_r of the ^3P-resonance in electron scattering by the ground-state Rb(5S) atom is considerably smaller than that for Na(3S) (see Fig. 5.2), this resonance should play an even greater role in the quenching of high-Rydberg Rb(nS) states in its own gas compared with the case of Na*+Na collisions, considered above. As follows from the impulse-approximation analysis [4.16], [5.75], the available experimental data [5.29] on the quenching process Rb(nS)+Rb(5S) can not be explained within the framework of the quasi-free electron model taking into account only the potential electron-perturber scattering. Simple estimates show that its contribution σ_{nS}^P to the total quenching cross section turns out to be appreciably smaller than the values measured [5.29] in the range of $34 \le n \le 43$.

This was confirmed by calculations [4.16], which indicate (see Fig. 6.14) that the experimental data for this process can be reasonably described by general formula (5.59) taking into account the ^3P-resonance scattering con-

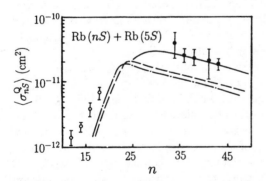

Fig. 6.14. The quenching cross sections $\langle \sigma_{ns}^Q \rangle$ of the Rb(nS) + Rb(5S) plotted against n. Full curve represents the impulse-approximation results [4.16] calculated at $T = 400$ K with semiempirical ^3P-resonance parameters $\epsilon_0 = 1.2 \cdot 10^{-3}$ a.u. and $\gamma = 18$ a.u.. Dashed-dotted curve is the result of similar calculation [5.75] at $T = 530$ K with the resonance parameters obtained using the modified effective range theory [5.19]. Dashed curve, the same for $T = 420$ K. The open and filled circles indicate the experimental data [5.80, 29] measured at $T = 520$ and 400 K, respectively

tribution (5.78) with the use of the semiempirical resonance parameters presented in Table 5.2. The dominant contribution to the quenching cross section $\sigma_{nS}^Q \equiv \sum_{n'} \sigma_{n',nS}$ at $n \sim 25$–50 is made by the $nS \to n-3$, $l' > 2$ transition with the minimum energy defect ($\delta_s^{Rb} \approx 3.13$), although all other $nS \to n'$ transitions have also been taken into account in [4.16]. A similar calculation of the quenching cross section σ_{nS}^Q for Rb(nS)+Rb($5S$) collisions based on the general formula (5.59) of the impulse approximation has been performed [5.75] using the data [5.19] for the resonance and potential amplitudes of electron-alkali atom scattering. In the range of $n > 25$, the curve of the quenching cross section obtained in [5.75] shows the same dependence on n as in [4.16]. However, the magnitudes of the former cross section are one half as large as those in [4.16] because the values of the ^3P-resonance parameters used in these papers slightly differ from each other (see Table 5.2).

All results for the quenching processes of Rydberg alkali-metal atoms by the ground-state parent atoms presented above was obtained for high enough values of n, which correspond to the range of weak coupling. Hence, the magnitudes of the calculated cross sections σ_{nl}^Q for both inelastic and quasi-elastic processes are considerably smaller than the geometrical cross section $\sigma_{\text{geom}} \propto \pi a_0^2 n_*^4$ of the Rydberg atom. Therefore, the use of the impulse approximation in calculations [4.16], [5.75] provides reasonable agreement with the experimental data [5.29] at high n. Nevertheless, a major part of experimental results on quenching of the Rydberg states by the ground-state alkali atoms have been obtained for low enough or intermediate n, where the quasi-free electron model and impulse approximation do not hold. For instance, there are experimental data on quenching of Rydberg states of rubidium (nS, nP, nD, nF with $n \leq 22$) and caesium (nS, nD with $n \leq 16$) in alkali-alkali collisions. Moreover, several authors have also measured cross sections for the fine-structure mixing ($n^2 D_{3/2} \to n^2 D_{5/2}$) collisions of Rb*+Rb and Cs*+Cs (see review [1.3] and references therein).

As follows from these experiments, the observed cross sections for both the quenching (Fig. 6.15, panel a) and the J-mixing (Fig. 6.15, panel b) processes increase rapidly with n growing in the studied range of $n \leq 16$–22. However, it is possible to distinguish two different cases. First, the total depopulation cross sections are of the same order of magnitude of the geometrical cross section $\sigma_{\text{geom}} \propto \pi a_0^2 n_*^4$ of the Rydberg atom, if the highly excited levels are close to some neighboring states (for example, nS, nF levels and nD_J sublevels of Rb and Cs). Second, the magnitudes of the depopulation cross sections for the well isolated Rydberg levels (such as nP and nD levels of Rb and Cs) turn out to be appreciably smaller than the value of σ_{geom}, in spite of the similar qualitative n-dependence [Fig. 6.15 (panel a)]. Naturally, this fact is accounted for by the influence of the inelasticity parameter $\lambda = n_* a_0 \omega_{n',nl}/V$ of the most important $nP \to n'$ and $nD \to n'$ transitions. On the other hand, the inelasticity parameter $\nu_{3/2,5/2}^{(d)} = n_*^2 a_0 \omega_{5/2,3/2}/V$ for

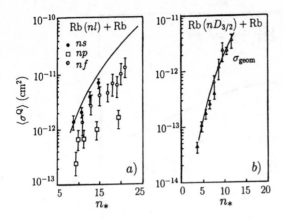

Fig. 6.15. The quenching cross sections $\left\langle \sigma_{nl}^{Q} \right\rangle$ and $\left\langle \sigma_{nD_{3/2}}^{Q} \right\rangle$ of the Rydberg nl-levels (panel a: nS, nP, and nF) and $n^2 D_{3/2}$-level (panel b) of rubidium by the ground-state Rb($5S$) atoms plotted against the effective principal quantum number $n_* = n - \delta$ ($\delta_s = 3.15$; $\delta_p = 2.65$; $\delta_d = 1.35$; and $\delta_f = 0.02$). Full circles, empty squares, and empty circles are the experimental data [5.80] and [5.66] for the nS-, nP-, and nF-levels, while full and empty triangles represent the experimental data for the $nD_{3/2}$ levels measured in [6.9, 5.80], respectively

the J-mixing transitions $n^2 D_{3/2} \rightarrow n^2 D_{5/2}$ is not sufficiently large in the experimentally studied range of $n \leq 16$, which corresponds to the region of close coupling in accordance with the simple estimate (6.11). Hence, it does not affect radically the J-mixing cross sections whose magnitudes increase roughly as the geometrical cross section $\sigma_{\text{geom}} \propto \pi a_0^2 n_*^4$ of the Rydberg atom (Fig. 6.15, panel b). Thus, in connection with the available experimental results, the depopulation cross sections of the fine-structure $n^2 D_J$ components in the studied range of low enough $n \leq 16$ is primarily due to the J-mixing transitions $n^2 D_{3/2} \rightarrow n^2 D_{5/2}$.

6.6 Quenching and Ionization of Rydberg States: Thermal Collisions with Molecules

6.6.1 Quasi-resonant Energy Exchange of Rydberg Electron with Rotational Motion of Molecule

Collisions of Rydberg atoms with molecules have some specific features due to the presence of internal rotational degrees of freedom and the long-range nature of electron-molecule interaction (Sect. 5.2.2). Their revealing is especially significant in the n-changing and ionization processes with large enough energy transfer to the Rydberg electron. The behavior of the depopulation cross sections for the rotationally elastic processes (in which the rotational

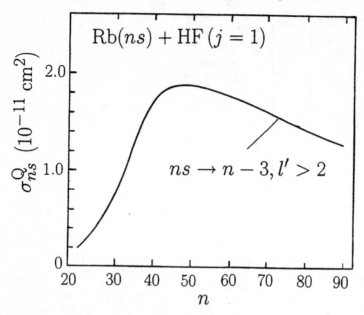

Fig. 6.16. The quenching cross sections σ_{ns}^Q of the Rb(nS) states by strongly polar molecules HF($j = 1$) via rotationally elastic ($j' = j$) transitions $ns \to n'l'$ to the nearest degenerate hydrogen-like sublevels with $n' = n - 3$ and $l' > 2$ ($\delta_s^{Rb} = 3.15$). Full curve represents the calculations [6.2] by semiclassical close-coupling method

energy of molecule remains unchanged) is qualitatively similar to the case of the atomic projectile. For instance, in quasi-elastic l- and J- mixing processes the nitrogen molecule as the perturber is similar to the rare gas atoms. However, for strongly polar molecules (such as HCl, HF, NH$_3$, etc.), the maximum magnitudes of the quasi-elastic l-mixing cross sections of Rydberg states turn out to be substantially greater than for neutral atoms, while the position of the maximum is shifted toward a higher principal quantum number $n \sim 50$–60 (Fig. 6.16). On the whole, the polar molecules in such processes are qualitatively similar to strongly polarizable alkali-metal atoms. This is the result of the major contribution of the second order term from the dipole potential, which falls off asymptotically as r^{-4}.

On the other hand, the inelastic n-changing transitions with large energy change ΔE_{fi} as well as the ionization of the Rydberg atom

$$A^*(nl) + B(vj) \to \begin{cases} A(n'l') + B(vj') \\ A^+(n'l') + B(vj') + e \end{cases},$$ (6.53)

may be a result of rotational $j \to j'$ (and sometimes vibrational $v \to v'$) de-excitation or excitation of the molecule. In spite of the large value of energy gained or released by the highly excited atom, these processes may occur with very large cross sections provided the quasi-resonant energy exchange between the Rydberg electron and internal rotational degrees of freedom of the molecule takes place. For quasi-resonant n-changing and ionizing collisions, the kinetic energy $\Delta \mathcal{E} = \Delta E_{jj'} - \Delta E_{\alpha'\alpha}$ transferred to the translational motion of the colliding particles is small in comparison with both the energy change $|\Delta E_{\alpha'\alpha}|$ of the Rydberg atom A^* and the rotational energy change $|\Delta E_{j'j}|$ of molecule B. In the opposite case, when the kinetic energy defect $\Delta \mathcal{E}$ is noticeable, the n-changing and ionization cross sections become small. The first results in this field were reviewed in [4.15], [6.10]. Further detailed calculations were made in [6.11, 12, 2].

In the impulse approximation the cross sections $\sigma_{\alpha'\alpha}^{j'j}$ of the inelastic $nl \rightarrow n'$ and $n \rightarrow n'$ transitions induced by slow collisions ($V \ll v_0/n$) with molecules may be evaluated by formulas (5.58) [or (5.67) combined with the binary-encounter expressions (3.54, 57) for the form factors]. The required expressions for the amplitude square $|f_{eB}^{j'j}(Q)|^2$ of electron scattering by the nonpolar and polar molecules are presented in Sect. 5.2.2. As is evident from these formulas, the cross sections of the n-changing transitions are determined by the value of the momentum transferred to the kinetic energy of colliding particles $A^* + B$. The quasi-resonant nature of these transitions and its dependence on the value of

$$Q_{\min} \approx |\Delta \mathcal{E}|/\hbar V = |Ry/(n'_*)^2 - Ry/n_*^2 - (E_j - E_{j'})|/\hbar V \quad (6.54)$$

was investigated in [5.56], [6.11]. It was shown that for a given magnitude of $|\Delta n|$, the resonant behavior of the cross sections $\sigma_{n',n}^{j\pm1,j}$ of the $n, j \rightarrow n', j\pm1$ transitions induced by collisions with dipolar molecules becomes more pronounced with n growing. For a given value of n, the form of the normalized function $\sigma_{n',n}^{j\pm1,j}(Q_{\min})/\sigma_{n',n}^{j\pm1,j}(0)$ becomes more broader as $|\Delta n|$ increases. In the resonant case the n-changing cross sections for the dipole-induced collisions are about several orders of magnitude higher than those for the quasi-elastic l-mixing process induced by the short-range electron-molecule interaction. Meanwhile, the quadrupole-induced n-changing collisions are much less efficient in comparison with the dipole case because of the shorter range of the Rydberg electron-molecule interaction.

Ionization of the Rydberg atom is most efficient, when it is accompanied by rotational de-excitation $j \rightarrow j'$ of the molecular projectile ($\Delta E_{jj'} = E_j - E_{j'} > 0$). Then, the ionization threshold ($\mathcal{E}_{\min} = \hbar^2 q_{\min}^2/2\mu = |E_{nl}| - \Delta E_{jj'} > 0$) of reaction (6.53) may exist only in the range of $n_* < n_*^{(0)}$ [where $n_*^{(0)} = (Ry/|\Delta E_{jj'}|)^{1/2}$]. In the opposite case, i.e. for high principal quantum numbers $n_* > n_*^{(0)}$ (when $|E_{nl}| - \Delta E_{jj'} < 0$ and $\mathcal{E}_{\min} = 0$), the resonant energy transfer to the Rydberg electron from the internal rotational motion of the molecule may occur for all possible values of the kinetic energy of heavy particles. For this range of $n_* > n_*^{(0)}$, Matsuzawa [5.60] has derived the following approximate expression for the ionization cross section:

$$\sigma_{nl}^{ion}(j', j) \approx \frac{4\pi v_0}{V} \int_0^\infty |f_{eB}^{j'j}(Q)|^2 \left[\frac{dF_{E,nl}(Q)}{d(E/2Ry)}\right]_{E=E_0} (Qa_0)^2 d(Qa_0) \qquad (6.55)$$

neglecting the energy exchange between the internal rotational motion of molecule and kinetic energy of the colliding particles relative motion. Here the energy of the ejected electron is taken to be equal to $E = E_0 = |\Delta E_{jj'}| - |E_{nl}|$, and the binding energy of the Rydberg electron is $|E_{nl}| = Ry/n_*^2$. According to the binary-encounter theory the form factor density $dF_{E,nl}(Q)/dE$ of the bound-free $nl \to E$ transition per unit energy interval is determined by (3.54, 58).

Using (6.55) and the Born approximation for the electron-perturber scattering amplitude square (5.15), the cross section $\sigma_{nl}^{ion}(j-2, j)$ of ionization of the Rydberg atom induced by rotational de-excitation $j \to j - 2$ by a quadrupolar molecule can be written as [4.15]

$$\sigma_{nl}^{ion}(j-2, j) \approx \frac{8\pi a_0^2 v_0}{15V}\left(\frac{Q}{ea_0^2}\right)^2 \frac{j(j-1)}{(2j+1)(2j-1)} I_q(nl, E_0), \quad n_* > n_*^{(0)},$$

$$(6.56)$$

$$I_q(nl, E_0) = \int_0^\infty \left[\frac{dF_{E,nl}(Q)}{d(E/2Ry)}\right]_{E=E_0} (Qa_0)^2 d(Qa_0).$$

A similar expression for the ionization cross section of the Rydberg atom by a diatomic polar molecule is given by

$$\sigma_{nl}^{ion}(j-1, j) \approx \frac{16\pi a_0^2 v_0}{3V}\left(\frac{D}{ea_0}\right)^2 \frac{j}{(2j+1)} I_d(nl, E_0), \quad n_* > n_*^{(0)},$$

$$(6.57)$$

$$I_d(nl, E_0) = \int_0^\infty \left[\frac{dF_{E,nl}(Q)}{d(E/2Ry)}\right]_{E=E_0} d(Qa_0).$$

A crude estimate, based on the comparison of (6.56, 57), show that the relative efficiency of the quadrupole- and dipole-induced ionization processes for a given velocity V of the colliding particles is determined by a factor [1.3]

$$\frac{(\sigma_n^{ion})_q}{(\sigma_n^{ion})_d} \sim \frac{(Q/ea_0^2)^2}{n^2(D/ea_0)^2} \ll 1, \qquad (6.58)$$

which becomes particularly important at high n. Thus, the ionization cross sections of the Rydberg atom by strongly polar molecules are about several order of magnitude greater than the corresponding values for nonpolar molecules.

To demonstrate typical behavior of the depopulation cross sections of Rydberg states by nonpolar, weakly polar, and strongly polar molecules we present in Figs. 6.17, 6.18 the experimental [6.11, 13] and corresponding theoretical results for the $Rb(nS)$ and $Rb(nD)$ atoms perturbed by N_2, CO, and HF. As can be seen from Fig. 6.17, the agreement between calculations based

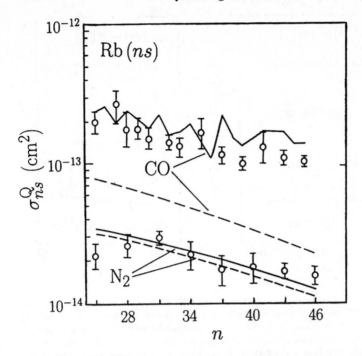

Fig. 6.17. Cross sections for collisional quenching of the Rb(nS) states by quadrupolar N_2 and weakly polar CO molecules ($T = 293$ K) versus n. Empty circles are the experimental data [6.11]. Full curves represent the impulse-approximation calculations [6.11] of the total depopulation cross sections σ_{ns}^Q including the quasi-elastic $nS \to n', l' > 2$ transitions to neighboring hydrogen-like levels with $n' = n - 3$ and $n' = n - 4$; inelastic n-changing $nS \to n', l' > 3$ transitions via rotational de-excitation and excitation of molecule ($j \to j \pm 2$ and $j \to j \pm 1$ for N_2 and CO, respectively); and ionization via rotational de-excitation. Dashed curves indicate the contribution $\sigma_{n-3,ns} + \sigma_{n-4,ns}$ of quasi-elastic transitions induced by the short-range part of electron-molecule interaction

on the impulse approximation and experimental data for N_2 and CO is quite satisfactory in the studied range of $20 < n < 50$. For thermal collisions ($T = 293$ K) with a nonpolar nitrogen molecule the total quenching cross section does not reveal the oscillatory behavior in dependence on n, since the n-changing and ionizing collisions via rotational $j \to j \pm 2$ transitions induced by the quadrupole part of the $e-N_2$ interaction are not efficient. The main contribution to σ_{ns}^Q is determined by the $ns \to n'l'$ transitions to neighboring degenerate sublevels with $n' = n - 3$ and $l' > 2$ ($\delta_s^{Rb} = 3.13$). These transitions are primarily due to the short-range part of the $e-N_2$ interaction.

However, the situation is changed radically even in the case of a weakly polar molecule such as CO. One can see from Fig. 6.17 that in the range of $20 < n < 50$ the total depopulation cross section of the Rb(ns) states by CO (at $T = 293$ K) turns out to be by one order of magnitude higher than for N_2

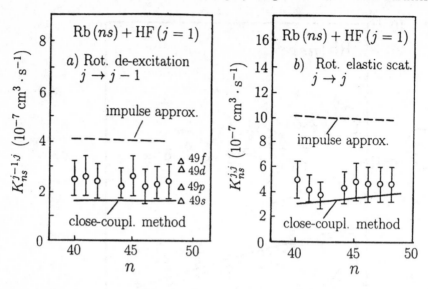

Fig. 6.18. Rate constants $K_{ns}^{j-1,j}$ (panel a) and $K_{ns}^{j,j}$ (panel b) of the inelastic n-changing upward transitions induced by rotational de-excitation $j \to j-1$ and of the quasi-elastic n, l-changing transitions via rotationally elastic electron-molecule scattering $(j \to j)$ in thermal collisions of $Rb(nS) + HF$ $(j = 1)$. Full curves are the semiclassical close-coupling calculations [6.2]. Dashed curves represent the corresponding impulse-approximation calculations. Empty triangles in *panel* a show the dependence of the n-upward rate constants on the initial orbital angular momentum l of the ns, np, nd, and nf states for $n = 49$. Empty circles are the experimental data [6.13]

as the perturber. In this case, the quenching process is predominantly determined by the quasi-resonant n-changing upward and downward transitions accompanied by the rotational dipole-de-excitation $(j \to j-1)$ and excitation $(j \to j+1)$ of CO [6.11]. The ionization process via rotational de-excitation $j \to j-1$ remains negligible in the studied range of n. The contribution of quasi-elastic $ns \to n-3$, $l' > 2$ transitions without change of the rotational energy of the CO molecule (the upper dashed curve in Fig. 6.17) is about 3 to 6 times less than the measured values of σ_{ns}^Q. Thus, the long-range electron-dipole interaction becomes dominant in the quenching of high-Rydberg levels even at small magnitudes of the dipole moment. The oscillatory structure of the quenching cross section of the $Rb(ns)$ by CO, observed in [6.11], reflects the resonant nature of the n-changing collisions via rotational $j \to j \pm 1$ transitions (i.e., strong dependence on the value of the energy defect of the reaction (6.53)).

For thermal collisions of high-Rydberg atoms $(n \sim 20 - 50)$ with strongly polar HF molecules $(\mathcal{D} = 0.72\ ea_0)$ the quenching cross sections become particularly large $\sigma_{nl}^Q \sim (1 - 2) \cdot 10^{-11}$ cm^2 (see Figs. 6.16 and [6.2, 11, 13, 14]).

The corresponding rate constants $K_{nl}^Q = \left\langle V \sigma_{nl}^Q \right\rangle_T$ may achieve the values up to $(1\text{–}5) \cdot 10^{-7}$ cm^3·s^{-1} (see Fig. 6.18). They are about two orders of magnitude greater than for the CO molecule as the perturber. This fact is in reasonable agreement with the impulse approximation (5.67) with the Born expression (5.17) for the differential cross section of rotational excitation (de-excitation) of polar molecule by slow electron impact, according to which σ_{nl}^Q is proportional to the square of the projectile dipole moment $\mathcal{D}^2$ (for example, $\mathcal{D}_{\mathrm{HF}}^2 / \mathcal{D}_{\mathrm{CO}}^2 \approx 266$). Nevertheless, the possibility of using the impulse approximation for quantitative calculations of collisions involving the Rydberg atoms and strongly polar molecules is not rigorously justified in the range of moderately high $n \sim 20\text{–}50$.

Some results of semiclassical close-coupling calculations by *Kimura* and *Lane* [6.2] of quasi-elastic n, l-changing and inelastic n-changing collisions between the Rydberg Rb(nl) atom and HF molecule are presented in Fig. 6.18 together with the corresponding experimental data and the impulse-approximation results [6.13]. One can see that the impulse approximation significantly overestimates the cross sections and rate constants when the energy defect of the reaction is small enough. In particular, the impulse approximation becomes inapplicable for describing the rotationally elastic $ns \to n - 3$, $l' > 2$ transitions (see Fig. 6.18) in thermal collisions of the Rydberg Rb(ns) atoms with HF molecules provided the principal quantum number is less than about 50–60. However, the qualitative agreement between the impulse and close coupling calculations at high n is reasonable. On the other hand, the semiclassical close-coupling method [6.2] provides satisfactory quantitative description of identified processes in a wide range of n and transition frequencies. Moreover, the semiclassical calculations [6.2] describe well the observed [6.14] increase of the n-upward rate constants with increasing the initial orbital moment l of the Rydberg atom ($10^7 \cdot K_{nl}^{n-\mathrm{ch}} = 1.6$, 2.1, 2.8 and 3.1 cm^3·s^{-1} for the s-, p-, d-, and f-states of Rb($49l$) in collision with HF($j = 1$). The rate constants of the inelastic n-changing downward transitions via rotational excitation $j \to j + 1$ of HF($j = 1$) turn out to be by one order of magnitude lower than for n-upward transitions induced by rotational de-excitation process $j \to j - 1$ (for $40 < n < 50$).

The recent experimental data and corresponding theoretical calculations [5.38] of the rate constants for destruction of very high-Rydberg states ($90 \leq n \leq 400$) and l-mixing process in collisions of K(nP)+HF are presented in Fig. 6.19. The destruction of high-Rydberg states is mainly a result of the collisional ionization process via a rotational de-excitation $j \to j - 1$ of the HF molecule and the corresponding rate constant increases with n-growing. The rate constant for the l-mixing collisions decreases with n increasing in accordance with the simple estimate by the Omont's formula [5.40]. Both theoretical dependencies of the rate constants on the principal quantum number are in good agreement with the measured values.

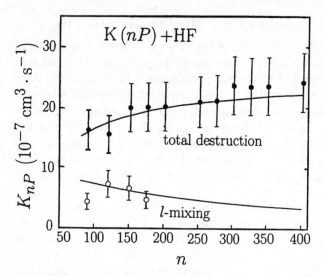

Fig. 6.19. Experimental and theoretical dependencies [5.38] of the rate constants for collisional destruction (full circles) and l-mixing process (empty circles) of high-Rydberg K(nP)-states by HF molecule

6.6.2 Ion-Pair Formation and Charge Transfer: Collision of Rydberg Atom with Electron-Attaching Molecule

Thermal collisions of the Rydberg atom with molecules (B = CD) that attach slow free electrons lead to efficient ionization processes of the two types:

$$A^*(nl) + CD \rightarrow \begin{cases} A^+ + CD^- \\ A^+ + C + D^- \end{cases}.$$

(6.59)

The first electron transfer process is accompanied by the formation of positive and negative ions in the final channel, while the second one corresponds to the electron transfer with simultaneous dissociative attachment of the electron to the molecule. At high n these processes can be described in terms of the quasi-free electron model, in which only the interaction between the Rydberg electron and perturbing molecule is important. Since the electron is in the bound state of the negative ion after collision, the rate coefficients $K_{nl}^{(1)}$ and $K_{nl}^{(2)}$ of processes (6.59) can be expressed through the cross sections of electron attachment $\sigma_a(k)$ [e + CD(β) $\rightarrow$ CD$^-$(γ)] and dissociative attachment $\sigma_{da}(k)$ [e + CD(β) $\rightarrow$ CD$^-$(γ) $\rightarrow$ C + D$^-$] of the free electron to the molecule. The final expressions are given by [4.15]

$$K_{nl}^{(1)} = \left\langle V\sigma_{nl}^{(1)} \right\rangle_T = \int_0^\infty (\hbar k/m)\sigma_a(k)W_{nl}(k)\,dk = \langle v\sigma_a(v)\rangle_{nl},$$

(6.60)

$$K_{nl}^{(2)} = \left\langle V\sigma_{nl}^{(2)} \right\rangle_T = \int_0^\infty (\hbar k/m)\sigma_{da}(k)W_{nl}(k)dk = \left\langle v\sigma_{da}(v) \right\rangle_{nl}. \qquad (6.61)$$

where $W_{nl}(k) = |\,g_{nl}(k)\,|^2\,k^2$ is the momentum distribution function of the Rydberg electron in the given state normalized to unity.

One of the most important example of this ionization mechanism is a collision of the Rydberg atom with a SF_6 molecule. This process was intensively studied experimentally for the Rydberg rare gas and alkali-metal atoms (see reviews [4.15], [6.10], [1.3]). The main contribution to collisional ionization results predominantly from the process $A^* + SF_6 \rightarrow A^+ + SF_6^-$, the rate constants of which are presented in Fig. 6.20. It can be seen that at high $n > 25$ (panel a) they are practically independent of n and does not reveal a significant dependence on the orbital angular momentum l of the Rydberg atom. Thus, the cross section for free electron attachment to SF_6 is inversely proportional to electron velocity v ($\sigma_a \propto \epsilon^{-1/2}$) at low energies $\epsilon \sim 1 - 10$ meV. Averaging over the results of different experimental works yields the following magnitude $K^{ion} = (4.3 \pm 0.4) \cdot 10^{-7}$ cm^{3}s^{-1} [6.15, 16] for the ionization rate constant of high-Rydberg atom by SF_6.

At low and intermediate n the quasi-free electron model does not hold since the molecular projectile interacts with the Rydberg electron and ionic core simultaneously. A significant reduction of the ionization rate constant in the range of at $n < 25$ with decreasing n (panel b in Fig. 6.20) occurs because many of the negative and positive ion-pairs initially formed remain electrostatically bound due to their Coulomb interaction in the final channel (see [6.15]). According to [6.16], the ionization rate constants of the reaction $A^* + SF_6 \rightarrow A^+ + SF_6^-$ averaged over the Maxwellian distribution of velocities in an atomic beam is described by the expression [6.16]

$$K^{ion} = K_0 \left\{ \frac{\sqrt{3}}{2} \exp\left[\frac{1.19}{n^2} \left(\frac{cv_0}{V_T} \right)^{2/3} \right] + 1 - \frac{\sqrt{3}}{2} \right\}^{-1}, \qquad (6.62)$$

which can be used both at high and intermediate n. Here $V_T = (2kT/\mu)^{1/2}$ is the relative velocity of the colliding particles (T is the temperature of the beam source); and c is the constant coefficient of the order of unity. Processing of the experimental data for the ionization rate constants of the reaction $Na(nP) + SF_6 \rightarrow Na^+ + SF_6^-$ at intermediate $n \sim 10-40$ leads to the following result: $K^{ion} = K_0 \exp(-n_0^2/n^2)$ with $K_0 = (5.5^{+1.4}_{-1.1}) \cdot 10^{-7}$ cm^{3}s^{-1} and $n_0 = 20.1 \pm 0.3$ [6.16].

The experimental data [5.37], [6.17] for the rate constant of the collisional destruction process

$$K(nl) + CCl_4 \rightarrow K^+ + \left(CCl_4^-\right)^* \rightarrow K^+ + CCl_3 + Cl^-$$

via the ionization of the Rydberg atom accompanied by the dissociative attachment to the CCl_4 molecule are shown in Fig. 6.21. The measurements

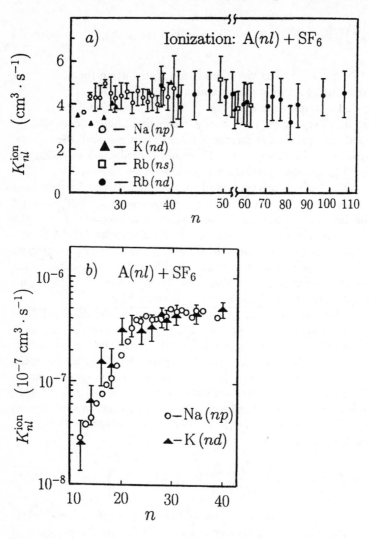

Fig. 6.20. The ionization rate constants K_{nl}^{ion} in collisions of the Rydberg atoms with electron-attaching molecule $A(nl) + SF_6 \rightarrow A^+ + SF_6^-$ at high (panel a) and intermediate (panel b) magnitudes of n. Experimental data [Na(np) – empty circles, K(nd) – full triangles, Rb(ns) – empty squares, and Rb(nd) – full circles] are those from [6.15, 16]

were carried out at very high principal quantum numbers n up to 1100. The full curve presents the rate constant calculated in the quasi-free electron model (6.61). The experimental data indicate that the rate constant of this process is practically independent of n and, hence, is independent of the specific form of the distribution function $W_{nl}(k) = | g_{nl}(k) |^2 k^2$ of the Rydberg

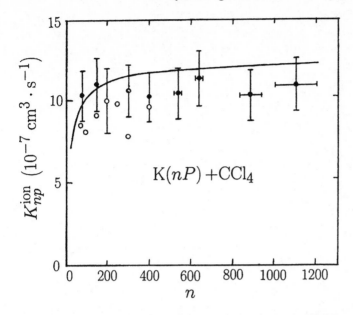

Fig. 6.21. Experimental and theoretical results for collisional destruction of Rydberg $K(nP)$-states by in collisions with CCl_4 molecule. Full circles are the measurments [5.37] at very high principal quantum numbers up to $n = 1100$). Empty circles are the measurments [6.17]. Full curve presents calculations [5.37] in quasi-free electron model

electron momenta. This means, that the cross section of dissociative attachment of the quasi-free electron to the CCl_4 molecule is inversely proportional to the electron momentum, i.e., $\sigma_{da}(k) \sim 1/k$. This fact is in full correspondence with the Wigner threshold law $\sigma_{da}(k) \sim 1/k^{2\ell-1}$ [where $\ell = 0$ for an inelastic scattering process with the incident partial s-wave [2.2].

7. Effects of Ion Core in Rydberg Atom-Neutral Collisions

In this chapter we shall continue to study the excitation (de-excitation), quenching, and ionization of highly excited atoms induced by collisions with neutral particles. However, our attention will be focused here on the analysis of the alternative physical mechanisms of these processes accounted for by the scattering of the perturbing neutral particle B on the parent core A^+ of the Rydberg atom A^*. This is, first of all, the most efficient mechanism of resonant energy exchange between the highly excited electron and inner electrons of the quasi-molecule $BA^+ + e$ temporally formed during the scattering of atom B on the ion core A^+. The identified mechanism plays usually the major role in the processes of associative and direct ionization, as well as, in the inelastic excitation (de-excitation) of the Rydberg atom in thermal collisions with the ground-state atoms. We shall also consider the dipole-induced n-changing transitions and ionization processes due to the mechanism of energy exchange between the Rydberg electron and translational motion of the colliding atoms. A theoretical description of the identified processes will be given within the framework of the quasi-molecular approach. Another separated atom-approach combined with the sudden perturbation theory will allow us to analyze transitions between highly excited states induced by the "shake up" of the Rydberg electron in the perturber-core scattering (the so-called noninertial mechanism).

7.1 Mechanisms of Perturber-Core Scattering

It is well known that the asymptotic behavior of the impact broadening for high-Rydberg levels is determined by the integral cross section for elastic scattering of perturbing atom B by the ionic core A^+ [3.8]. For quasi-elastic and inelastic collisions of Rydberg atoms with neutral particles several authors investigated the noninertial mechanism of the l-mixing [7.1–6] and n-changing transitions [7.3, 4, 7], proposed by *V.Smirnov* [7.8]. These transitions are due to the inertial force acting on the Rydberg electron as a result of acceleration of the Coulomb center A^+ upon collision with neutral perturber B. The identified mechanism does not make, as a rule, any substantial contribution to the quasi-elastic $nl \rightarrow nl'$ transitions in the experimentally studied range of n. Such transitions are primarily due to the quasi-free-electron–perturber scattering. The influence of the perturber-core scattering on the behavior of the l-mixing process becomes important only in the range of very high n [7.4, 6] or for low and intermediate n, where the quasi-free electron model does not hold. As was shown by *Hahn* [5.71] and *de Prunelé* [7.9], at low enough values of n the distortion corrections to the impulse approximation due to perturber-core interaction are large, so that the B-e and B-A$^+$ collisions can not be considered independently. However, for the inelastic n-changing transitions the predominant contribution to the cross section $\sigma_{n'n}$ in a wide range of n can be a result of the perturber-core scattering alone [7.4], while the interaction between the neutral particle B and the outer electron e of the Rydberg atom A* can be neglected.

The mechanisms of the perturber-core scattering become particularly important for the ionization [7.10–15] and n-changing transitions [7.4, 7, 16, 17] with large energy transfer ΔE_{fi} to the Rydberg electron. That is why the cross sections of inelastic processes induced by scattering of the quasi-free electron by a perturber reveal strong reduction with an increase of ΔE_{fi}. During the scattering of the perturbing atom B on the ionic core A$^+$ of the Rydberg atom A* the quasi-molecular BA$^+$ + e system is temporally formed with large orbital radius $r_n \sim n_*^2 a_0$ of the external electron e as compared to the internuclear distance R. As a result, the different types of inelastic transitions and ionization in such a system can occur because of the interaction between the Rydberg electron e and the quasi-molecular ion BA$^+$.

The excitation and ionization processes induced by the perturber-core scattering are particularly efficient when there is a resonant (or quasi-resonant) exchange of the Rydberg electron energy $\Delta E_{fi} = \hbar\omega$ with the energy of the electronic shell of quasi-molecular ion BA$^+$. Such processes occur near the crossing point R_ω of electronic terms of quasi-molecule BA$^+$ + e, in which the energy defect ΔE_{fi} of the $|i\rangle \rightarrow |f\rangle$ transition becomes equal to the energy splitting $\Delta U_{fi}(R_\omega)$ of electronic terms of quasi-molecular ion BA$^+$. This mechanism of excitation and ionization takes place, for example, in thermal collisions of Rydberg atom A* with the ground-state parent atom A. It is well known that in this case the n-changing transitions with large enough energy transfer as well as direct and associative ionization are due to the dipole transitions between the symmetrical and antisymmetrical terms of the homonuclear A$_2^+$ ion. Such a mechanism (proposed in [7.16, 17]) has been studied by *Devdariani* et al. [7.10], *Janev* and *Mihajlov* [7.11], *Duman* and *Shmatov* [7.12] for the ionization processes. The detailed experimental studies of associative and direct ionization have been carried out for the Rydberg alkali-metal atoms in symmetrical A(nl) + A and nonsymmetrical A(nl) + B thermal collisions with ground-state alkali atoms (see [7.18]).

The resonant energy exchange of external and inner electrons of quasi-molecular ion BA$^+$ also has been investigated for the inelastic quenching [7.19] and ionization [7.15] of the Rydberg rare-gas A$\left[n_0 p^5\left(^2P_{3/2}\right)nl\right]$

atoms in thermal collisions with the ground-state $B(^1S_0)$ atoms of the buffer rare gas. In this case, the excitation (de-excitation) or ionization of the Rydberg electron is accompanied by the $A_1 |j_i = 3/2, \Omega_i = 3/2\rangle \rightleftarrows X |j_f = 3/2, \Omega_f = 1/2\rangle$ transition between two split terms of the heteronuclear BA^+ ion with different $\Omega_i = 3/2$ and $\Omega_f = 1/2$ projections of the electronic angular $j = 3/2$ moment on the internuclear $\mathbf{R}$ axis. Due to the selection rules the main role in the mechanism involved is played by the quadrupole and short-range parts of the Coulomb interaction between the Rydberg electron and inner $n_0 p^5$ electrons of the atomic core A^+.

Another mechanism of inelastic $n \rightarrow n'$ transitions [7.4] and ionization [7.14, 15] associated with the scattering of perturbing B atom on the ionic core A^+ has been proposed by *Lebedev* et al. [7.3]. In this mechanism, the Rydberg electron transition is accompanied by the translational transition of colliding particles A^+ and B within one electronic $U(R)$ term of the quasi-molecular BA^+ ion. The identified processes are due to the interaction of the Rydberg electron of the heteronuclear quasi-molecular $BA^+ + e$ system with the dipole moment of the BA^+ ion in the given electronic state. This mechanism can play an essential role for collisions of atoms $A^* - B$ with small reduced mass, provided no resonance transfer of energy from the inner electronic shell of the BA^+ system to the outer electron occurs [for example, for $H(n) + He$].

7.2 Separated-Atoms Approach: Shake-Up Model

Consider first the processes of excitation (de-excitation) and ionization of the Rydberg electron e due to the transitions within one electronic $U(R)$ term of the quasi-molecular BA^+ ion. They are the result of direct exchange of the Rydberg electron energy with the kinetic energy of relative motion of colliding A^+ and B particles. The magnitudes and the behavior of the cross sections for such processes are qualitatively different in the range of small $\omega \ll V/\rho_{cap}$ and large $\omega \gtrsim V/\rho_{cap}$ transition frequencies. Here ρ_{cap} is the impact parameter for the capture of the perturbing atom B by the ionic core A^+ of the Rydberg atom, which determines the characteristic dimension of internuclear separation R_{BA^+}, and V is the relative velocity of the colliding particles.

7.2.1 General Tréatment

a) Basic Formulas of Sudden Perturbation Theory. In the range of low frequencies $\omega \ll V/\rho_{cap}$ quasi-elastic and inelastic transitions of the Rydberg electron occur in the range of internuclear separations $\rho_{cap} \lesssim R_\omega \ll n^2 a_0$. Thus, for calculating the transition cross sections it is natural to use the separated-atoms approach based on the sudden perturbation theory [4.19, 7.4] or on the impulse approximation [7.20, 6]. The shake-up model of electron, proposed by *Migdal* [2.6] for ionization of atoms by neutrons, was applied for calculating the cross sections of the n-changing, l-mixing, quenching, and ionization processes involving a highly excited atom and neutral projectile in [5.21, 7.4, 14]. In its general form the shake-up model of quantum system was formulated in [7.21]. Here we outline the basic idea of the shake-up model for the Rydberg arom-neutral particle collisions.

Let the velocity of the ion core A^+ be changed by a value ΔV_{A^+} as a result of a binary encounter with projectile B. For small enough transition frequencies ω_{fi} the collision time $\tau_{A+B} \sim \rho_{cap}/V$ of these particles is small compared to the characteristic period $1/\omega_{fi}$ of the Rydberg electron motion, which can be estimated as $T_n \sim 2\pi n^3 (a_0/v_0)$. Therefore, the interaction between the ion core and neutral projectile can be regarded as a sudden perturbation of the Rydberg atom Hamiltonian:

$$H(t) = \begin{cases} H_1 \equiv H_A(\mathbf{r}) = -\dfrac{\hbar^2}{2m} \Delta_\mathbf{r} + U(r) & t < 0, \\[2mm] H_2 \equiv H_A(\mathbf{r}') = -\dfrac{\hbar^2}{2m} \Delta_{\mathbf{r}'} + U(r') & t \gg \tau_{A+B}. \end{cases} \qquad (7.1)$$

Here $\mathbf{r}$ and $\mathbf{r}'$ are the radius vectors of a highly excited electron in the frames of reference moving with the ion core before and after collision so that $\mathbf{r}' = \mathbf{r} - (\Delta \mathbf{V}_{A^+})t$. Since the orbital radius of a highly excited electron $r_n \sim n^2 a_0$ is large $(r_n \gg \rho_{cap})$, the displacement $\Delta R_{A^+} \sim (\Delta V_{A^+})(\tau_{A+B}) \sim (\mu/M_A)\rho_{cap}$ of the ion core during the time interval τ_{A+B} of the core-projectile interaction is small $\Delta R_{A^+} \ll r_n$ compared to the characteristic size of the Rydberg atom $\Delta R_{A^+} \ll r_n$, and hence can be neglected. As a result, one can assume that the electron coordinates r' and r directly before and after a collision practically coincide.

Further, the perturber-core encounter leads to the $\alpha \to \alpha'$ transition between two eigenstates of the Rydberg atom. It is important to stress that the initial eigenwave function $\psi_\alpha(\mathbf{r})$ of the Hamiltonian H_1 is not the eigenfunction of the Hamiltonian H_2. The initial wave function of the Rydberg atom in the frame of reference moving with the ion core directly after collision is $\psi_i^{(0)}(\mathbf{r}', t = 0) = \psi_\alpha(\mathbf{r}') \exp(-i\mathbf{K} \cdot \mathbf{r}')$, where $\mathbf{K} = (m\Delta\mathbf{V}_{A^+})/\hbar$ is the electron wave vector corresponding to the change of the parent core velocity $\Delta\mathbf{V}_{A^+} = (\mathbf{q}' - \mathbf{q})/M_A$ in its collision with the perturbing particle B. Hence, using the basic formula of sudden perturbation theory for the transition amplitude $a_{fi}(t = \infty) = \left\langle \psi_f(\mathbf{r}') \middle| \psi_i^{(0)}(\mathbf{r}', t = 0) \right\rangle$ from the initial α state to the final eigenstate $f = \alpha'$ and for the corresponding transition probability W_{fi}, we have

$$a_{\alpha'\alpha}(t = \infty) = \langle \psi_{\alpha'}(\mathbf{r}) | \exp(-i\mathbf{K} \cdot \mathbf{r}) | \psi_\alpha(\mathbf{r}) \rangle, \qquad W_{\alpha'\alpha} = |a_{\alpha'\alpha}(t = \infty)|^2,$$

$$(7.2)$$

where we additionally put $\mathbf{r}' = \mathbf{r}$ in accordance with the discussion presented above.

Thus, in the shake-up model the probability of the $nl \rightarrow n'l'$ transition between the Rydberg levels with the given values of the principal and orbital quantum numbers caused by the perturber-core scattering is given by

$$W_{n'l',nl}(K) = \frac{1}{2l+1} \sum_{mm'} |\langle \psi_{n'l'm'}(\mathbf{r})| \exp(i\mathbf{K} \cdot \mathbf{r}) |\psi_{nlm}(\mathbf{r})\rangle|^2. \quad (7.3)$$

Here $\mathbf{r} = (r, \theta, \varphi)$ is the radius vector of the highly excited electron relative to the parent core A^+, $\psi_{nlm}(\mathbf{r}) = R_{nl}(r) Y_{lm}(\theta, \varphi)$ is the atomic wave function. The electron wave vector $\mathbf{K}$ corresponding to the change of the parent core velocity $\Delta\mathbf{V}_{A^+}$ in its collision with the perturbing particle B can be expressed in terms of the momentum transfer vector $\mathbf{Q}$. The values of $\mathbf{K} = (m/M_A)\mathbf{Q}$ and $\mathbf{Q} = \mathbf{q}' - \mathbf{q}$ in the Rydberg atom-neutral collision are given through their scattering angle $\theta_{\mathbf{q}'\mathbf{q}}$ in the center of mass system by the relation

$$K^2 = \left(\frac{m}{M_A}\right)^2 (q^2 + q'^2 - 2qq' \cos\theta_{\mathbf{q}'\mathbf{q}}) \approx 2a_0^{-2} \left(\frac{\mu V}{v_0 M_A}\right)^2 (1 - \cos\theta_{\mathbf{q}'\mathbf{q}}).$$

$$(7.4)$$

Here $q = \mu V/\hbar$ and $q' = \mu V'/\hbar$ are the relative wave numbers of Rydberg atom A^* and perturber B (μ is their reduced mass), whereas $q' = [q^2 + 2\mu\Delta E_{nl,n'l'}/\hbar^2]^{1/2} \approx q$ at $|\Delta E_{n'l',nl}| \ll \hbar^2 q^2/2\mu$.

The summed probability $W_{nl}^{in} = 1 - W_{nl}^{el}$ of all inelastic bound-bound and bound-free transitions of the Rydberg atom (excluding only the contribution of elastic scattering $W_{nl}^{el} \equiv W_{nl,nl}$) can be written as [5.21]

$$W_{nl}^{in}(K) = 1 - (2l+1) \sum_{\varkappa=0}^{2l} (2\varkappa + 1) \begin{pmatrix} l & \varkappa & l \\ 0 & 0 & 0 \end{pmatrix}^2 |\langle nl |j_\varkappa (Kr)| nl\rangle|^2. \quad (7.5)$$

This result proceeds directly from (7.3) at $|n'l'\rangle = |nl\rangle$ using the expansion in spherical Bessel functions for the $\exp(i\mathbf{K} \cdot \mathbf{r})$ factor.

The cross section of the $nl \rightarrow n'l'$ transition can be expressed in the shake-up model through the probability $W_{n'l',nl}[K(\theta_{\mathbf{q}'\mathbf{q}})]$ and the differential cross section $d\sigma_{A+B}/d\Omega_{\mathbf{q}'\mathbf{q}} = |f_{A+B}(q, \theta_{\mathbf{q}'\mathbf{q}})|^2$ for the elastic scattering of perturbing particle B by the ionic core A^+ [4.19]

$$\sigma_{n'l',nl} = \int W_{n'l',nl} |f_{A+B}|^2 d\Omega_{\mathbf{q}'\mathbf{q}}. \quad (7.6)$$

Here f_{A+B} is the elastic perturber-core scattering amplitude. Then, using (7.3–6) and replacing the integration with respect to the solid angle $d\Omega_{\mathbf{q}'\mathbf{q}}$ by integration over the momentum transfer dQ, we obtain [5.21]

$$\sigma_{n'l',nl} = \frac{2\pi\hbar^2}{\mu^2 V^2 (2l+1)} \sum_{mm'} \int_{Q_{min}}^{Q_{max}} \left| \left\langle n'l'm' \left| \exp\left[i\left(\frac{m}{M_A}\right) \mathbf{Q} \cdot \mathbf{r} \right] \right| nlm \right\rangle \right|^2$$
$$\times \left| f_{A+B}(Q) \right|^2 Q dQ, \tag{7.7}$$

where $Q_{min} = |q' - q| \approx |\Delta E_{n'l',nl}|/\hbar V$ and $Q_{max} = q' + q \approx 2\mu V/\hbar$.

b) Equivalence of Shake-Up Model and Impulse Approximation.
This result for the contribution of the core-perturber scattering mechanism to
the transition cross section can also be derived in the impulse approximation
similar to the case of transitions induced by the electron-perturber scattering
considered previously in Sect.4.5.1. Indeed, in accordance with [7.20, 6] the
basic expression for the matrix elements of the scattering T_{A+B}-operator is
expressed in terms of the two-body t_{A+B}-operator over the plane waves for
the core-perturber relative motion

$$\langle \mathbf{q}', f | T_{A+B}(E) | \mathbf{q}, i \rangle = \int G^*_{\alpha'}(\kappa') \langle \mathbf{q}' | t_{A+B}(\mathcal{E}) | \mathbf{q} \rangle G_\alpha(\kappa) d\kappa, \tag{7.8}$$

$$t_{A+B}(\mathbf{q}', \mathbf{q}; \mathcal{E}) = \left\langle \mathbf{q}' \left| V_{A+B} + V_{A+B} \left(\frac{1}{\mathcal{E} - K_{A+B} - V_{A+B} + i0} \right) V_{A+B} \right| \mathbf{q} \right\rangle \tag{7.9}$$

in full correspondence with the analogous result (4.104). Here we have additionally used the mass disparity relations $(m \ll M_A)$ according to which
$\kappa' \approx \kappa + (m/M_A) \mathbf{Q}$, where $\mathbf{Q} = \mathbf{q}' - \mathbf{q}$. In the coordinate representation,
expression (7.8) can be rewritten as

$$\langle \mathbf{q}', f | T_{A+B}(E) | \mathbf{q}, i \rangle = \left\langle \psi_f(\mathbf{r}) \left| \exp\left(-\frac{m}{M_A} \mathbf{Q} \cdot \mathbf{r} \right) \right| \psi_i(\mathbf{r}) \right\rangle$$
$$\times \langle \mathbf{q}' | t_{A+B}(\mathcal{E} = \hbar^2 q^2/2\mu) | \mathbf{q} \rangle. \tag{7.10}$$

Here the matrix elements of the two-body (core-perturber) operator over the
plane waves are related with the amplitude $f_{A+B}(Q)$ for elastic scattering of
the A^+B-particles

$$f_{A+B}(Q) = -\frac{\mu}{2\pi\hbar^2} \langle \mathbf{q}' | t_{A+B}(\mathcal{E} = \hbar^2 q^2/2\mu) | \mathbf{q} \rangle. \tag{7.11}$$

Thus, the final formula of the impulse approximation for the contribution
of the core-perturber scattering to the amplitude of the inelastic transition
between Rydberg states of an atom takes the following form:

$$f_{fi}^{A^+B} = \left\langle f \left| \exp\left(-\frac{m}{M_A} \mathbf{Q} \cdot \mathbf{r} \right) \right| i \right\rangle f_{A+B}(Q). \tag{7.12}$$

It is evident that integration of the amplitude square $\left| f_{fi}^{A^+B}(Q) \right|^2$ over
the momentum transfer $Q dQ$ leads directly to (7.7). Therefore, for the

perturber-core scattering mechanism, the shake-up model used by *Lebedev* and *Marchenko* [7.4] for transitions between Rydberg states and the impulse approximation used by *Gounand* and *Petitjean* [7.6] should lead to completely equivalent results. Comparison of this expression with the basic formula of the impulse approximation for the electron-perturber scattering shows that usually the core-perturber scattering mechanism yields a small contribution to the cross sections of the inelastic transitions $i \neq f$ due to the presence of a small factor (m/M_A) in the momentum transferred to the Rydberg atom $\mathbf{K} = (m/M_A)\mathbf{Q}$. However, we shall show in this section that in some cases it becomes important.

c) **Limiting and Asymptotic Behavior of the Cross Sections.** *Flannery* [4.19] has shown that the total cross section σ_i^{tot} for all inelastic and elastic transitions caused by the mechanism under consideration is determined by the integral cross section for elastic scattering of the neutral projectile B on the ionic core A^+ of the Rydberg atom, i.e.,

$$\sigma_i^{tot} = \sum_f \sigma_{fi} \equiv \sigma_i^{el} + \sigma_i^{in} = \sigma_{A+B}^{el} , \qquad \sigma_{A+B}^{el}(V) = \int |f_{A+B}|^2 \, d\Omega_{q'q} \qquad (7.13)$$

Formula (7.13) follows from the basic equations (7.3, 6) taking into account the summation rule $W_i^{tot} = \sum_f W_{fi} = 1$ for the total transition probability.

Thus in accordance with [4.19], the cross section σ_{fi} of any bound-bound or bound-free transitions induced by the perturber-core scattering mechanism is limited by the magnitude of the elastic scattering cross section σ_{A+B}^{el}.

In the asymptotic region of very high principal quantum numbers $n \gg v_0 M_A/V\mu$ ($n \gg 10^3$–10^4 at thermal velocities) the cross section σ_{nl}^{in} of all inelastic transitions tends asymptotically to the integral cross section for elastic scattering of perturbing particle B by the parent core A^+ of the Rydberg atom A^*. Correspondingly, contribution $\sigma_{nl}^{el}(V)$ of elastic scattering of perturber B on the Rydberg atom becomes very small and can be neglected, so that

$$\sigma_{nl}^{in}(V) \xrightarrow[n \to \infty]{} \sigma_{A+B}^{el}(V) , \qquad \sigma_{nl}^{el}(V) \xrightarrow[n \to \infty]{} 0. \qquad (7.14)$$

As has been shown in [5.44], in this range of $n \gg v_0 M_A/\mu V$ the main contribution to the cross section $\sigma_i^{in} = \sum_{f \neq i} \sigma_{fi}$ makes the bound-free transitions of the highly excited electron. This means that at very high magnitudes of n the cross section of all inelastic transitions is primarily determined by the ionization of the Rydberg atom, i.e., $\sigma_{nl}^{in}(V) \approx \sigma_{nl}^{ion}(V)$.

7.2.2 Dipole Approximation

Simple expressions for the cross sections of inelastic and quasi-elastic transitions between the Rydberg states can be derived in the shake-up model for the range of $n \ll (v_0 M_A/V\mu)^{1/2}$. At thermal velocities V this condition leads usually to the following restriction $n \overset{\sim}{<} 30 - 100$ on the principal quantum numbers (depending on the kind of colliding particles). Due to the relation $Kn^2 a_0 \ll 1$, the exponential factor in (7.4) can be expanded in series of (Kr). Then, in the dipole approximation, the transition probability $W_{n'l',nl}$ of the $nl \to n'l'$ transitions (for which $l' \neq l$ at $n' = n$) is given by

$$W_{n'l',nl}(K) = \frac{K^2}{3(2l+1)} \sum_{mm'} \left| \langle n'l'm' | \mathbf{r} | nlm \rangle^2 \right|. \tag{7.15}$$

Substitution of this expression (7.15) into the basic formula (7.6) yields the following result for the corresponding cross section:

$$\sigma_{n'l',nl} = \frac{2}{3} \left(\frac{\mu V}{v_0 M_A} \right)^2 M_{n'l',nl} \sigma_{A+B}^{tr}, \quad M_{n'l',nl} = \frac{l_>}{(2l+1)a_0^2} \left| R_{nl}^{n'l'} \right|^2. \tag{7.16}$$

Here the dimensionless quantity $M_{n'l',nl}$ is determined by the radial matrix element $R_{nl}^{n'l'} = \langle sn'l' | r | nl \rangle$ of the Rydberg coordinate $[l' = l \pm 1, \, l_> = \max\{l, l'\}]$, and

$$\sigma_{A+B}^{tr}(V) = \int |f_{A+B}|^2 \left(1 - \cos\theta_{q'q} \right) d\Omega_{q'q} \tag{7.17}$$

is the momentum transfer cross section for the elastic scattering of neutral particle B by the ionic core A^+.

Formula for the total cross section σ_{nl}^{in} of all inelastic $nl \to n'l'$ transitions (i.e. the quenching cross section of the nl-level) caused by the perturber-core scattering [5.21]

$$\sigma_{nl}^{in}(V) = \sum_{n'l'}' \sigma_{n'l',nl} = \frac{2}{3} \left\langle \left(\frac{r}{a_0} \right)^2 \right\rangle_{nl} \left(\frac{\mu V}{v_0 M_A} \right)^2 \sigma_{A+B}^{tr}(V) \tag{7.18}$$

directly follows in the dipole approximation from (7.5). Here $\langle r^2 \rangle_{nl}$ is determined by a simple analytic expression [2.3] in the case of hydrogen-like degenerate nl-levels, whereas $\langle r^2 \rangle_{nl} \approx 5n^4 a_0^2/2$ at $l \ll n$.

As is evident from (7.16) and expressions of Sect. 3.1 for the dipole matrix elements $R_{nl}^{n,l\pm1}$, the contribution of pure quasi-elastic $nl \to n, l\pm1$ transitions $\Delta E_{n'l',nl} = 0$ (considered analytically in [7.4, 5]) to the total quenching cross section is predominant. The contribution of all other inelastic $nl \to n', l \pm 1$ transitions with $n' \neq n$ (n-changing processes) is appreciably smaller for quenching of degenerate hydrogenlike nl-levels. The corresponding M_{nl}^{l-mix} and $M_{nl}^{n,l-ch}$ terms in (7.16) determining the l-mixing and n-changing ($n' \neq n$) cross sections σ_{nl}^{l-mix} and $\sigma_{nl}^{n,l-ch}$ are given by (at $l \ll n$)

$$M_{nl}^{l-mix} = \sum_{l'=l\pm1} M_{nl',nl} \approx \frac{9n^4}{4}, \quad M_{nl}^{n,l-ch} = \sum_{n' \neq n} \sum_{l'=l\pm1} M_{n'l',nl} \approx \frac{n^4}{4}. \tag{7.19}$$

Provided there are equally populated lm-sublevels within a given principal quantum number n, the shake-up model leads to a simple analytic expression [7.4]

$$\sigma_{n'n}(V) = \sum_{ll'} \left(\frac{2l+1}{n^2}\right) \sigma_{n'l',nl} = \frac{4\mathcal{G}(\Delta n)}{3^{3/2}\pi} \left(\frac{n}{\Delta n}\right)^4 \left(\frac{\mu V}{v_0 M_A}\right)^2 \sigma_{A+B}^{tr}(V)$$

(7.20)

for the cross section of the inelastic $n \to n'$ transitions in the dipole range of $n \ll (v_0 M_A/\mu V)^{1/2}$. Here $\mathcal{G}(\Delta n)$ is the Gaunt factor ($\mathcal{G} = 1$ in the Kramers approximation). Expression (7.20) is valid for small enough transition frequencies $\omega = 2Ry|\Delta n|/\hbar n^3 \ll V/\rho_{cap}$.

Now we will concern with the behavior of the elastic scattering cross section $\sigma_{nl}^{el} \equiv \sigma_{nl,nl}$ of neutral particle B by the Rydberg atom $A(nl)$ for the perturber-core scattering mechanism. In the dipole range of $n \ll (v_0 M_A/\mu V)^{1/2}$ the probability of elastic scattering $W_{nl}^{el} = 1 - W_{nl}^{in}$ in the shake-up model is approximately equal to $W_{nl}^{el}(K) = 1 - O\left[(Kn^2 a_0)^2\right]$ [i.e., equal to unity in the zero approximation over the $(Kn^2 a_0)$ parameter, see (7.5)]. Thus, in this approximation ($\sigma_{nl}^{el} \gg \sigma_{nl}^{in}$) the elastic scattering cross section is independent of the principal quantum number n and is determined by the integral cross section σ_{A+B}^{el} for perturber-core scattering, so that

$$\sigma_{nl}^{el}(V) = \sigma_{A+B}^{el}(V) - \sigma_{nl}^{in}(V) \approx \sigma_{A+B}^{el}(V), \qquad n \ll (v_0 M_A/\mu V)^{1/2} . \quad (7.21)$$

At thermal energies, the major contribution to the integral σ_{A+B}^{el} and the momentum transfer σ_{A+B}^{tr} cross sections is determined by the long-range part of the perturber-core interaction $U(R)$. For the power approximation $U^{l.r}(R) = -C_\nu/R^\nu$ the semiclassical expression for the integral cross section (7.22) is given by [2.2]

$$\sigma_{A+B}^{el}(V) = 2\pi^{\nu/(\nu-1)} \sin\left[\frac{\pi}{2}\left(\frac{\nu-3}{\nu-1}\right)\right] \Gamma\left(\frac{\nu-3}{\nu-1}\right)$$

$$\times \left[\frac{\Gamma\left(\frac{\nu-1}{2}\right)}{\Gamma\left(\frac{\nu}{2}\right)}\right]^{2/(\nu-1)} \left(\frac{C_\nu}{\hbar V}\right)^{2/(\nu-1)} .$$

(7.22)

For a given relative velocity V of colliding particles, magnitudes of the momentum transfer cross section $\sigma_{A+B}^{tr}(V)$ (7.17) are significantly smaller than $\sigma_{A+B}^{el}(V)$, since the contribution of small scattering angles $\theta_{q'q}$ turns out to be partially compensated by a factor $(1 - \cos\theta_{q'q})$. As a result, the value of σ_{A+B}^{tr} is closer to that of the cross section σ_{A+B}^{cap} for the capture of perturbing atom B by the ionic core A^+. The capture cross section is given by the simple expression [2.1]

$$\sigma_{A^+B}^{cap}(V) = \pi \rho_{cap}^2(V) = \pi \left(\frac{\nu}{\nu-2}\right)^{\nu/(\nu-2)} \left(\frac{\nu C_\nu}{2\mathcal{E}}\right)^{2/\nu}, \qquad (7.23)$$

where $\mathcal{E} = \mu V^2/2$, and ρ_{cap} is the capture impact parameter. In the special case of polarization interaction $U_{pol} = -\alpha e^2/2R^4$ (when $\nu = 4$ and $C_4 = \alpha e^2/2$) between the B and A^+ particles, expressions, for the capture cross section, as well as, the momentum transfer and integral elastic scattering cross sections can be written as

$$\sigma_{A^+B}^{el} = \frac{\pi^{5/3}}{2^{4/3}} \Gamma\left(\frac{1}{3}\right) \left(\frac{\alpha e^2}{\hbar V}\right)^{2/3} \approx 7.16 \left(\frac{\alpha v_0}{V}\right)^{2/3},$$

$$\sigma_{A^+B}^{tr} \approx 1.12\sigma_{A^+B}^{cap}, \qquad \sigma_{A^+B}^{cap} = \pi \left(\frac{2\alpha e^2}{\mathcal{E}}\right)^{1/2}, \qquad (7.24)$$

where $\mathcal{E} = \mu V^2/2$. It is seen that the $\sigma_{A^+B}^{el}/\sigma_{A^+B}^{tr}$ ratio is approximately determined by the parameter

$$\sigma_{A^+B}^{el}/\sigma_{A^+B}^{tr} \sim \left[(\mu/m)^2 (\mathcal{E}/2Ry) (\alpha/a_0^3)\right]^{1/6} \qquad (7.25)$$

whose value turns out to be of the order of 5–50 at thermal energies depending on the polarizability α of neutral perturber B and the reduced mass μ of the colliding partners.

Thus, as follows from the results of the shake-up model of Rydberg atom A^* by neutral particle B, the quenching cross section $\sigma_{nl}^Q \equiv \sigma_{nl}^{in}$ of a highly excited nl-level increase rapidly with principal quantum number ($\sigma_{nl}^Q \propto n^4$) in the dipole range of $n \ll (v_0 M_A/\mu V)^{1/2}$. It becomes of the order of the momentum transfer cross section $\sigma_{A^+B}^{tr}$ for the perturber-core scattering at $n \sim (v_0 M_A/\mu V)^{1/2}$. Further increase of n leads to a slower growing of the total inelastic cross section. The limiting value $\sigma_{A^+B}^{el}$ of the total inelastic cross section $\sigma_{nl}^{in} \approx \sigma_{nl}^{ion}$ for the perturber-core scattering mechanism is achieved only in the range of very high principal quantum numbers $n \gg v_0 M_A/\mu V$, which are of no importance for the experimental applications at thermal energies. Therefore, the ratio of the total inelastic cross section magnitudes at very high n and at $n \sim (v_0 M_A/\mu V)^{1/2}$ may be estimated by relation (7.25).

All results presented above describe the case of collisions between the Rydberg atom A^* and the perturbing atom or molecule B of the buffer gas. Some comments should be made on symmetrical collisions of highly excited atom A^* with the ground-state parent atom A. In this case the scattering of atom A on the ionic core A^+ leads to charge exchange of the valence electron of the projectile. Hence, the velocity of the Coulomb center, in which the outer Rydberg electron e is localized, is periodically changing from V_1 to V_2, where V_1 and V_2 are velocities of the heavy particles A^+ and A. Therefore, as in the case of nonsymmetrical B$\rightarrow A^*$ collisions, transitions of the Rydberg electron may be a result of an inertial force acting on this electron due to the change of the coordinate frame of reference connecting with one or another

Coulomb center. However, for the symmetrical Rydberg atom-neutral $A \rightarrow A^*$ thermal collisions, such transitions are primarily accounted for by charge exchange of the inner electron of the composite $A_2^+ + e$ system (i.e., of the valence electron of the ground-state atom A) but not the momentum transfer to the Coulomb center A^+ in its collision with projectile A. This is due to the small contribution of the perturber-core scattering $A \rightarrow A^+$ to the momentum transfer cross section σ_{A+A}^{tr} as compared to the charge exchange effect at thermal energies, so that $\sigma_{A+A}^{tr} \approx 2\sigma_{A+A}^{ex}$. The cross section σ_{A+A}^{ex} for resonance charge exchange of atomic ion A^+ by the parent atom A is determined by the well-known expression [2.2]

$$\sigma_{A+A}^{ex}(V) = \int\limits_0^\infty \sin^2\left[\eta\left(\rho, V\right)\right] 2\pi\rho d\rho, \quad \eta(\rho, V) = \frac{1}{2\hbar} \int\limits_{-\infty}^\infty \Delta U_{gu}\left[R\left(t\right)\right] dt, \quad (7.26)$$

where $\Delta U_{gu}(R) = |U_g(R) - U_u(R)|$ is the exchange splitting of even and odd lower terms of quasi-molecular ion A_2^+, and $V = |\mathbf{V}_1 - \mathbf{V}_2|$ is the relative velocity of the colliding A^+ and A particles.

As has been shown in [7.7, 5], analytic formulas of the dipole approximation (7.16, 20) presented above for nonsymmetrical collisions with small enough transition frequencies remain the same also in the symmetrical case. However, it should be taken into account that the reduced mass of the colliding atoms is $\mu = M_A/2$ and the momentum transfer cross section $\sigma_{A+A}^{tr} \approx 2\sigma_{A+A}^{ex}$ is given by expression (7.26).

7.3 Quasi-molecular Approach: Basic Assumptions

For large transition frequencies $\omega > V/\rho_{cap}$ the shake model of the Rydberg electron based on the sudden perturbation approximation becomes inapplicable. For symmetrical collisions this occur at $\omega > V/R_{ex}$, where the characteristic internuclear distance is determined by the cross section of the resonance exchange process $R_{ex} \sim \left(\sigma_{A+A}^{ex}/\pi\right)^{1/2}$, see (7.26). In this range of ω the inelastic transitions of a highly excited electron, induced by the perturber-core scattering, occur at small enough internuclear distances $R_\omega < \rho_{cap}$ ($R_\omega < R_{ex}$) as compared to the orbital radius $r_n \sim n^2 a_0$ of the Rydberg electron. Thus, it is natural to use the quasi-molecular approach in theoretical analysis of both transitions within one and between different electronic terms of the BA^+ ion. In this approach, a simple description of excitation (de-excitation) and ionization processes may be given on the basis of stationary perturbation theory over interaction $\mathcal{V}$ between Rydberg electron e and quasi-molecular BA^+ ion.

The total Hamiltonian H of the quasi-molecular $BA^+ + e$ system can be presented as

$$H = -\frac{\hbar^2}{2\mu}\Delta_{\mathbf{R}} + H_{BA^+}(\mathbf{r}_\kappa, \mathbf{R}) + H_e(\mathbf{r}) + \mathcal{V},$$

$$H_e = -\frac{\hbar^2}{2m}\Delta_{\mathbf{r}} - \frac{e^2}{r},$$

$$\text{(7.27)}$$

$$H_e\psi_{nlm}(\mathbf{r}) = E_{nl}\psi_{nlm}(\mathbf{r}), \qquad H_e\psi_{\mathbf{k}}^\pm(\mathbf{r}) = E\psi_{\mathbf{k}}^\pm(\mathbf{r}), \tag{7.28}$$

$$H_{BA^+}\varphi_i(\mathbf{r}_\kappa, \mathbf{R}) = U_i(R)\varphi_i(\mathbf{r}_\kappa, \mathbf{R}),$$

$$H_{BA^+}\varphi_f(\mathbf{r}_\kappa, \mathbf{R}) = U_f(R)\varphi_f(\mathbf{r}_\kappa, \mathbf{R}),$$

$$\text{(7.29)}$$

$$\mathcal{V} = \sum_{\kappa=1}^{N}\frac{e^2}{|\mathbf{r}-\mathbf{r}_\kappa|} - \frac{Z_A e^2}{|\mathbf{r}-\mathbf{R}_A|} - \frac{Z_B e^2}{|\mathbf{r}-\mathbf{R}_B|} + \frac{e^2}{r}. \tag{7.30}$$

H_e is the Hamiltonian of the outer electron in the Coulomb field of the atomic core A^+, $\psi_{nlm}(\mathbf{r})$ is the atomic wave function with the given magnitudes of quantum numbers n, l, and m and with energy $E_{nl} = -Ry/n_*^2$, while $\psi_{\mathbf{k}}^\pm(\mathbf{r})$ are the Coulomb wave functions of a continuous spectrum with wave vector $\mathbf{k} = (k, \theta_{\mathbf{k}}, \varphi_{\mathbf{k}})$ and energy $E = \hbar^2 k^2/2m$ of the ejected electron. H_{BA^+} is the Hamiltonian of the electronic shell of the BA^+ ion, $\varphi_i(\mathbf{r}_\kappa, \mathbf{R})$ and $\varphi_f(\mathbf{r}_\kappa, \mathbf{R})$ and $U_i(R)$ and $U_f(R)$ are its wave eigenfunctions and the corresponding electronic terms in the initial $|i\rangle$ and final $|f\rangle$ states ($\mathbf{R} = \mathbf{R}_B - \mathbf{R}_A$ is the radius vector joining the nuclei A and B; $\mathcal{V}$ is the operator of the Coulomb interaction of the outer electron ($\mathbf{r}$) with all inner electrons ($\mathbf{r}_\kappa$, $\kappa = 1, 2, \ldots, N$) and nuclei (Z_A and Z_B) of the BA^++e system. Note that the Coulomb term (e^2/r) is to be included in (7.30) to compensate the corresponding $(-e^2/r)$ term in the Hamiltonian of the zeroth approximation

$$H_0 = -\hbar^2\Delta_{\mathbf{R}}/2\mu + H_{BA^+} + H_e.$$

Within the framework of the first-order of stationary perturbation theory and impact-parameter approach, the cross section $\sigma_{n'l',nl}^{f,i}$ of the inelastic $|i, nl\rangle \rightarrow |f, n'l'\rangle$ transition can be presented as

$$\sigma_{n'l',nl}^{f,i}(\mathcal{E}) = \frac{4\pi^3 g_f}{(2l+1)g_i q^2}\sum_{mm'}\int_0^\infty\left|\left\langle\chi_{\mathcal{E}'J}^{(f)}(R)\left|\mathcal{V}_{i,nlm}^{f,n'l'm'}(R)\right|\chi_{\mathcal{E}J}^{(i)}(R)\right\rangle\right|^2 2J dJ, \tag{7.31}$$

$$\mathcal{V}_{i,nlm}^{f,n'l'm'}(R) = \langle\psi_{n'l'm'}(\mathbf{r})|\langle\varphi_f(\mathbf{r}_\kappa, \mathbf{R})|\mathcal{V}|\varphi_i(\mathbf{r}_\kappa, \mathbf{R})\rangle|\psi_{nlm}(\mathbf{r})\rangle. \tag{7.32}$$

Here $\mathcal{V}_{i,nlm}^{f,n'l'm'}$ is the electronic matrix element of perturbation potential (7.30) for transitions between different terms $U_{i,nl}(R) = U_i(R) + E_{nl}$ and $U_{f,n'l'}(R) = U_f(R) + E_{n'l'}$ of quasi-molecule BA^++e; g_i and g_f are the statistical weights of the BA^+ ion. Radial nuclear wave functions $\chi_{\mathcal{E}J}^{(i)}(R)$ and $\chi_{\mathcal{E}'J}^{(f)}(R)$ of continuous spectrum with definite magnitudes of the kinetic energy $(\mathcal{E} = \hbar^2 q^2/2\mu$ and $\mathcal{E}' = \hbar^2 q'^2/2\mu)$ and orbital angular momentum $J = q\rho$ for the perturber-core relative motion, correspond to the initial $|i\rangle$

and final $|f\rangle$ electronic terms of the quasi-molecular BA^+ ion, respectively; ρ is the impact-parameter of the colliding particles B and A^+.

For the cross section $\sigma_{nl}^{\mathrm{d.i.}}(\mathcal{E})$ of the direct ionization process $A(nl) + B \to A^+ + B + e$, caused by the perturber-core scattering mechanism, a similar result can be written as [7.15]

$$\sigma_{nl}^{\mathrm{d.i.}} = \frac{4\pi^3 g_f}{(2l+1)g_i q^2} \sum_{l'} \sum_{mm'} \int_0^{E_{\max}^{\mathrm{d.i.}}} dE \int_0^\infty 2J dJ \left| \left\langle \chi_{\mathcal{E}'J}^{(f)}(R) \left| \mathcal{V}_{i,nlm}^{f,El'm'}(R) \right| \chi_{\mathcal{E}J}^{(i)}(R) \right\rangle \right|^2,$$

(7.33)

where integration over dE is taken over all possible values of the ejected electron energy. Here the electronic matrix element $\mathcal{V}_{i,nlm}^{f,El'm'}(R) = \langle \psi_{El'm'} | \langle \varphi_f | \mathcal{V} | \varphi_i \rangle | \psi_{nlm} \rangle$ corresponds to the bound-free transition of the outer electron, so that its final wave function in (7.33) should be replaced by the Coulomb wave function $\psi_{El'm'}(\mathbf{r}) = \mathcal{R}_{El'}(r) Y_{l'm'}(\theta, \varphi)$ of the continuous spectrum normalized to the δ-function of energy.

The semiclassical expression for the cross section $\sigma_{nl}^{\mathrm{a.i.}}(\mathcal{E}) = \sum_{vJ} \sigma_{nl}^{fi}(\mathcal{E} \to vJ)$ of the associative ionization $A(nl) + B \to BA^+(vJ) + e$, summed over all possible values of vibrational v and rotational J quantum numbers of the molecular $BA^+(vJ)$ ion in the final electronic term $U_f(R)$, is given by [7.15]

$$\sigma_{nl}^{\mathrm{a.i.}} = \frac{4\pi^3 g_f}{(2l+1) g_i q^2} \sum_{l'} \sum_{mm'} \int_0^{v_{\max}} dv \int_0^\infty 2J dJ \left| \left\langle \chi_{vJ}^{(f)}(R) \left| \mathcal{V}_{i,nlm}^{f,El'm'}(R) \right| \chi_{\mathcal{E}J}^{(i)}(R) \right\rangle \right|^2.$$

(7.34)

Here summation over all possible values of v and J is replaced by integration over dv and dJ. This procedure corresponds to approximation of the quasi-discrete spectrum for energy levels $\mathcal{E}_{vJ}$ of the molecular BA^+ ion.

In the JWKB-approximation, the radial nuclear wave functions $\chi_{\mathcal{E}J}^{(i)}(R)$ and $\chi_{\mathcal{E}'J}^{(f)}(R)$ in the continuous or discrete $\chi_{vJ}^{(f)}(R)$ spectra have a form

$$\chi(R) = \frac{C}{R\sqrt{V(R)}} \cos \left[\int_b^R q(R) dR - \pi/4 \right],$$

(7.35)

$$q(R) = \frac{\mu V(R)}{\hbar} = \frac{1}{\hbar} \left\{ 2\mu \left[\mathcal{E} - U(R) - \frac{\hbar^2 (J + 1/2)^2}{2\mu R^2} \right] \right\}^{1/2}.$$

(7.36)

Here $q_i(R)$ and $q_f(R)$ are the wave numbers, $V_i(R)$ and $V_f(R)$ are relative velocities of the nuclei, b_i and b_f are the left turning points for the motion of A^+ and B particles in the initial and final terms of the quasi-molecular (or molecular) BA^+ ion, respectively; C is the normalizing constant equal to $C_{\mathcal{E}} = (2/\pi\hbar)^{1/2}$ in the continuous $(\mathcal{E} > 0)$ and $C_{vJ} = 2/\left(T_{vJ}^{(f)}\right)^{1/2}$ in the discrete $(\mathcal{E}_{vJ} < 0)$ spectra, and $T_{vJ}^{(f)}$ is the classical period of vibrational-rotational motion of the nuclei in the final $U_f(R)$ term with energy $\mathcal{E}_{vJ}$. Wave functions $\chi_{\mathcal{E}J}^{(i)}(R)$ and $\chi_{\mathcal{E}'J}^{(f)}(R)$ are normalized to the delta function of energy $\delta(\mathcal{E} - \mathcal{E}')$, while $\chi_{vJ}^{(f)}(R)$ is normalized to unity.

7.4 Exchange of Rydberg Electron Energy with Translational Motion of Atoms

One of the possible mechanism of dipole-induced transitions associated with the scattering of perturbing atom B on the ion core A^+ corresponds to excitation and ionization processes of Rydberg atom A^*, which do not affect the variation of electronic states of quasi-molecular ion BA^+ (i.e., when $U_i(R) = U_f(R)$ and $\varphi_i = \varphi_f$ in the basic equations of Sect. 7.3). In contrast to the case of small transition frequencies (see Sect. 7.2), here our attention will be focused on the inelastic n-changing transitions [7.4] and ionization [7.15] of a highly excited atom with large enough values of ω ($\omega \underset{\sim}{>} V/\rho_{\mathrm{cap}}$ and $R_\omega < \rho_{\mathrm{cap}} \ll n^2 a_0$). A simple description of such processes may be given on the basis of the general semiclassical formulas of stationary perturbation theory presented above.

7.4.1 Matrix Elements of Transitions Within One Electronic Term of Quasi-molecular Ion

Due to a large enough electron orbital radius $r_n \sim n^2 a_0$ compared with the internuclear distances R_ω of the quasi-molecular BA^+ ion, the main contribution to the excitation and ionization cross section is determined by the interaction

$$\langle \varphi_i(\mathbf{r}_\kappa, \mathbf{R})| V |\varphi_i(\mathbf{r}_\kappa, \mathbf{R})\rangle = -\frac{e\mathbf{D}(R) \cdot \mathbf{r}}{r^3} + O(r^{-3})$$

$$= -\frac{e\mathbf{D}_1(R) \cdot \mathbf{r}}{r^3} - \frac{e\mathbf{D}_2(R) \cdot \mathbf{r}}{r^3} + O(r^{-3})$$

(7.37)

of a Rydberg electron with the total dipole moment $\mathbf{D}(R) = \langle \varphi_i(\mathbf{r}_\kappa, \mathbf{R})| \mathbf{D}(\mathbf{r}_\kappa, \mathbf{R}) |\varphi_i(\mathbf{r}_\kappa, \mathbf{R})\rangle$ of the BA^+ ion (relative to its center of mass, $\mu = M_B M_A / (M_B + M_A)$ is the reduced mass) in the electronic state $|\varphi_i\rangle$ under consideration. The dipole moment $\mathbf{D}(R) = \mathbf{D}_1(R) + \mathbf{D}_2(R)$ involves two terms. The first one $\mathbf{D}_1 = e\mathbf{R}_A = (\mu/M_A)e\mathbf{R}$ determines the contribution of the positive Coulomb A^+ center. As has been shown in [7.4], its contribution to the transition cross section corresponds to the noninertial effect. The second term $\mathbf{D}_2(R)$ (which tends to zero as $R \to \infty$) is due to the displacement (polarization) of the inner electrons of the BA^+ ion with respect to its nuclei induced by the scattering of perturbing atom B on the ionic core A^+.

The transition matrix elements of the dipole interaction (7.37) can be written as (see review [1.3])

$$\frac{1}{2l+1} \sum_{l'} \sum_{mm'} \left| \left\langle \chi_{\mathcal{E}'J}^{(i)} \left| V_{i,nlm}^{f,n'l'm'} \right| \chi_{\mathcal{E}J}^{(i)} \right\rangle \right|^2 = \frac{m^2 \omega^4}{3e^2} \left| D_{\mathcal{E}'\mathcal{E}} \right|^2$$

$$\times \left(\frac{l}{2l+1} \left| R_{nl}^{n',l-1} \right|^2 + \frac{l+1}{2l+1} \left| R_{nl}^{n',l+1} \right|^2 \right). \tag{7.38}$$

Here $D_{\mathcal{E}'\mathcal{E}}(J) \equiv \left\langle \chi_{\mathcal{E}'J}^{(i)} \left| D(R) \right| \chi_{\mathcal{E}J}^{(i)} \right\rangle$ is the radial matrix element of the dipole moment of the quasi-molecular ion for the free-free $(\mathcal{E} \to \mathcal{E}')$ transition of nuclei within one electronic term $U(R)$; and $R_{nl}^{n',l\pm1} = \langle n', l \pm 1 | r | nl \rangle$ is the radial matrix element of the Rydberg electron coordinate. Expressions for the free-bound $(\mathcal{E} \to v)$ transition of nuclei B and A$^+$ and bound-free $(n \to E)$ electron transitions have the same form.

In thermal collisions of Rydberg atom A* with ground-state B atoms of the buffer gas, the processes considered here prove to be most efficient, provided the electronic term $U(R)$ has a deep well $\mathcal{E}_0 \equiv |U(R_e)|$ and a large value ω_e of lower vibrational quantum. For this case the dipole matrix elements $D_{\mathcal{E}'\mathcal{E}}(J)$ and $D_{v\mathcal{E}}(J)$ for translational transitions of nuclei within one electronic term of the heteronuclear BA$^+$ ion near its dissociation limit $\mathcal{E}, \mathcal{E}' \ll \mathcal{E}_0$ (or $|\mathcal{E}_{vJ}|, |\mathcal{E}_{v'J}| \ll \mathcal{E}_0$) have been calculated in [7.4]. These semiclassical calculations are based on the JWKB-approximation for the wave functions (7.35) of the perturber-core relative motion and the method of the Fourier components (the correspondence principle) for the transition matrix elements (at $\Delta E = \hbar\omega \ll \mathcal{E}_0$). The final result for the free-free $\mathcal{E} \to \mathcal{E}'$ transitions can be presented as

$$D_{\mathcal{E}',\mathcal{E}}(J) = \begin{cases} D_{\mathcal{E}',\mathcal{E}}(0) = \dfrac{\pi}{2} C_{\mathcal{E}'} C_{\mathcal{E}} \dfrac{d(\omega)}{\omega} \exp(-\omega\tau), & J \leq J_{\text{cap}}, \\ \propto D_{\mathcal{E}',\mathcal{E}}(0) \exp(-\omega\rho/V), & J \geq J_{\text{cap}}. \end{cases} \tag{7.39}$$

The matrix elements $D_{v,\mathcal{E}}(J)$ of the free-bound $\mathcal{E} \to v$ transitions are determined by the same expression, but with the other value of the normalizing constant C_v of the semiclassical wave function in the final channel (7.35). Here $J_{\text{cap}} = q\rho_{\text{cap}}$ is the angular momentum corresponding to the capture of the perturbing particle B by ionic core A$^+$; and τ is the collision time of B and A$^+$ particles in the left turning point b_i on the repulsive branch of the term $U(R)$. The value $d(\omega) \sim (dD/dR)|_{R_\omega} \Delta R_\omega$ is determined by the increment of the dipole moment of the BA$^+$ ion over the characteristic length $\Delta R_\omega = |U'(R_\omega)/U''(R_\omega)|$ of the $U(R)$ term in the vicinity of the distance R_ω, which makes a major contribution to the transition with frequency ω. The particular form of the $d(\omega)$ function for various approximations of the $U(R)$ term is presented in [7.4]. For instance, for the Morse potential

$$U(R) = \mathcal{E}_0 \left[\exp\left(-2\alpha \frac{R-R_e}{R_e}\right) - 2\exp\left(-\alpha \frac{R-R_e}{R_e}\right) \right] \tag{7.40}$$

with linear approximation of the dipole moment in the vicinity of the well bottom $R = R_e$

$$D(R) = D_e + \left(\frac{dD}{dR}\bigg|_{R_e}\right)(R - R_e) \ . \tag{7.41}$$

calculation of the $d(\omega)$ function and the collision time τ of B and A^+ particles in a zero approximation with respect to the parameter $|\mathcal{E}|/\mathcal{E}_0 \ll 1$ yields

$$d(\omega) = \left(\frac{dD}{dR}\bigg|_{R_e}\right)\frac{R_e}{\alpha} \ , \qquad \tau = \omega_e^{-1} = \frac{R_e}{\alpha(2\mathcal{E}_0/\mu)^{1/2}} \ . \tag{7.42}$$

Thus, in this special case the $d(\omega)$ quantity is independent of the transition frequency, i.e., $d(\omega) = const$. Moreover, the quantity reciprocal to τ coincides with the value of the lowest vibrational quantum ω_e of the molecular BA^+ ion, which separates the regions of adiabatic $\omega \gg \tau^{-1}$ and nonadiabatic $\omega \ll \tau^{-1}$ transition frequencies. Hence the use of approximations (7.40) and (7.41) allows us to express the d and τ quantities in terms of several parameters of the potential $U(R)$ and the dipole moment $D(R)$ alone.

7.4.2 Cross Sections of de-excitation and Ionization

As is evident from expression (7.39), in the most interesting case of small kinetic energy $\mathcal{E} = \mu V^2/2$ of colliding atoms as compared to the depth of the potential well $\mathcal{E}_0$, the transition matrix elements $D_{\mathcal{E}'\mathcal{E}}(J)$ [or $D_{v\mathcal{E}}(J)$] with impact parameters $\rho \leq \rho_{cap}$ are practically independent of ρ and are determined only by transition frequency ω. This range of ρ corresponds to the capture of perturbing atom B by the ionic core A^+ in the region of small internuclear distances $R_\omega < \rho_{cap}$. In the opposite case $\rho \gg \rho_{cap}$, they reveal sharp exponential reduction with ρ. Hence the contribution of large values of ρ to the excitation and ionization cross sections can be neglected. Substituting (7.38, 39) into the general formula (7.31) and performing integration over the $2\pi\rho d\rho$, we obtain the final analytic expressions for the cross section $\sigma_{n'n}$ and rate constant $K_{n'n} = \langle V\sigma_{n'n}\rangle_T$ of de-excitation $n \to n'$ $(n > n')$ of the Rydberg atom [7.4]

$$\sigma_{n'n}(\mathcal{E}) = \frac{8\pi\mathcal{G}(\Delta n)}{3^{3/2}n^5(n')^3}\left(\frac{2Ry}{\hbar\omega_{n'n}}\right)^2\left(\frac{d(\omega_{n'n})}{ea_0}\right)^2\exp(-2\omega_{n'n}\tau)\,\sigma_{A^+B}^{cap}(\mathcal{E}) \ , \tag{7.43}$$

$$K_{n'n}(T) = \Gamma\left(2 - \frac{2}{\nu}\right)\langle V\rangle_T\,\sigma_{n'n}(V_T) \ . \tag{7.44}$$

Here $\langle V\rangle_T = (8kT/\pi\mu)^{1/2}$ and $\sigma_{nn'}(V_T)$ is the cross section at $\mathcal{E} = kT$ (and $\mathcal{G} = 1$ in the Kramers approximation). These expressions are valid provided the following restrictions on the transition frequency and principal quantum number

$$V/\rho_{\text{cap}} \ll \omega \ll \mathcal{E}_0/\hbar, \qquad n^2 a_0 \gg \rho_{\text{cap}} \tag{7.45}$$

are satisfied. The excitation cross section and rate constant can be determined from the detailed balance relations.

To explain the physical meaning of the results presented above we take into account only the major noninertial effect (corresponding to the contribution $\mathbf{D}_1 = (\mu/M_A)\, e\mathbf{R}$ of the Coulomb center A^+ to the dipole moment) and neglect completely the induced dipole interaction. This allows rewriting the result (7.43) in the simple form

$$\sigma_{n'n} \propto \left(\frac{n}{\Delta n}\right)^4 \left(\frac{\mu V (R_\omega)}{v_0 M_A}\right)^2 \exp\left(-2\omega_{n'n}\tau\right) \sigma_{\text{A}^+\text{B}}^{\text{cap}} (\mathcal{E}) , \tag{7.46}$$

which clarifies the dependence of the de-excitation cross section $\sigma_{n'n}$ on the principal quantum number n and transition frequency $\omega_{n'n}$, on the mass M_A, M_B and relative velocity $V (R_\omega)$ of the colliding atoms. It is interesting to compare this expression with the result (7.20) of the shake-up model. One can see that the magnitudes of the transition cross section $\sigma_{n'n}$ turns out to be proportional to a factor $n^4 V^2 (R_\omega)$ for the entire considered region of $n \lesssim (v_0 M_A/\mu V)^{1/2}$. However, in the shake-up region of small frequencies $(\omega_{n'n} \lesssim V/\rho_{\text{cap}})$ they fall with decreasing n like n^4, since major contribution to the inelastic or quasi-elastic transition is made by large distances $R_\omega \gtrsim \rho_{\text{cap}}$ [where relative velocity of the colliding A^+ and B particles is determined by $V (R_\omega) \sim V = (2\mathcal{E}/\mu)^{1/2}$]. In the opposite limiting case $\omega_{n'n} \gg V/\rho_{\text{cap}}$, the behavior of inelastic transitions is changed radically. Such transitions take place in the region of small internuclear distances $R_\omega \ll \rho_{\text{cap}}$, in which the relative velocity $V (R_\omega) \approx (2\left|U (R_\omega)\right|/\mu)^{1/2}$ of perturbing atom B and ionic core A^+ increases significantly with increasing $\omega_{n'n}$, owing to their acceleration in the potential well. In particular, for transition frequencies $\omega_{n'n} \sim \tau^{-1} \approx \omega_e$ (when R_ω corresponds to the region in the vicinity of the potential well R_e) the relative velocity reaches the value of $V_0 = (2\mathcal{E}_0/\mu)^{1/2}$ determined by the dissociation energy $\mathcal{E}_0 = |U (R_e)|$ of the BA^+ ion, so that $V_0 \gg V$ at thermal energies. This increase in the velocity $V (R_\omega)$ causes a slight enough change of the cross section $\sigma_{n'n}$ with decreasing n in the frequency region $V/\rho_{\text{cap}} \lesssim \omega_{n'n} \lesssim \tau^{-1}$. Strong reduction of the inelastic cross sections $\sigma_{n'n} \propto \exp\left(-2\omega_{n'n}\tau\right)$ appears only in the adiabatic region of frequencies $\omega_{n'n} \gg \tau^{-1} \approx \omega_e$, i.e., for low enough principal quantum numbers $n \ll (2Ry\tau/\hbar)^{1/3}$ at $\Delta n \sim 1$.

All these effects were also taken into account in calculations [7.15] of the ionization processes of Rydberg atom A^* induced by the scattering of atom B on the ionic core A^+. A particularly simple formula has been obtained for the total cross section $\sigma_{nl}^{\text{ion}} = \sigma_{nl}^{\text{d.i.}} + \sigma_{nl}^{\text{a.i.}}$ of direct and associative ionization

$$\sigma_{nl}^{\text{ion}} = \frac{1}{\alpha} \left(\frac{m}{e\hbar}\right)^2 \sigma_{\text{A}^+\text{B}}^{\text{cap}} (\mathcal{E}) \int\limits_{\omega_{nl}}^{\omega_{\text{max}}} d^2 (\omega)\, \sigma_{nl}^{\text{ph}} (\omega) \exp\left(-2\omega\tau\right) \omega d\omega . \tag{7.47}$$

Here $\alpha = e^2/\hbar c = 1/137$, and the lower and upper limits of integration are given by $\omega_{nl} = |E_{nl}|/\hbar = Ry/\hbar n_*^2$ and $\omega_{\max} = (\mathcal{E}_0 + \mathcal{E})/\hbar$, respectively. The use of simple analytic expressions for the photoionization cross section

$$\sigma_n^{\rm ph}(\omega) = \sum_l \left[(2l+1)/n^2\right] \sigma_{nl}^{\rm ph}(\omega)$$

averaged over all degenerate lm-sublevels (see Sect. 3.2) leads to the following final result for ionization rate constant $K_n^{\rm ion} = \langle V\sigma_n^{\rm ion}\rangle_T$ [7.15]:

$$K_n^{\rm ion}(T) = \frac{32\pi^{1/2}\Gamma(2-2/\nu)}{3^{3/2}n^3}\left(\frac{d(\omega_n)}{ea_0}\right)^2 V_T\sigma_{\rm A+B}^{\rm cap}(T) E_2(2\omega_n\tau), \qquad (7.48)$$

where $V_T = (2kT/\mu)^{1/2}$, and $E_2(z) = \int\limits_1^\infty t^{-2}e^{-zt}dt$ is the exponential integral of the second order. This formula is valid for the Rydberg levels $n \gg (Ry/\mathcal{E}_0)^{1/2}$ with small binding energy $|E_n|$ compared with the depth $\mathcal{E}_0 = |U(R_e)|$ of the potential well. Another restriction for the permissible values of the principal quantum number in (7.48) follows from the condition (7.45), which yields $n \ll (\rho_{\rm cap}v_0/a_0V)^{1/2}$.

7.4.3 Collisions of Highly Excited Hydrogen with Helium

Efficiency of the considered mechanism of excitation and ionization of Rydberg states by neutral particles can be illustrated for thermal collisions between highly excited hydrogen $H(nl)$ and the ground-state helium $He(1s^2)$ atoms. The results of calculations [7.4] of the cross sections for inelastic n-changing processes induced by the scattering of He atom on the ionic core H^+ of the highly excited hydrogen atom are presented in Fig. 7.1 by the full curve for the kinetic energy $\mathcal{E} = 0.026$ eV. The dashed curve in Fig. 7.1 indicates the contribution of the traditional perturber-quasi-free electron scattering mechanism.

One can see, for the inelastic de-excitation of $H(n)$ atoms with change in the principal quantum number $n \to n-1$ the shake-up region with $\omega_{n-1,n} < V/\rho_{\rm cap}$ (where $\sigma_{n-1,n} \propto n^4$) is realized only for $n > 15$. In the opposite case, with $n < 15$ the behavior of the inelastic cross section $\sigma_{n-1,n}(He - H^+)$ with large enough transition frequencies $\omega_{n-1,n} > V/\rho_{\rm cap}$ is described within the framework of the electron-dipole mechanism. Formula (7.43) leads to practically n independent magnitudes of the de-excitation cross section $\sigma_{n-1,n}$ for $6 < n < 15$ and to its sharp exponential decrease $\sigma_{n-1,n} \propto \exp(-2\omega\tau)$ only for low principal quantum numbers $n < 6$. As a result, for the inelastic $n \to n-1$ transitions with change in the principal quantum number the perturber-core scattering mechanism plays the main role at intermediate and sufficiently low $n < 10$. As follows from calculations by formula (7.20), the scattering of the perturbing atom He on the ionic core H^+ becomes also predominant in the range of high enough principal quantum numbers $n > 35$–40.

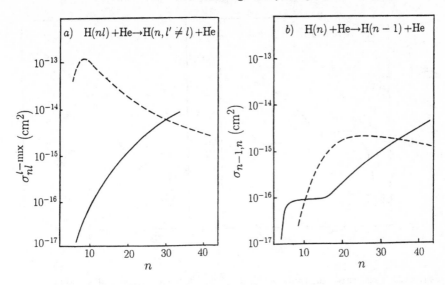

Fig. 7.1. Cross sections for quasi-elastic l-mixing collisions ($\sigma_{nl}^{l-\text{mix}}$ -panel a) and inelastic n-changing de-excitation ($\sigma_{n-1,n}$ -panel b) of highly excited hydrogen atoms by He ($T = 300$ K) calculated in [7.4]. Full curves represent the contributions of the core-perturber scattering mechanism. Dashed curves indicate the contributions of the quasi-free electron-perturber scattering mechanism

The corresponding magnitude of the quenching cross section turns out to be of the order of $\sigma_{\text{tr}}^{\text{H}^+,\text{He}}$ at $n \sim (v_0/V)^{1/2} \sim 40$ ($\sigma_{\text{tr}}^{\text{H}^+,\text{He}} = 6 \cdot 10^{-15}$ cm^2 is the momentum transfer cross section for $\mathcal{E} = 0.026$ eV).

The results of the calculations [7.15] of the total ionization rate constant $K_n(T)$ for thermal collisions of H(n) + He are presented in Fig. 7.2. One can see that the values of $K_n(T)$ in the perturber-core scattering mechanism (full curve) are practically independent of the gas temperature T and have the maximum $\sim 10^{-12}$ cm$^3 \cdot$s^{-1} at $n \sim 10$. Thus, for the ionization processes of the Rydberg atoms H(n)+He, the perturber-core scattering mechanism is more effective than the traditional Fermi mechanism (dashed curves) in the range of $n < 20$–30 at $T = 300$ K.

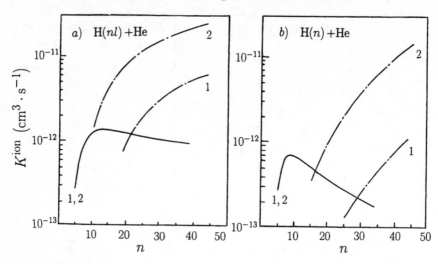

Fig. 7.2. Total rate constants $K^{\text{ion}} = K^{\text{d.i.}} + K^{\text{a.i.}}$ of direct and associative ionization of the Rydberg hydrogen atom in thermal collisions $T = 300$ K (curves 1) and $T = 1000$ K (curves 2) with He atoms calculated in [7.4]. Full curves 1 and 2 (coinciding for $T = 300$ K and $T = 1000$ K) are the contributions of the core-perturber scattering. Dashed curves 1 and 2 are the contributions of the quasi-free electron-perturber scattering. Panels a and b represent calculations for selectively excited nl levels (with $l \ll n$) and for n-levels (with equally populated sublevels), respectively

7.5 Resonant Excitation and Ionization

7.5.1 Exchange of Electron Energy in Quasi-molecule

The processes of resonant excitation and ionization of the Rydberg atom induced by scattering of the perturbing atom by the ionic core take place in the vicinity of the crossing point R_ω of two different electronic terms of the quasi-molecular $BA^+ + e$ system. Hence, the nuclear matrix elements for such transitions in the general formulas of Sect. 7.3 may be calculated within the framework of the semiclassical theory of Landau-Zener [2.2]

$$\left| \left\langle \chi^{(f)}_{\mathcal{E}'J}(R) \left| V^{f,n'l'm'}_{i,nlm}(R) \right| \chi^{(i)}_{\mathcal{E}J}(R) \right\rangle \right|^2$$

$$= \frac{\pi C^2_{\mathcal{E}} C^2_{\mathcal{E}'} \left| V^{f,n'l'm'}_{i,nlm}(R_\omega) \right|^2 \cos^2 S_0(J)}{2V \Delta F_{fi}(R_\omega) \left[1 - U_i(R_\omega)/\mathcal{E} - (J + 1/2)^2/q^2 R_\omega^2 \right]^{1/2}} , \tag{7.49}$$

$$S_0(J) = \int_{b_i}^{R_\omega} q_i(R)\, dR - \int_{b_f}^{R_\omega} q_f(R)\, dR + \pi/4 . \tag{7.50}$$

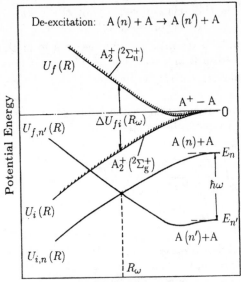

De-excitation: $A(n) + A \rightarrow A(n') + A$

Potential Energy

$U_f(R)$

$A_2^+ \left({}^2\Sigma_u^+ \right)$

$A^+ - A$

$U_{f,n'}(R)$ $\Delta U_{fi}(R_\omega)$ 0

$A(n) + A$ E_n

$A_2^+ \left({}^2\Sigma_g^+ \right)$ $\hbar\omega$

$U_i(R)$ $E_{n'}$

$A(n') + A$

$U_{i,n}(R)$ R_ω

Internuclear Separation

Fig. 7.3. Schematic diagram of the potential energy curves $U_i(R)$ and $U_f(R)$ for the lower ${}^2\Sigma_g^+$ and ${}^2\Sigma_u^+$ terms of the homonuclear A_2^+ ion in the asymptotic region of internuclear separation R. $U_{i,n}(R) = U_i(R) + E_n$ and $U_{f,n'}(R) = U_f(R) + E_{n'}$ are the potential energy curves of the compose $A_2^+ + e$ system correlating with the initial $A(n) + A$ and final $A(n') + A$ states of separated atoms

Here $\Delta F_{fi}(R_\omega) = |dU_f/dR - dU_i/dR|_{R=R_\omega}$ is the difference of the derivatives of the quasi-molecule $BA^+ + e$ at the crossing point R_ω defined by relation $U_{i,nl}(R_\omega) = U_{f,n'l'}(R_\omega)$ [or $U_{i,nl}(R_\omega) = U_{f,E}(R_\omega)$ for the ionization processes]. Thus, at $R = R_\omega$, the energy $\Delta E_{fi} = \hbar\omega$ transferred to the outer electron in the bound-bound ($\Delta E_{fi} = E_{n'l'} - E_{nl}$) or bound-free ($\Delta E_{fi} = E + |E_{nl}|$) transition becomes equal to the energy splitting of the electronic terms of the quasi-molecular BA^+ ion, i.e.,

$$\Delta E_{fi} \equiv \hbar\omega = U_i(R_\omega) - U_f(R_\omega) \equiv \Delta U_{if}(R_\omega) , \qquad (7.51)$$

where $\hbar\omega = \mathcal{E} - \mathcal{E}'$ for the excitation (de-excitation) and direct ionization, and $\hbar\omega = \mathcal{E} + |\mathcal{E}_{vJ}|$ for the associative ionization. Figure 7.3 illustrates the schematic diagram of the potential energy curves of homonuclear quasi-molecule $A_2^+ + e$ and quasi-molecular ion A_2^+ (temporally formed during the scattering of projectile atom A by the ionic core A^+) and clarifies the main idea of resonant de-excitation mechanism of Rydberg states $n \rightarrow n'$.

The fast oscillating $\cos^2 S_0(J)$ function in (7.49) can be replaced by its average value equal to $1/2$. Then, integration over dJ in the general semiclassical formula (7.31) of the first-order perturbation theory (within the limits of $0 \le J \le J_{\max}(R_\omega) = qR_\omega[1 - U_i(R_\omega)/\mathcal{E}]^{1/2}$) leads to the following expression for the cross section of the inelastic $nl \rightarrow n'l'$ transition:

$$\sigma_{n'l',nl}^{f,i}(\mathcal{E}) = \frac{8\pi^2 g_f R_\omega^2 [1 - U_i(R_\omega)/\mathcal{E}]^{1/2}}{(2l+1) g_i \hbar V \Delta F_{fi}(R_\omega)} \sum_{mm'} \left| \mathcal{V}_{i,nlm}^{f,n'l'm'}(R_\omega) \right|^2 , \qquad (7.52)$$

where $V = \hbar q/\mu = (2\mathcal{E}/\mu)^{1/2}$ is the relative velocity of the colliding $A(nl)$ and B atoms as $R \to \infty$. Due to the presence of the $V(R_\omega)/V = [1 - U_i(R_\omega)/\mathcal{E}]^{1/2}$ term this formula takes into account the change in the relative velocity of the perturbing atom B and ionic core A^+ in the vicinity of the transition point R_ω in comparison with its value $V \equiv V_\infty = (2\mathcal{E}/\mu)^{1/2}$ as $R \to \infty$.

A similar semiclassical expression for the total cross section $\sigma_{nl}^{ion} = \sigma_{nl}^{d.i.} + \sigma_{nl}^{a.i.}$ of direct and associative ionization can be written as [7.15]

$$\sigma_{nl}^{ion}(\mathcal{E}) = \frac{8\pi^2 g_f}{(2l+1)g_i \hbar V} \int_{R_{min}}^{R_{max}} \sum_{l'} \sum_{mm'} \left| \mathcal{V}_{i,nlm}^{f,El'm'}(R_\omega) \right|^2$$

$$\left(1 - \frac{U_i(R_\omega)}{\mathcal{E}}\right)^{1/2} R_\omega^2 dR_\omega. \tag{7.53}$$

Thus, the final result for the total ionization cross section can be expressed as an integral over the transition frequencies $\hbar d\omega = d[\Delta U_{fi}(R_\omega)] = \Delta F_{fi}(R_\omega)dR_\omega$ or over the internuclear distances. The lower R_{min} and upper R_{max} limits of integration in (7.53) correspond to the maximum ω_{max} and minimum ω_{min} possible values of the energy splitting terms of the BA^+ ion. Averaging (7.53) over the Maxwellian distribution of relative velocities of colliding atoms, yields

$$K_{nl}^{ion}(T) = \frac{8\pi^2 g_f}{(2l+1)g_i \hbar} \int_0^{R_{nl}} \sum_{l'} \sum_{mm'} \left| \mathcal{V}_{i,nlm}^{f,El'm'}(R_\omega) \right|^2$$

$$\exp\left(-\frac{U_i(R)}{kT}\right) \mathcal{H}(R) R^2 dR. \tag{7.54}$$

Here R_{nl} is defined by the condition $\Delta U_{fi}(R_{nl}) = |E_{nl}| = Ry/n_*^2$, and function $\mathcal{H}(R)$ assumes a different form in the repulsive ($R < R_0^{(i)}$ and $U_i(R) > 0$) and attractive ($R \geq R_0^{(i)}$ and $U_i(R) < 0$) regions of the initial electronic term of the quasi-molecular BA^+ ion, i.e.,

$$\mathcal{H}(R) = \left\{ \begin{array}{ll} 1, & R < R_0^{(i)}, \\ \dfrac{\Gamma(3/2, |U_i(R)|/kT)}{\Gamma(3/2)}, & R \geq R_0^{(i)}, \end{array} \right\} \tag{7.55}$$

where $\Gamma(3/2, z)$ is the incomplete gamma function, $\Gamma(3/2) = \pi^{1/2}/2$, and $U_i\left(R_0^{(i)}\right) = 0$.

Semiclassical formulas (7.52–54) of first-order perturbation theory are valid to an arbitrary form and symmetry of the initial U_i and final U_f terms of heteronuclear BA^+ or homonuclear A_2^+ quasi-molecular ion converging to one dissociation limit [i.e., $U_i(\infty) = U_f(\infty)$]. However, in some cases the perturbation theory becomes inapplicable. Thus, it is necessary to have simple estimates for the maximum possible values of the excitation and ionization rate constants. They can be obtained by assuming that the probability of the bound-bound or bound-free transition is equal to $W_{n'l',nl} = 1$ (or $W_{nl}^{\text{ion}} = 1$) for all impact parameters $0 \leq \rho \leq \rho_{\max}(\mathcal{E})$. Hence, for the resonant mechanism associated with the crossing of various electronic terms, the maximum value of the cross section can be written as

$$\sigma_{\max} = \pi \rho_{\max}^2(\mathcal{E}), \qquad \rho_{\max}(\mathcal{E}) = R_\omega \left[1 - U_i(R_\omega)/\mathcal{E}\right]^{1/2}, \qquad (7.56)$$

where $\Delta U_{if}(R_\omega) = |\Delta E_{n'l',nl}| = 2Ry \left|n'_* - n_*\right|/n_*^3$ and $\Delta U_{if}(R_{nl}) = |E_{nl}| = Ry/n_*^2$ (i.e. $R_\omega = R_{nl}$) for the bound-bound and bound-free transitions of the Rydberg electron, respectively. A simple estimate for the maximum value of the ionization rate constant $K_{\max}^{\text{ion}}(T) = \langle V \sigma_{\max} \rangle$ [7.15]

$$K_{\max}^{\text{ion}}(T) = \pi R_{nl}^2 \langle V \rangle_T P_{nl}(T), \qquad (7.57)$$

$$P_{nl}(T) = \begin{cases} \exp\left[U_i(R_{nl})/kT\right], & R_{nl} < R_0^{(i)}, \\ 1 + |U_i(R_{nl})|/kT, & R_{nl} \geq R_0^{(i)} \end{cases}$$

proceeds directly from (7.56) after integrating over the Maxwellian distribution of velocities $V = (2\mathcal{E}/\mu)^{1/2}$ [over the range of $\mathcal{E}_{\min} \leq \mathcal{E} < \infty$, where $\mathcal{E}_{\min} = \max\{0, U_i(R_{nl})\}$].

7.5.2 Dipole Transitions Between Symmetrical and Antisymmetrical Terms

The formulas presented above can be applied to the analysis of the excitation and ionization processes in thermal collisions of the Rydberg atoms $A^*(nl)$ with the ground-state parent atoms A. In this case they are due to the dipole transition between the symmetrical $U_g(R)$ and antisymmetrical $U_u(R)$ terms of the quasi-molecular A_2^+ ion (see Fig. 7.1). Due to the large value of the exchange splitting $\Delta(R) \equiv |\Delta U_{gu}(R)|$ of these terms the inelastic transitions $n \to n'$ and ionization $n \to E$ of the Rydberg electron occur in the asymptotic range of internuclear distances $R_\omega \gg a_0$, in which the $U_g(R)$ and $U_u(R)$ magnitudes are given by simple relations [5.74]

$$U_g(R) = -\frac{\Delta(R)}{2} - \frac{\alpha_A e^2}{2R^4}, \qquad U_u(R) = \frac{\Delta(R)}{2} - \frac{\alpha_A e^2}{2R^4}, \qquad (7.58)$$

where α_A is the polarizability; $\Delta(R) = CR^\nu \exp(-\gamma R)$, and C, ν, γ are the constant coefficients determined by parameters of the ground-state atom A.

The electronic matrix element for the dipole part $V = e\mathbf{r} \cdot \mathbf{D}(\mathbf{r}_\kappa)/r^3$ of the Coulomb interaction (7.30) between the outer electron $(\mathbf{r})$ and the inner electrons $(\mathbf{r}_\kappa)$ of the $A_2^+ + e$ system may be written as

$$V_{i,nlm}^{f,n'l'm'}(R) = \left\langle \psi_{n'l'm'}(\mathbf{r}) \left| \left\langle \varphi_{A_2^+}^{(f)}(\mathbf{r}_\kappa, \mathbf{R}) \left| \frac{e\mathbf{r} \cdot \mathbf{D}(\mathbf{r}_\kappa)}{r^3} \right| \varphi_{A_2^+}^{(i)}(\mathbf{r}_\kappa, \mathbf{R}) \right\rangle \right| \psi_{nlm}(\mathbf{r}) \right\rangle$$

$$= -a_0^{-3} \left(\frac{\hbar \omega_{fi}}{2Ry}\right)^2 \langle n'l'm'| e\mathbf{r} \cdot \mathbf{D}_{fi}(R) |nlm\rangle. \qquad (7.59)$$

Here $\mathbf{D}_{fi}(R) \equiv \left\langle \varphi_{A_2^+}^{(f)}(\mathbf{r}_\kappa, \mathbf{R}) \middle| \mathbf{D} \middle| \varphi_{A_2^+}^{(i)}(\mathbf{r}_\kappa, \mathbf{R}) \right\rangle$ is the matrix element of the dipole moment $\mathbf{D} = -e \sum_\kappa \mathbf{r}_\kappa$ of the inner electrons over the initial and final electronic wave functions of the A_2^+ ion (the radius vectors of $\mathbf{r}$ and $\mathbf{r}_\kappa$ are measured from its center of mass). Using (7.52,59) the cross section $\sigma_{n'n}$ of the inelastic n-changing process may be described by the simple expression

$$\sigma_{n'n}(V) = \frac{8\pi g_f \mathcal{G}(\Delta n) R_\omega^2}{3^{3/2} g_i n^5 (n')^3 (V/v_0)}$$

$$\left(1 - \frac{U_i(R_\omega)}{\mathcal{E}}\right)^{1/2} \left(\frac{D_{fi}(R_\omega)}{ea_0}\right)^2 \left(\frac{2Ry}{a_0 \Delta F_{fi}(R_\omega)}\right) . \tag{7.60}$$

For the inelastic transitions between the neighboring Rydberg levels n and $n \pm \Delta n$ (with $\Delta n \ll n$) the values of the dipole matrix elements are usually large compared with the transition energy $|\Delta E_{n'n}| = 2Ry|\Delta n|/n^3$. Therefore, the perturbation theory becomes, as a rule, inapplicable for the description of the resonant excitation or de-excitation of the Rydberg atom at $\Delta n \ll n$. Hence, the simple expression (7.60) may be used only for a qualitative estimate of the cross section at high enough n.

For the total rate constant $K_{nl}^{ion}(T)$ of direct and associative ionization of Rydberg atom $A^*(nl)$ by the ground-state parent A atom, the use of first-order perturbation theory and dipole approximation for the electronic matrix element $V_{i,nlm}^{f,El'm'}(R_\omega)$ of the bound-free transition yields

$$K_{nl}^{ion}(T) = \frac{2g_f}{g_i \alpha v_0^2} \int_0^{R_{nl}} \left(\frac{D_{fi}(R_\omega)}{ea_0}\right)^2 \omega^3 \sigma_{nl}^{ph}(\omega)$$

$$\exp\left(-\frac{U_i(R)}{kT}\right) \mathcal{H}(R_\omega) R_\omega^2 dR_\omega , \tag{7.61}$$

where $\omega \equiv |\Delta U_{fi}(R_\omega)|$, and the R_{nl} value is determined by the relation $|\Delta U_{fi}(R_{nl})| = |E_{nl}| = Ry/n_*^2$ and $\alpha = e^2/\hbar c \approx 1/137$. It should be emphasized that the dipole matrix element square of the bound-free electron transition in this formula is expressed in terms of the photoionization cross section $\sigma_{nl}^{ph}(\omega)$. Formula (7.61) involves contributions of both the repulsive $[U_i(R) > 0$ at $R < R_0^{(i)}]$ and attractive $[U_i(R) \leq 0$ at $R \geq R_0^{(i)}]$ regions of the upper $U_i(R)$ term [see (7.55)]. For the case of pure repulsive upper term $U_i(R)$ of quasi-molecular A_2^+ ion (when $\mathcal{H}(R) = 1$ for all internuclear distances $0 < R < \infty$) (7.61) is equivalent to the expressions obtained by *Janev* and *Mihajlov* [7.11] and *Duman* and *Shmatov* [7.12]. It should be noted that for the matrix element of dipole moment one can use the following asymptotic approximation $D_{fi}(R_\omega) = -eR_\omega/2$, since the major contribution to the excitation (de-excitation) and ionization cross sections is determined, as a rule, by large enough internuclear distances $R_\omega \gg a_0$.

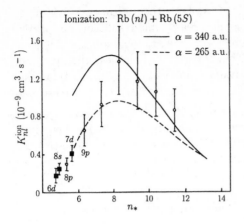

Fig. 7.4. Ionization rate constant $K_{nl}^{ion}(T)$ for the Rb(nl) + Rb($5S$) thermal collisions ($T = 520$ K) plotted against the effective principal quantum number $n_* = n - \delta_l$. Full and dashed curves are the results [7.11] calculated for two values ($\alpha = 340$ a.u., $\alpha = 265$ a.u.) of polarizability of the Rb(5S). Empty circles – (nP), and full squares – (nS and nD) are the experimental data from [7.22, 23], respectively

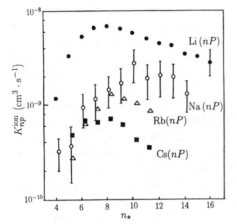

Fig. 7.5. Comparison of experimental data for the ionization rate constants $K_{np}^{ion}(T)$ of the Rydberg np-states of alkali-metal atoms by the ground-state parent atoms plotted against the effective principal quantum number $n_* = n - \delta_p$ ($\delta_p^{Li} = 0.053$, $\delta_p^{Na} = 0.85$, $\delta_p^{Rb} = 2.65$, $\delta_p^{Cs} = 3.57$). Filled circles: Li(np) + Li($2s$) ([7.24], $T = 1100$ K); open circles: Na(np)+Na($3s$) ([7.24], $T = 600$ K); open triangles: Rb(np) + Rb($5s$) ([7.22], $T = 520$ K); filled squares: Cs(np) + Cs($6s$) ([7.10], $T = 500$ K)

Formulas presented above are used in several papers for calculations of ionization [7.11, 12] and quenching [7.7] of the Rydberg states of alkali-metal atoms in thermal collisions with the ground-state parent atoms Li, Na, K, Rb, and Cs. As follows from the comparison with the available experimental data [see [7.18] and references therein] they are, on the whole, in reasonable agreement with the measured rate constants of ionization. For instance, Fig. 7.4 shows the experimental data [7.22, 23] and the results of calculations [7.11] for the total rate constant of direct and associative ionization of the Rydberg ns, np, and nd levels of rubidium perturbed by Rb($5s$) atoms of the own gas. It should be noted that at thermal energies major contribution to the total ionization rate constant $K_{nl}^{ion} = K_{nl}^{d.i.} + K_{nl}^{a.i.}$ is determined by the process A*(nl) +A→A$_2^+$+e of associative ionization, while the role of direct ionization A*(nl) +A→A$^+$+A+e becomes important at high energies of the colliding particles. Experimental dependencies [7.22–25] of the ionization rate constants on the kind of the alkali-metal atoms is shown in Fig. 7.5.

As follows from the detailed analysis of available theoretical and experimental data, the perturbation theory provides reasonable results in the range of high enough principal quantum numbers $n \gg n_{max}$ (where $n_{max} \sim 8-15$ at thermal energies $T \sim 300-1000$ K), but it does not hold near the maximum $n \sim n_{max}$ and at low principal quantum numbers $n \ll n_{max}$. In this range $n \underset{\sim}{<} n_{max}$ some more realistic magnitudes may be obtained by simple estimate (7.57). Note also that the values of the ionization rate constants, measured [7.26–28] for the case of nonsymmetrical collisions between the Rydberg alkali atoms A^* (nl) and the ground-state alkali B atoms of the buffer gas, turn out to be of the same order of magnitude as those for symmetrical A^* (nl) +A collisions.

7.5.3 Quadrupole Transitions in Nonsymmetrical Collisions of Rare Gas Atoms

Another example of resonant processes of inelastic $n \rightarrow n'$ transitions [7.19] and ionization [7.15] is thermal collisions of the Rydberg rare gas A^* (nl) atoms with the ground-state $B(^1S_0)$ atoms of the buffer rare gas. In this case, the excitation and ionization of the highly excited electron e induced by scattering of atom B on the atomic core A^+ $[n_0 p^5 \, (^2P_{3/2})]$ are due to the transition $A_1 \, |j = 3/2, \Omega = 3/2\rangle \rightarrow X \, |j = 3/2, \Omega = 1/2\rangle$ between the first excited $U_i(R)$ and lower $U_f(R)$ electronic terms of the heteronuclear quasi-molecular BA^+ ion (while the de-excitation process is being accompanied by the inverse transition). These terms, split by electrostatic interaction, correlate with the ground-states of the separated $B(^1S_0)$ atom and atomic core A^+ $[n_0 p^5 \, (^2P_{3/2})]$. They correspond to the different $\Omega_i = 3/2$ and $\Omega_f = 1/2$ projections of the total angular moment $\mathbf{j} = \mathbf{L} + \mathbf{S}$ $(j_i = j_f = 3/2)$ of the quasi-molecular BA^+ ion on the internuclear $\mathbf{R}$ axis. The second excited $A_2 \, |j' = 1/2, \Omega' = 1/2\rangle$ term does not take part in the excitation and ionization processes involved (for the case of heavy rare gas Rydberg Ar^*, Kr^*, and Xe^* atoms under consideration). This is due to the large value $\Delta_{j'j}$ of the spin-orbit splitting for the $^2P_{3/2}$ and $^2P_{1/2}$ states of the ground-electronic p^5-shell of the A^+ ion in comparison with thermal energy $kT \sim 0.03-0.1$ eV of colliding atoms and energy splitting $\Delta U_{if}(R)$ of the lower electronic terms. In the zero-approximation the final result of the calculation for the sum of matrix elements determining the values of the quenching and ionization rate constants can be written as

$$\sum_{mm'} \left| \mathcal{V}_{i,nlm}^{f,n'l'm'} \right|^2 = (2Ry)^2 \, \frac{2\gamma_{l'l}}{25 n_*^3 \, (n_*')^3} \, ,$$

$$\mathcal{V}_{i,nlm}^{f,El'm'} = \left(\frac{(n_*')^3}{2Ry} \right)^{1/2} \mathcal{V}_{i,nlm}^{f,n'l'm'}$$

(7.62)

Here the n-independent value $\gamma_{l'l}$ characterizes the coupling constant of different Rydberg states with different magnitudes l and l' of the orbital moment.

Its is determined by the short range $(r < r_\kappa)$ and the long-range $(r > r_\kappa)$ parts of the Coulomb interaction (7.30) between the Rydberg electron $(\mathbf{r})$ and the electrons $(\mathbf{r}_\kappa)$ of the ionic core A^+ from its ground $n_0 p^5$-shell (interaction with the electrons of the filled $^1S_0$ shell of the rare gas B atom can be disregarded). As has been shown in [7.19], the magnitude of $\gamma_{l'l}$ can be expressed in terms of the angular coefficient and the radial matrix elements of the quadrupole kind $r_<^2/r_>^3$ over the Rydberg electron wave function $\mathcal{R}_{nl}(r)$ and the Hartree-Fock wave functions $\mathcal{R}_{n_0 p}(r_\kappa)$ of the rare gas ion A^+ $(n_0 p^5)$ with the principal quantum number n_0 and orbital moment $L = 1$. Here $r_< = \min(r, r_\kappa)$ and $r_> = \max(r, r_\kappa)$, while $n_0 = 3, 4$ and 5 for Ar^+, Kr^+ and Xe^+, respectively. The dipole part of the interaction does not make any contribution to the transition matrix elements due to the selection rules $\Delta l = 0, \pm 2$, $\Delta m = -1$ for transitions with a change $\Delta\Omega = 1$ of the total angular moment projections on the internuclear axis.

With the use of the general expression of first order perturbation theory (7.52, 62), the cross section of the inelastic $nl \to n'l$ transition in the Rydberg rare gas atom $A[n_0 p^5 (^2P_{3/2}) nl]$ induced by the scattering of the rare gas atom $B(^1S_0)$ on its ionic core $A^+[n_0 p^5 (^2P_{3/2})]$ can be described by the simple analytic formula [7.19]

$$\sigma_{nl',nl} = \frac{8\pi^2 a_0^2 \gamma_{l'l} (A/2Ry)^{3/\nu} [1 - U_i(R_\omega)/\mathcal{E}]^{1/2}}{25(2l+1)\nu(V/v_0) n_*^{3(1-3/\nu)} |n'_* - n_*|^{1+3/\nu}}. \tag{7.63}$$

This formula pertains to the case of a power-law approximation of the energy splitting $|\Delta U_{fi}(R)| = A(a_0/R)^\nu$, when the crossing point R_ω of the Rydberg $U_{i,nl}(R_\omega) = U_i(R_\omega) + E_{nl}$ and $U_{f,n'l'}(R_\omega) = U_f(R_\omega) + E_{n'l'}$ terms of the quasi-molecule $BA^+ + e$ is given by the simple relation

$$R_\omega = a_0(A/\hbar\omega)^{1/\nu}, \qquad \hbar\omega = |\Delta E_{n'l',nl}| = |\Delta U_{fi}(R_\omega)|. \tag{7.64}$$

A detailed analysis shows, that the most effective transitions in this mechanism are $np \to n'p$ with $l' = l = 1$. The cross sections $\sigma_{n'l',nl}$ of the inelastic $nl \to n'l'$ transitions with $l' \neq 1$ and $l \neq 1$ decreases rapidly with increase of the orbital quantum number l and l'. The $ns \to n's$ transitions accompanied by the change of the angular momentum projection $\Delta\Omega_{3/2,1/2} = 1$ of the quasi-molecular BA^+ ion are forbidden in first order perturbation theory.

Thus, the total rate constant of collisional de-excitation

$$K_n(T) = \sum_{n'<n} \sum_{l'} [(2l+1)/n^2] \langle V\sigma_{nl',nl}\rangle$$

of a given n level of the Rydberg rare gas $A[n_0 p^5 (^2P_{3/2}) n]$ atom by the ground-state rare gas $B(^1S_0)$ atoms of the buffer gas may be approximately presented as [7.19]

$$K_n(T) \approx (v_0 a_0^2) \frac{8\pi^2 \gamma (A/2Ry)^{3/\nu}}{25\nu n^{5-9/\nu}} \zeta(1+3/\nu) A\left(\overline{U_i(R_\omega)/kT}\right). \tag{7.65}$$

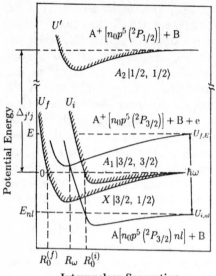

Fig. 7.6. Schematic diagram of the potential energy curves $U_f(R)$, $U_i(R)$ and $U'(R)$ for the lower electronic states $X\,|j_f = 3/2$, $\Omega_f = 1/2\rangle$, $A_1\,|j_i = 3/2$, $\Omega_i = 3/2\rangle$ and $A_2\,|j' = 1/2$, $\Omega' = 1/2\rangle$ of heteronuclear rare gas BA^+ ion. $U_{i,nl}(R) = U_i(R) + E_{nl}$ and $U_{f,E}(R) = U_f(R) + E$ are the potential energy curves of the compose $BA^+ + e$ system correlating with the initial $A\left[n_0p^5(^2P_{3/2})nl\right] + B$ and final $A^+\left[n_0p^5(^2P_{3/2})\right] + B + e$ states of separated particles

Here $\gamma = \sum_{ll'} \gamma_{l'l} \approx \gamma_{11}$, $\zeta(z) = \sum_{k=1}^{\infty} k^{-z}$ is the Riemann zeta function, $U_i(R)$ is the lower $X\,|j_i = 3/2$, $\Omega_i = 1/2\rangle$ term of the quasi-molecular BA^+ term, and the value of

$$\mathcal{A}\left(|U_i(R_\omega)|/kT\right) = \exp\left[-U_i(R_\omega)/kT\right]\mathcal{H}(R_\omega)$$

is expressed through the $\mathcal{H}(R_\omega)$ function [see (7.55)]. A simple estimate shows that in the case of $Xe\left[5p^5(^2P_{3/2})n\right] + He(^1S_0)$ collisions, total rate constant of inelastic de-excitation of the Rydberg n-level is weakly dependent on the gas temperature in the range of $T \sim 300\text{–}600$ K. Its value can be approximately estimated as $K_n \approx 3 \cdot 10^{-7}/n_*^{3.85}$ cm^{3}s^{-1} if the principal quantum number is not too small [when the perturbation theory does not hold and the limiting magnitude of K_n is determined by (7.56)].

As a result, the de-excitation of highly excited $Xe\left[5p^5(^2P_{3/2})n\right]$ atoms in inelastic n-changing collisions with He atoms via the resonant mechanism proposed in [7.19] is substantially more effective than that via the traditional Fermi mechanism (dashed-dotted curve) in the region of $n < 15\text{–}20$. Moreover, the probabilities $W_n^e = \langle v\sigma_n^e\rangle N_e$ for inelastic de-excitation of Rydberg levels of $Xe\left[5p^5(^2P_{3/2})n\right]$ atoms by electron impact also turn out to be lower

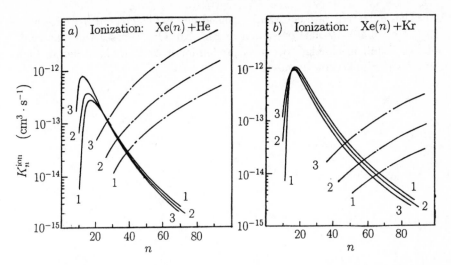

Fig. 7.7. Total rate constants $K_n^{ion} = K_{nl}^{d.i.} + K_{nl}^{a.i.}$ of direct and associative ioniza-tion of the Rydberg $Xe\left[5p^5(^2P_{3/2})n\right]$ atoms in thermal collisions with ground-state He (panel a) and Kr (panel b) atoms calculated in [7.15]. The present results cor-respond to the case of equally populated nlm-sublevels within a given n-level

than the probabilities $W_n^{He} = \langle V\sigma_n^{He}\rangle N_{He}$ of de-excitation of these levels in collisions with He atoms for $n < 8$, 12, and 16 at plasma ionization degrees $\alpha = N_e/N_{He} = 10^{-6}$, 10^{-7} and 10^{-8}, respectively. Hence, the mechanism considered here for inelastic quenching of highly excited atomic $Xe(n)$ states by atoms of the buffer rare gas He play a significant role in the process of three-particle electron-ion recombination (see [7.19] fore more details).

The ionization rate constants of the Rydberg rare gas $A^*\left[n_0p^5\left(^2P_{3/2}\right)nl\right]$ atoms in thermal collisions with lighter rare gas $B(^1S_0)$ atoms in the ground state have been calculated in [7.15]. In the case of selectively excited nl levels with small orbital moment $l \ll n$, the final formula describing the contribu-tion of the perturber-core scattering mechanism can be presented as

$$K_{nl}^{ion}(T) = \frac{16\pi^2\gamma_l\,(v_0/a_0)\,g_f}{25(2l+1)n_*^3 g_i} \int\limits_0^{R_{nl}} \exp\left(-\frac{U_i(R)}{kT}\right) \mathcal{H}(R)R^2 dR \ . \qquad (7.66)$$

where $\gamma_l = \sum_{l'}\gamma_{l'l}$; and the initial term $U_i(R)$ corresponds to the first ex-cited $A_1\,|j_i = 3/2, \Omega_i = 3/2\rangle$ state of the quasi-molecular BA^+ ion, while the final term $U_f(R)$ is its ground state $X\,|j_f = 3/2, \Omega_f = 1/2\rangle$. The ionization rate constants for the Rydberg states of Xe atoms in thermal collisions with $Kr(^1S_0)$ are presented in Fig. 7.6 by the full curves. It can be seen that the values of the rate constants are equal to $K_{np}^{ion} \sim 10^{-11}$–$10^{-10}$ $cm^3 s^{-1}$ for the case of selectively excited np-levels in the range of $n \sim 20$–60. For comparison, the ionization rate constants (dashed curves) calculated in [7.15] with the use of (6.46) are also presented in Fig. 7.6. One can see that the competitive Fermi mechanism becomes predominant only in the range of high enough principal quantum numbers.

8. Inelastic Transitions Induced by Collisions of Rydberg Atom (Ion) with Charged Particles

The theory of collisions of Rydberg atoms and ions is reviewed. Classical, semiclassical and quantum approaches for transitions induced by collisions with charged particles are considered in detail. We paid special attention to the comparison with available experimental data and Born approximation as the limiting case of any theory.

The final formulas are written for any incident charged particle (M, Ze, V, E are mass, charge, velocity and energy of the particle, respectively, μ is the reduced mass of the particle and Rydberg atom). Some formulas are valid for incident electrons only. In this case we use designations: m, e, v, E for mass, charge, velocity and energy of the incident electron.

8.1 Basic Problems

The first theoretical investigations of collisions of charged particles with atoms in the framework of the binary approximation of classical mechanics had begun in twenties (see [8.1] and [1.3]). In 1962 *Seaton* [4.3] built his theory of transitions between Rydberg states induced by charged particles collisions using the perturbation theory with the dipole approximation for the interaction potential. In 1964, 66 two clear papers by *Stabler* [8.2] and *Gerjuoy* [8.3] were published, in which the classical impulse approximation for energy transfer to Rydberg atoms by charged particles was practically completed.

The next studies (late sixties – early seventies) were stimulated, both by applications to astrophysics and by internal requirements of the theory. The discovery of the recombination radio lines between highly excited states in the nebulas opened new perspectives for diagnostics of the planetary nebulae, interstellar medium, HII zones, etc. Need was felt to get the rate coefficients of the transitions between Rydberg states induced by electron impact.

The classical approach gives the cross sections of the $n - n'$ transition between nearby levels to be proportional to $\pi a_0^2 \cdot n^4/(\Delta n)^3$, $(\Delta n = n'-n,\ \Delta n \ll n)$. The corresponding dipole perturbation theory gives the cross sections to be proportional to $\pi a_0^2 \cdot n^4/(\Delta n)^4$ and depends on energy logarithmically at large energies (or velocities) of the incident charged particles. Which one is correct?

Naturally, the classical approach is valid and fruitful for the states with high quantum numbers. However, if the initial condition that the motion of the Rydberg electron in a classical orbit satisfies the Bohr quantization condition is set, then after collision we usually obtain the classical orbit, which does not satisfy the Bohr quantization condition. What should be the interpretation of the results?

Third problem is as follows: It is clear, that the perturbation theory is not valid at the incident particle velocities, which are about the velocity of the atomic highly excited electron. Besides, methods developed for describing the low level excitation more sophisticated than the first order perturbation approximation become difficult to apply to Rydberg states. So, new methods are needed.

It was shown by *Beigman, Shevelko and Urnov* [3.16] that the consistent quantum perturbation theory (Born approximation) includes both the classical impulse approximation and the dipole approximation. Three approaches for the collision theory of Rydberg atoms were developed independently: "Consistent semiclassical description of the transitions between Rydberg states and classical perturbation theory in the action variables" by *Beigman, Vainshtein* and *Sobelman* [4.6], "Model of the equidistant levels" by *Presnyakov* and *Urnov*[4.8], and the transition theory based on the "Correspondence principles" by *Percival* and *Richards* [8.4]. They have answered the second and the third questions mentioned above. The equivalence of all these approaches was established in [4.7]. In the following decade many calculations have been performed, and the approximation formulas convenient for applications were obtained (see [1.3] for more detailes). Here we consider mainly the approaches based on the ideas of [3.16, 4.6]. The approach based on the "Correspondence principles" has been highlighted in detail in reviews [3.15, 8.5]. The theory for the spin-changing transitions of the Rydberg atoms has been developed in [3.24, 8.6, 7].

In 1980 the experiment of MacAdam's group on collisions of the Na Rydberg atom's beam with a charged particle beam [8.8] was reported. This was the first time that the cross sections of the transitions between Rydberg states induced by charged particles were measured directly. The sudden approximation theory [8.5, 9] and various normalized perturbation theories [8.10] were in good agreement with the experimental total l-changing cross section (from Na nd-state into all $l' > 2$) up to velocities about the average velocity of the highly excited electron. The region of lower velocities requires a more refined theory. Close coupling methods [8.10] allow describing the region of small velocities and predict the redistribution over final l-states. The cross sections of large angular momentum transfer transitions ($\Delta l > 1$) appear to be larger than those of the allowed transitions ($\Delta l = 1$). This prediction has been confirmed by experiment [8.11]. Recent developments of this approach is presented in [8.12].

Thus, the present status of the theoretical and experimental investigations appears as follows:

1. We have consistent theory of the transitions with energy transfer (or n-changing).
2. Calculated inelastic broadening induced by electrons in a plasma is in satisfactory agreement with the laboratory experiment [8.13, 14]. The dependence of the theoretical inelastic broadening induced by electrons in a plasma on the principal quantum number is in good agreement with many observations of the radio recombination lines (see, for example, [1.12]). Transitions between the neighboring levels ($\Delta n = 1$) dominate in the inelastic broadening. In reality, we cannot check the theory for the transitions with large energy transfer which are of relevance for Rydberg atom kinetics.
3. Until recently direct measurements of the n-changing transition cross sections were absent for the Rydberg atoms. Now the experiment on electron-collision induced $30s - 30p$ and $30s - 29p$ transitions in Na is carried out in [8.15]. The experimental data related to large electron energies are in agreement with the Born approximation.
4. The modern theory of the l-changing transitions is in satisfactory agreement with experiment.

8.2 n-Changing Transitions

8.2.1 Classical Approach

From the point of view of classical mechanics, the collision problem of atoms with charged particles is the Coulomb problem of the N-particles system. This system allows the similarity transformation (for example, see [2.2]) if $r_i(t)$ are the solutions of the corresponding differential equation set, and E is the total energy of the system, then

$$\mathbf{r}'_i(t') = \lambda \cdot \mathbf{r}_i(t), \ t' = \lambda^{3/2} \cdot t, \ E' = \lambda^{-1} \cdot E, \ \mathbf{v}' = \lambda^{-1/2} \cdot \mathbf{v}$$

are the solutions of the same set with the corresponding energy of the system and velocities of particles. The similarity properties allow expressing the energy transfer cross section from any level in terms of the cross section from the single n level, i.e. (scale factor λ is equal to n^2):

$$\sigma = \pi a_0^2 \cdot n^4 \cdot f(E/E_n, \Delta E/E_n), \tag{8.1}$$

where E is the energy of the incident charged particle, E_n is the energy of the "classical" atom, and ΔE is the energy transfer. Representation of the cross section in the form (8.1) is a necessary condition for applicability of classical methods. Function f is usually calculated by numerical methods, with the help of various analytical approximations, for example, the classical perturbation theory, or impulse approximation. Here we consider the impulse or binary encounter approach.

The approach is based on the assumption that the atomic core influence on the process of collision of the incident particle with the highly excited electron may be neglected. This means that we deal with interactions of the two free particles: the incident one with energy or velocity given, and the atomic electron with the same velocity distribution as that of the classical atom. It is important to take into account the atomic electron distribution even if the projectile velocity is much larger than that of the atomic electron. The well-known Rutherford's formula for the electron at rest is valid, if the transfer momentum appreciably exceeds the average atomic electron momentum.

A detailed history has been presented in review [3.15]. We would like to draw the readers attention to paper [8.2] in which the simple expression for the rate of the transfer energy from the incident electron to the electron gas with monokinetic isotropic velocity distribution was obtained. We give an expression from [3.15]:

$$\frac{d\sigma}{d(\Delta E)} = \pi \frac{M}{m} \frac{Z^2 e^4}{E} \left[\frac{1}{(\Delta E)^2} + \frac{4}{3} \frac{T_n}{(\Delta E)^3} \right] \qquad (8.2)$$

or

$$\frac{d\sigma}{d(\Delta E/T_n)} = 4Z^2 \frac{\pi a_0^2 n^4}{Z^4} \frac{M}{m} \frac{T_n}{E} \left[\left(\frac{T_n}{\Delta E} \right)^2 + \frac{4}{3} \left(\frac{T_n}{\Delta E} \right)^3 \right], \qquad (8.3)$$

where M, Ze, E are the mass, charge and the energy of the incident particle, respectively, $T_n = Z^2 Ry/n^2$ is the kinetic energy of the electron gas corresponding to the Rydberg state with spectroscopical symbol Z and principal quantum number n, ΔE is the energy transfer. Formula (8.3) is the same as (8.2), but in the form (8.1).

The ionization cross section may be obtained from (8.2,3) by integrating over all possible energy transfer. For electron collisions ($|E_n| \ll \Delta E \ll E_1$, where $|E_n| = T_n$ is the ionization potential) the formula

$$\sigma_i = \frac{\pi e^4}{3E^3 |E_n|} \frac{5E + 2|E_n|}{E - |E_n|},$$

$$\sigma_i = \frac{4}{3}\pi a_0^2 \left(\frac{Ry}{|E_n|} \right)^2 \frac{(u-1)(2+5u)}{u^3}, \qquad u = E/|E_n| \qquad (8.4)$$

is presented, for example, in paper [3.15]. For ion collisions the upper integral limit is practically equal to infinity, and the ionization cross section is

$$\sigma_i = \frac{\pi Z^2 e^4}{E |E_n|} \left(\frac{5}{3} - \frac{|E_n|}{4(E - |E_n|)} \right), \quad E = m\mathcal{E}/M,$$

$$\sigma_i = 4\pi a_0^2 Z^2 \left(\frac{Ry}{|E_n|} \right)^2 \frac{1}{u} \left(\frac{5}{3} - \frac{1}{4(u-1)} \right), \quad u = E/|E_n|, \quad u > 1.46. \qquad (8.5)$$

For applications to the Rydberg states the case of $u \gg 1$ is of most interest. In this case the formulas (8.4,5) yield the same result:

$$\sigma_i = \left(\frac{20}{3}\right) \cdot \pi a_0^2 Z^2 \frac{Ry}{|E_n|} \frac{M Ry}{m\mathcal{E}} = \frac{20}{3} \cdot \pi a_0^2 Z^2 \frac{Ry}{|E_n|} \left(\frac{v_0}{V}\right)^2, \qquad (8.6)$$

where V is the velocity of the incident particle, and v_0 is, as usual, the atomic velocity unit.

8.2.2 Born Approximation

In the modern theory of atomic collisions any reasonable method with rather wide application must include the Born approximation as the limit case. Most methods of the cross section calculations include the Born amplitude (or the cross section) as a necessary element.

The quantum Born method is valid only for the case of weak interactions; however, this case is considered without any suggestions or approximations of the interaction potential. It should be noted that the cross sections averaged over orbital quantum numbers are of special interest both from theoretical and applied points of view.

Approach to the Born cross sections based on classical distribution functions (see Sect. 2.4) is considered in [8.16]. The references to analytical formulas for some cases are presented in review [1.3]. Numerical methods for the transitions between hydrogen states with principal quantum numbers that are not large were developed by many authors (see, for example, [3.8]). The extension of the numerical methods of this kind to highly excited states is practically impossible. In this section we consider application of the above mentioned (see Sect. 3.3) analytical expressions of the form factors to the Born cross sections.

The representation of the momentum transfer, in which the transition amplitude is expressed directly in terms of the form factor, is most convenient for the Born cross sections. However, in this representation it is very difficult to formulate the approach beyond the limits of the perturbation theory. The impact parameter representation is more convenient for such approaches. The Born transition amplitude in the impact parameter representation for the dipole potential was obtained in [4.3]. We use a general relation between amplitudes in both representations and obtain the approximation of the total Born amplitude, which takes into account not only the long-range dipole interaction but the short-range "impulsive" part of the interaction as well.

a) **Momentum Representation.** The cross section of the $n - n'$ transition of the ion with a spectroscopic symbol Z induced by electron impact may be expressed in the Born approximation in terms of the form factor $\mathcal{F}_{n'n}(Q)$ (see Chap. 3):

$$\sigma_{n'n} = \frac{8\pi}{q^2} \frac{1}{n^2} \int\limits_{|q-q'|}^{q+q'} \mathcal{F}_{n'n}(Q) \cdot \left(\frac{Z}{Qa_0}\right)^3 d\left(\frac{Qa_0}{Z}\right), \quad \mathbf{Q} = \mathbf{q}' - \mathbf{q}, \qquad (8.7)$$

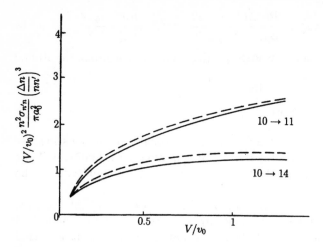

Fig. 8.1. The cross sections of the hydrogen transitions 10–11, 10–14. Full curves are the momentum transfer representation; dashed curves are the impact parameter representation

where $\hbar q$, $\hbar q'$ are the initial and final momenta of the external electron, $\mathcal{F}_{n'n}(Q)$ is defined by formula (3.32). We substitute asymptotic expressions (3.32) into (8.7) and integrate by parts:

$$\sigma_{n'n} = \frac{8\pi}{3q^2 Z^2} \left(\frac{nn'}{\Delta n}\right)^2 \left[\int_{\vartheta_-}^{\vartheta_+} \frac{A(\vartheta)\,\vartheta d\vartheta}{(\vartheta^2 - 1)} + \left(\vartheta_-^2 - 1\right)^{-3/2} \int_{\vartheta_-}^{\vartheta_+} \frac{A(\vartheta)\,\vartheta d\vartheta}{(\vartheta^2 - 1)^{1/2}}\right],$$

$$\vartheta = \left[1 + \left(a_0 \frac{Qnn'}{\Delta n}\right)^2\right]^{1/2}, \quad \vartheta_\pm = \left[1 + \left(a_0 \frac{(q \pm q')nn'}{\Delta n}\right)^2\right]^{1/2}, \quad \Delta n = |n - n'|$$

$$(8.8)$$

In Sect. 3.3 we have seen, that the asymptotic formulas are satisfactory even for principal quantum numbers that are not large. We expect the same result for the cross sections, too. Figure 8.1 shows the comparison of the cross sections of the 2–3, 5–6, 4–7 transitions obtained from (8.8) with the usual numerical Born ones.

Formulas (8.8) may be significantly simplified, if in the expansion (8.8) over $(\Delta n/n)$ only the first term is being kept. Besides, we factor out the integral singularities and rewrite (8.8) in the form:

$$\sigma_{nn'} = \frac{\pi}{q^2}\frac{1}{Z^2 n^2}\left(\frac{nn'}{\Delta n}\right)^3 \Omega_{nn'},$$

$$\Omega_{n'n} = \frac{16}{9}\left[\int_{\vartheta_-}^{\vartheta_+} d\vartheta\left(\frac{\mathcal{U}(\vartheta)-\mathcal{U}(1)}{\vartheta^2(\vartheta^2-1)} + \Delta n\frac{\mathcal{V}(\vartheta)}{\vartheta} + \frac{\mathcal{U}(1)}{\vartheta^2(\vartheta^2-1)}\right)+\right.\tag{8.9}$$

$$\left.+(\vartheta_-^2-1)^{-3/2}\int_1^{\vartheta_+} d\vartheta\left(\frac{(\vartheta^2-1)\mathcal{U}(\vartheta)}{\vartheta^2} + \Delta n\frac{(\vartheta^2-1)^{3/2}\mathcal{U}(\vartheta)}{\vartheta^2}\right)\right],$$

where

$$\mathcal{U}(\vartheta) = (4/\vartheta^2-1)J_{\Delta n}(\Delta n\vartheta)J'_{\Delta n}(\Delta n\vartheta),$$
$$\mathcal{V}(\vartheta) = [(\vartheta^2-1)/\vartheta^2](1+12/\vartheta^2)J_{\Delta n}^2(\Delta n\vartheta) - (1-12/\vartheta^2)[J'_{\Delta n}(\Delta n\vartheta)]^2.$$

Now we consider the cross sections at large energies of the external electrons $E \gg |E_n|$. It is clear, that $\vartheta_-^2 - 1 \approx (k_a/q)^2 \ll 1$, where $k_a = (Z/a_0)\cdot(1/n + 1/n')/2$, and ϑ_+ tend to infinity. We keep only the first term of the series over $\vartheta_-^2 - 1$, and have for $\Omega_{n'n}$:

$$\Omega_{n'n} = A[\ln(2q/k_a) - 2/3 - B_1] + (4/3)B,\tag{8.10}$$

where A, B, B_1 depend only on Δn:

$$A = (16/9)\cdot\mathcal{U}(1) = \frac{16}{3}\cdot J_{\Delta n}(\Delta n)J'_{\Delta n}(\Delta n),$$

$$B = \frac{4}{3}\Delta n\int_1^\infty \frac{\Delta\vartheta}{\vartheta^3}\left\{\frac{\vartheta^2-1}{\vartheta^2}(1+\frac{12}{\vartheta^2})J_{\Delta n}^2(\Delta n\vartheta) - (1-\frac{12}{\vartheta^2})[J'_{\Delta n}(\Delta n\vartheta)]^2\right\},$$

$$B_1 = \int_1^\infty \frac{\Delta\vartheta}{\vartheta^2(\vartheta^2-1)}\left(\frac{(1-4/\vartheta^2)J_{\Delta n}(\Delta n\vartheta)J'_{\Delta n}(\Delta n\vartheta)}{3J_{\Delta n}(\Delta n)} + 1\right)\tag{8.11}$$

We consider the behavior of A, B, B_1 in the limit case of $\Delta n \to \infty$. The known Nicholson formulas for Bessel functions give the asymptotic expressions (see, for example, [2.4]):

$$J_p(px) = \begin{cases} \frac{\Gamma(1/3)}{2\pi\sqrt{3}}\cdot(6/px)^{1/3}, & 0 < x-1 \ll c/p^{2/3},\\[2mm] \left(\frac{2}{x(x-1)}\right)^{1/4}(3\pi p)^{-1/2}\cdot\cos\left(\frac{p[2(x-1)]^{3/2}}{3\sqrt{x}} - \frac{\pi}{4}\right), & x-1 \gg c/p^{2/3}, \end{cases}$$

$$J'_p(px) = \begin{cases} \frac{\Gamma(2/3)}{2\pi\sqrt{3}}(6/p)^{2/3}\left[\frac{2}{3}x^{-1/3} + \frac{1}{3}x^{-4/3}\right], & 0 < x-1 \ll c/p^{2/3},\\[2mm] -\left(\frac{8(x-1)}{x^7}\right)^{1/4}(3\pi p)^{-1/2}\frac{2x+1}{3}\sin\left(\frac{p[2(x-1)]^{3/2}}{3\sqrt{x}} - \frac{\pi}{4}\right), & x-1 \gg c/p^{2/3} \end{cases}\tag{8.12}$$

It follows from (8.11) with the help of (8.12), that as $\Delta n \to \infty$ $A(\Delta n) \to 0.98/\Delta n$, $B \to 0.98$, and B_1 is proportional to $\ln(\Delta n)$. The quantities A, B, B_1 may be fitted with error less 5% by formulas

$$A = (1 - 0.25/\Delta n)/\Delta n, \quad B = (1 - 1/(3\Delta n)), \quad B_1 = 0.27\ln(5 + \Delta n). \quad (8.13)$$

It should be noted, that the classical impulse approximation (see Sect. 8.1.1, and, for example, paper [8.2]) gives the B value equal to 1.

Thus, formulas (8.10, 13) show that the dipole interaction dominates the Born cross section at small Δn, and the classical impulse approximation describes the cross section at large Δn. The interpolation formula including both limiting cases may be written in the form:

$$\sigma = \left(\pi a_0^2 \mathcal{Z}^2/Z^4\right) \frac{1}{n^2}\frac{1}{\varepsilon_e} 8\left\{\left(1 - \frac{0.25}{\Delta n}\right)\frac{(\varepsilon\varepsilon')^{3/2}}{(\Delta\varepsilon)^4}\ln(1 + \varepsilon_e/\varepsilon)\right.$$

$$\left. + \left(1 - \frac{0.6}{\Delta n}\right)\frac{\varepsilon_e/\varepsilon}{1 + \varepsilon_e/\varepsilon}\frac{(\varepsilon')^{3/2}}{(\Delta\varepsilon)^2}\left(\frac{4}{3\Delta n} + \frac{1}{\varepsilon}\right)\right\}, \quad \varepsilon_e = \frac{mE}{MZ^2 R_y}, \quad (8.14)$$

where $\varepsilon = 1/n^2$, $\varepsilon' = 1/(n')^2$, $\Delta\varepsilon = \varepsilon - \varepsilon'$, $\Delta\varepsilon > 0$, $\mathcal{Z}e$, M, E are charge, mass and energy of the projectile, respectively. In the limit case of $\Delta n \ll n$, formula (8.14) reduces to (8.10, 13), and at $\Delta n \approx n$ (8.14) goes to the classical impulse approximation. Formula (8.14) was derived for the electron - Rydberg ion collisions. However, since extension to collisions with heavy particles is obviously we presented it in a form applicable also to transitions induced by heavy particle collisions.

b) **Impact Parameter Representation.** A number of more sophisticated, perturbation theory methods are formulated in the framework of the impact parameter representation. In this representation, the incident charged particle motion is described as a classical particle by a rectilinear trajectory with impact parameter ρ, and the problem is to determine the transition probability of atom from n state to n' state for the time of the particle motion. Below we consider the Born transition amplitude in the impact parameter representation, which is the base of the more complicated methods.

We have seen in the previous section, that the momentum transfer probability consists of two terms. The first term dominating at small momentum transfer corresponds to the long range dipole interaction of the atom and incident particle. The second term gives the main contribution at the large momentum transfer, and corresponds to the classical impulse approximation, which neglects the atomic core influence on the collisional process.

First we consider the dipole part. The nondiagonal interaction potential matrix element of the atom and charged particle at large distance is

$$\left\langle nlm \left| \frac{\mathcal{Z}e^2}{r - R} \right| n'l'm' \right\rangle \approx [\mathcal{Z}e^2/R^3]\langle nlm|\mathbf{r}\cdot\mathbf{R}|n'l'm'\rangle, \quad (8.15)$$

where $\mathcal{Z}e$ is the charge of the incident particle, $\mathbf{r}$ and $\mathbf{R}$ are the position vectors of the atomic electron and incident particle, respectively. In the Born approximation we have for the transition probability:

$$W_{n'n}(\rho) = \sum_{l,m,l',m'} \left| \frac{i}{\hbar} \int_{-\infty}^{+\infty} \mathcal{V}_{n'l'm',nlm}(t) \exp(i\omega_{n'n}t)dt \right|^2 ,$$

$$\mathcal{V}_{n'l'm',nlm}(t) = [\mathcal{Z}e^2/R^3(t)]\langle nlm|\mathbf{r} \cdot \mathbf{R}|n'l'm'\rangle , \quad \mathbf{R}(t) = \rho + \mathbf{V}t,$$

(8.16)

where $\omega_{n'n}$ is the transition frequency, $\mathbf{V}$ is the velocity of the incident particle, ρ is the impact parameter, and we assume the rectilinear trajectory of the incident particle. The integrals in (8.16) were expressed in terms of the MacDonald functions K_0, K_1 in paper [4.3], and $W_{n'n}$ may be written in the form ($f_{n'n}$ is the oscillator strength):

$$n^2 W_{n'n}(\rho) = |A_d(\rho)|^2 = \left(\frac{2\mathcal{Z}a_0v_0n^2}{\rho V} \right)^2 \left(\frac{n^2 f_{n'n}\Delta n}{(n+n')(nn')Z^2} \right) \cdot \xi(\beta),$$

$$\xi(\beta) = \beta^2[K_0^2(\beta) + K_1^2(\beta)], \quad \beta = \omega_{n'n}\rho/V.$$

(8.17)

Below, we call the $A_d(\rho)$ by the averaged Born amplitude of the dipole interaction. It should be noted the function $\xi(\beta)$ can be fitted with an error less than 10% by the simple expression

$$\xi(\beta) = (1 + \pi\beta)e^{-2\beta},$$

(8.18)

which is asymptotically exact in both limiting cases $\beta \to 0$ and $\beta \to \infty$.

$A_d(\rho)$ in (8.17) is singular at $\rho \to 0$. The pole at $\rho \to 0$ arises due to nonapplicability of the dipole approximation at small distances. We change (8.17) to regularize it by means of the substitution $\rho^2 \to \rho^2 + \rho_0^2$:

$$A_d(\rho) = \frac{\mathcal{Z}L_d\sqrt{1+\pi\beta}e^{-\beta}n^2a_0v_0}{V(\Delta n)^2(\rho^2 + \rho_0^2)^{1/2}}, \quad \beta = \frac{\omega_{n'n}}{V}(\rho^2 + \rho_0^2)^{1/2},$$

$$L_d = 2f_{n'n}\frac{n^2(\Delta n)^3}{(n+n')(nn')}.$$

(8.19)

In the Kramers approximation, L_d is equal to 0.7 for $\Delta n = 1$. The exact value is 0.6.

Now we consider the "nondipole", "impulse" part. We use the general relation between the amplitude in the impact parameter representation and the momentum transfer one (4.56). Let us introduce function $\mathsf{F}(Q)$:

$$\left(\frac{nn'}{\Delta n} \right)^3 \mathsf{F}^2(Q) = \mathcal{F}_{n'n}^{(i)}(Q)/Q^2,$$

$$\mathcal{F}_{n'n}^{(i)}(Q) = \frac{1}{Q} \int_0^Q dQ \left\{ \frac{2}{3} \left(\frac{nn'}{\Delta n} \right) \exp \left(-\frac{(\Delta n)^2\varepsilon^2}{2(n+n')} \right) \mathfrak{Q}(\Delta n/n) \cdot a_2(\varepsilon) \right\},$$

(8.20)

where function $a_2(\varepsilon)$ is defined by formula (3.43). According to (4.56) the impulse part of the amplitude in the impact parameter representation is given by the Fourier transformation taken over the plane defined by the condition $\omega + \mathbf{Q} \cdot \mathbf{V} = 0$. Let $\mathbf{Q}_\parallel$ and $\mathbf{Q}_\perp$ be the parallel and perpendicular $\mathbf{Q}$-components, respectively. Performing integration over the angle formed by the $\mathbf{Q}$ and ρ vectors, we obtain

$$A_i(\rho) = \frac{2Zv_0}{Vn} \left(\frac{nn'}{\Delta n}\right)^{3/2} \int_0^\infty \frac{J_0(Q_\perp \rho)F(Q)}{Q} Q_\perp dQ_\perp,$$

$$Q^2 = Q_\parallel^2 + \mathbf{Q}_\perp^2, \quad Q_\parallel = \omega/V. \tag{8.21}$$

There is no analytical expression for the integral (8.21). Below we fit $F(Q)$ by the formula which allows us to obtain integral (8.21) in closed form. The analysis of $\mathcal{F}^{(i)}(Q)$ shows, that there is a maximum at $a_0 Qnn'/Z\Delta n \approx 1$, and the main the argument is in the form $(\Delta n)^2 + (a_0 Qnn'/Z)^2$. Therefore, we write the fitting formula as follows:

$$F(x) = F_m \left(\frac{x}{q_m \Delta n}\right) \left(\frac{2\alpha}{(2\alpha + 1) + (x/q_m \Delta n)^2}\right)^\alpha, \quad x = a_0 Qnn'/Z. \tag{8.22}$$

Function (8.22) is maximal at $a_0 Qnn'/Z \approx (\Delta n)q_m$, and the maximum value is equal to F_m. α is the fitting parameter. The substitution of (8.22) into (8.21) gives

$$A_i(\tilde{\rho}) = 2ZF_m \frac{(2\alpha q_m^2)^\alpha v_0}{V\sqrt{\Delta nq_m}} \cdot \frac{x^{\alpha-1}K_{\alpha-1}(y)}{q_0^{2(\alpha-1)}2^{\alpha-1}\Gamma(\alpha - 1)}, \tag{8.23}$$

where $q_0 = q_m(2\alpha - 1) + (k_a a_0 v_0/V)$, $y = q_0 \Delta n\tilde{\rho}$, $\tilde{\rho} = Z\rho/nn'a_0$. If α is a half-integer number, (4.29) may be expressed in terms of elementary functions:

$$A_i(\tilde{\rho}) = \frac{2ZF_m}{VZ} \left(\frac{2\alpha q_m^2}{q_0^2}\right)^\alpha \frac{q_0^2}{q_m} \frac{e^{-y}}{2(\alpha - 1)} \sum_{i=0}^{j} \frac{(2y)^i (2j - i)!j!}{i!(2j)!(j - i)!}, \quad j = \alpha - 3/2. \tag{8.24}$$

A comparison with direct calculation gives the parameters F_m and q_m equal to 0.33 and 1.1, respectively. The function $A_i(\tilde{\rho})$ shows a weak dependence on α. The value $\alpha = 3.5$ was accepted. A more detailed analysis of the fitting formula (8.24) is presented in [1.3]. Now we have for the Born cross section in the impact parameter representation:

$$\sigma_{n'n} = \int_0^\infty 2\pi\rho d\rho |A_{n'n}|^2, \quad A_{n'n}^2 = A_d^2(\rho) + A_i^2(\rho). \tag{8.25}$$

The formula (8.19) for $A_d(\rho)$ involves the regularization parameter ρ_0. We determine this parameter from the comparison of the calculations of (8.25) with the Born cross sections in the momentum transfer representation [formulas (8.7,8)]. The agreement is better than 20% at $\tilde{\rho}_0 = 3$ ($\tilde{\rho}_0 = Z\rho_0/nn'a_0$).

8.2.3 Semiclassical Approach

The semiclassical approach is based on determining the action increment of the Rydberg electron as result of a collision. The corresponding general formulas are given in Sect. 4.5.

a) **Transition Probabilities and Cross sections.** Calculating the probabilities of $n - n'$ transitions requires averaging over l, m and summing over $l'm'$. In the framework of the semiclassical approach it is equivalent to averaging the expression for transition probability in term of the action increment (4.65) over $\Delta l = l' - l$, and $\Delta m = m' - m$. The quantum perturbation theory allows us to carry out these operations analytically before the final calculation. The analytical averaging of (4.65) is impossible, and numerical methods are too cumbersome. Calculations may be simplified by using the averaged A_κ coefficients. The averaged A_κ coefficients are independent of angular momenta and therefore the one-dimensional formula (4.78) is applicable. Finally, the approximate averaging procedure is given by formulas:

$$W_{n+\kappa,n} = \left| \int_0^{2\pi} e^{iS/\hbar - i\kappa u} du \right|^2 , \quad S(n,u) = \sum_\kappa A_\kappa \exp(i\kappa u),$$

$$A_\kappa^2 = (1/n^2) \sum_{l,m,l',m'} \left| \int_{-\infty}^{\infty} dt \, \langle n+\kappa, l'm' | \mathcal{V}(\rho, vt) | nlm \rangle \right|^2 = A_d^2 + A_i^2,$$

$$(8.26)$$

where the amplitudes A_d, A_i are given by quantum formulas $(8.19, 23)$ from the Sect. 8.2 with

$$j = \alpha - 3/2, \ z = q_0 \Delta n (Z\rho/nn'a_0), \ q_0^2 = q_m^2(2\alpha - 1) + (k_a a_0 v_0/v)^2,$$
$$k_a = (Z/a_0)(1/n + 1/n')/2, \ q_m = 1.1, \ f_m = 0.33, \ \alpha = 3.5, \ \rho_0 = 3(nn'a_0/Z).$$

The use of the quantum amplitudes is possible due to the relation between the Born amplitudes and the Fourier coefficients of the action increment. Of course, it is possible to determine these coefficients in the framework of classical mechanics, but this however, has not been done so far.

Figure 8.2 shows the probabilities of the transitions: $100 - 101$, $100 - 102$, and $100 - 104$ in neutral Rydberg atom as functions of the impact parameter ρ. In the case of $\Delta n = 1$ the transition probability is maximal at impact parameter ρ exceeding the characteristic orbit dimension by the factor of 6–8. The maximum is about 0.3. It should be noted that the transition probabilities with $\Delta n > 1$ exceeds the Born one in a wide region of the impact parameter. This suggests the significant contribution of the "step by step" excitation. Function $W(\rho)$ has oscillations at small ρ, but the contribution of this region to the total cross section is small.

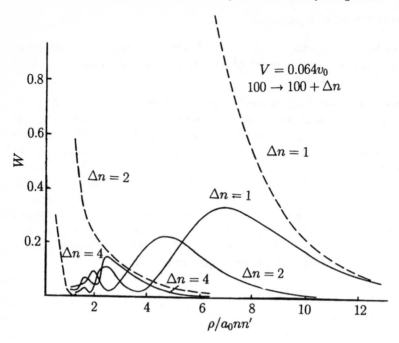

Fig. 8.2. The transition probability in neutral Rydberg atom versus the impact parameter, induced by collision with ion ($Z = 1$). Full curves are the perturbation theory for action increment (semiclassical perturbation theory); dashed curves are the Born approximation

The cross section may be obtained from (8.26) by integration over the impact parameter. The Born cross section is proportional to the factor $Z^{-4}(nn'/\Delta n)^3$, therefore, we write the cross section in the form:

$$n^2\sigma_{n'n} = (\pi a_0^2/Z^4)\left(\frac{nn'}{\Delta n}\right)^3 \cdot s_{nn'}, \quad s_{nn'} = 2(\Delta n)^3 Z^2 \int\limits_0^\infty W_{n'n}(\tilde\rho)\tilde\rho d\tilde\rho, \quad (8.27)$$

where probability $W_{n'n}(\tilde\rho)$ is given by formula (8.26), and $\tilde\rho = \rho/Za_0 n^2$.

Figure 8.3 shows the s value as a function of the velocity for the transitions: $10 - 11$, $10 - 12$, $10 - 14$. In the case of $\Delta n = 1$, the cross section is maximum at velocity $V = 0.3v_0$ and is significantly smaller than the Born one. These results are in agreement with data of [4.3], where the perturbation theory was used for large impact parameters, and the transition probability is assumed $1/2$ for small ρ. Thus, we see once more, that the dipole interaction region of large impact parameters dominates for the $\Delta n = 1$ transitions. The situation changes drastically for the $\Delta n > 1$ transitions, for which the "impulsive" part is significant.

The numerical results for the semiclassical cross section σ^{Sc} may be written in the form:

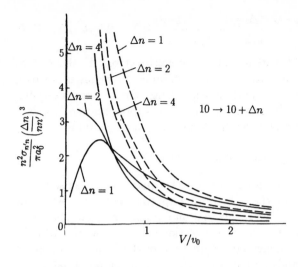

Fig. 8.3. The cross sections of the $10 \to 10 + \Delta n$ transitions, induced by collision with ion ($Z = 1$). Full curves are the semiclassical perturbation theory; dashed curves are the Born approximation

$$\sigma^{\text{Sc}}_{n'n} = \sigma^{\text{B}}_{n'n} \cdot f(n, Z \cdot \Delta n), \tag{8.28}$$

where $\sigma^{\text{B}}_{n'n}$ is the Born cross section of the $n \to n'$ transition [see (8.14) from the Sect. 8.1.2]. In the limiting case of large energies or large Z function f tends to unity. In [8.17] it was shown the function $f(n, Z \cdot \Delta n)$ may be fitted by the expression

$$f(n, Z \cdot \Delta n) = \frac{\ln\{1 + x/(1 + (1.25n/(xy))^{3/2})\}}{\ln\{1 + x/y\}}, \quad x = V/v_n, \ y = Z \cdot \Delta n, \tag{8.29}$$

which reduces to 1 in the previously mentioned limiting cases and differ from direct calculations not more than 10–20%.

b) **Rate Coefficient.** The rate coefficient, i.e., $v\sigma$ value averaged over the Maxwellian distribution of electrons, is required for a number of applications. Using the formulas (8.14, 28, 29), fitting formulas for the rate coefficient of the $n - n'$ transitions may be written in the form (see also [1.3], $n < n'$ is assumed):

$$\langle v\sigma_{n',n} \rangle = \langle v\sigma^{\text{B}}_{n',n} \rangle \cdot f_\theta(Z \cdot \Delta n, n),$$

$$\langle v\sigma^{\text{B}}_{n',n} \rangle = \frac{16}{\sqrt{\pi}} \left(\frac{\pi a_0^2 v_0}{Z^3} \right) \left(\frac{Z^2 Ry}{kT} \right)^{1/2} \frac{e^{-\Delta E/kT}}{n^2}$$

$$\times \left\{ \left(1 - \frac{0.25}{\Delta n} \right) \frac{(\varepsilon\varepsilon')^{3/2}}{(\Delta\varepsilon)^4} \varphi(|E_n|/kT) \right.$$

$$\left. + \left(1 - \frac{0.6}{\Delta n} \right) \frac{(\varepsilon')^{3/2}}{(\Delta\varepsilon)^2} \left(\frac{4}{3\Delta n} + \frac{1}{\varepsilon} \right) \left[1 - \frac{|E_n|}{kT} \cdot \varphi \left(\frac{|E_n|}{kT} \right) \right] \right\}, \tag{8.30}$$

where $\varepsilon = 1/n^2$, $\varepsilon' = 1/(n')^2$, $\Delta\varepsilon = \varepsilon - \varepsilon'$, $\Delta\varepsilon > 0$, $\varphi(x) = -e^x \mathbf{E}_i(-x)$

$$f_\theta(Z \cdot \Delta n, n) = \frac{\ln[1 + (n\sqrt{\theta}/Z \cdot \Delta n)/1 + c/Z\Delta n\sqrt{\theta}]}{\ln[1 + (n\sqrt{\theta}/Z \cdot \Delta n)]}, \quad c = 2.5, \ \theta = kT/Z^2 Ry.$$

c) **Dipole Potential in Rotating Axis Approximation.** In this section we consider the dipole model, which allows obtaining the result in a closed analytical form. Let the incident particle with charge Ze fly upon a rectilinear trajectory with velocity V. In the rotating axis approximation (z-axis is directed to the incident particle) the dipole interaction potential is

$$V = \frac{2}{\pi} Zed_z/R^2 = \frac{2}{\pi} Zed_z/(\rho^2 + \rho_0^2 + (Vt)^2). \tag{8.31}$$

Here d_z is the dipole moment projection on the quantization axis. Factor $(2/\pi)$ is introduced to provide the right Born amplitude at large ρ, when the perturbation theory is valid. The regularization parameter ρ_0 may be determined from comparison with the Born cross section in the momentum transfer representation. Using the Kramers approximation for the Fourier component of the dipole moment we find the Fourier component of the potential V_κ:

$$V_\kappa(t) = \frac{2Z \cdot e^2 a_0 n^2}{\pi 3^{3/4} \kappa^2 [\rho^2 + \rho^2 + (Vt)^2]}, \tag{8.32}$$

and

$$A_\kappa = \hbar \frac{2\alpha \exp[-\kappa(\omega\rho_1/v)]}{\kappa^2 V \rho_1}, \quad \alpha = 0.7Z \cdot v_0 a_0 n^2, \ \rho_1 = (\rho^2 + \rho_0^2)^{1/2}$$

$$S = \sum_{\kappa \neq 0} A_\kappa e^{i\kappa u} = \hbar \frac{2\alpha}{V\rho_1} \sum_{\kappa \neq 0} \frac{\exp(-|\kappa|\beta + i\kappa u)}{\kappa^2}, \ \beta = \omega\rho_1/V. \tag{8.33}$$

We used the one-dimensional expression (4.78) according to the symmetry of the problem.

Exclusive simplicity of the Born amplitudes and the corresponding potential Fourier components in this model allows us to obtain the closed expression for the action increment. Taking into account

$$\sum_{\kappa \neq 0} \frac{\exp(-|\kappa|\beta + i\kappa u)}{\kappa^2} = i \int_0^u du \left\{ \sum_{\kappa=1}^{\infty} \frac{(e^{-\beta+iu})^\kappa}{\kappa} - \sum_{\kappa=1}^{\infty} \frac{(e^{-\beta-iu})^\kappa}{\kappa} \right\}$$

$$= i \int_0^u du \ln \frac{1 - e^{-\beta-iu}}{1 - e^{-\beta+iu}},$$

we find

$$S(n, u) = i\hbar \frac{2\alpha}{V\rho_1} \int_0^u du \ln \frac{1 - e^{-\beta-iu}}{1 - e^{-\beta+iu}} = \hbar \frac{4\alpha}{V\rho_1} \int_0^u du \, \text{arctg} \left(\frac{e^{-\beta} \sin u}{1 - e^{-\beta} \cos u} \right). \tag{8.34}$$

If $\rho_1 \gg \alpha/V$, the perturbation theory condition $S \ll \hbar$ is valid, and the transition probability is

$$W_{n'n} = |A_\kappa/\hbar|^2 = 4\alpha^2/[\kappa^4 V^2 \rho_1^2]. \tag{8.35}$$

At rather small ρ_1 we can substitute unity for $\exp(-\omega\rho_1/V)$ and taking into account equality $(1 - e^{-iu})/(1 - e^{iu}) = -e^{-iu}$ we obtain

$$S = \hbar\frac{2\alpha}{V\rho_1}\frac{u}{2}(u - 2\pi) = \hbar\frac{\alpha}{V\rho_1}[(u - \pi)^2 - \pi^2]; \tag{8.36}$$

$$W_{n'n} = \left| \frac{1}{2\pi}\int_0^{2\pi} du \exp\left[-i\kappa u + i\frac{\alpha}{v\rho_1}(u - \pi)^2\right] \right|^2 \tag{8.37}$$

$$= \frac{1}{8\pi}(V\rho_1/\alpha)\{[C(x_+) - C(x_-)]^2 + [S(x_+) - S(x_-)]^2\},$$

$$x_\pm = (\pi\kappa/2)^{1/2} \cdot (y^{1/2} \mp y^{-1/2}), \quad y = \rho_1/\rho_\kappa, \quad \rho_\kappa = 2\pi\alpha/v\kappa,$$

where C and S are the Fresnel integrals.

These expressions are valid, if $\rho_1 < V/\omega\kappa$. Condition $\rho_1 < V/\omega$ is not enough, since the transition probability is dependent on the Fourier component of the quantity $e^{iS/\hbar}$ with number κ, therefore the corresponding condition must be valid for the first κ terms of the Fourier series.

Now we can illustrate the relation between the classical cross section $d\sigma/dE$ and the quantum cross section in this model. If the action increment is given by (8.36), condition $\Delta = 2\pi\hbar\kappa$ may be satisfied only in the region of $\rho_1 < \rho_\kappa$. In this case, formulas (4.68, 49) yield

$$W_{n'n}(\rho) = \rho_1/2\kappa\rho_\kappa. \tag{8.38}$$

At small ρ_1, formula (8.38) leads to the same result as formula (8.36). At $\rho_1 > \rho_\kappa$ the stationary phase point is absent, therefore no classical expression can in principle be described in this region. The result of formula (8.36) at $\rho = \rho_\kappa$ with $\kappa \gg 1$ is $W \approx (8\kappa)^{-1}$. This value is one-forth of the corresponding value from (8.38)

d) Applicability Conditions. It is clear that the "rotating axis approximation" is suitable for applicability estimates. According to (8.34) the action increment is

$$S(n, u) = \hbar\frac{4ZL_dv_0 \cdot a_0 n^2}{V}\frac{a_0 n^2}{Z\rho_1}\int_0^u du \arctan\left(\frac{e^{-\beta}\sin u}{1 - e^{-\beta}\cos u}\right), \tag{8.39}$$

where $\rho_1 = (\rho_0^2 + \rho^2)^{1/2}$, $\rho_0 = 3a_0 n^2/Z$, and $L_d = 0.7$. Since the integrand in (8.39) is less than $\pi/2$, we have

$$|S| < \pi^2 \hbar|Z|(v_0/V). \tag{8.40}$$

From here we see that for collisions with electrons or one charged ions the validity condition of the classical perturbation theory $|S| \ll S_0 \sim 2\pi n\hbar$

reduces to condition $V \ll v_0/n$ or inequality $E \ll (1/Z^2)|E_n|$ for electrons and $E \ll (1/Z^2) \cdot (M/m) \cdot E_n$ for ions. The quantum perturbation theory is valid $(S \ll \hbar)$ if $V \ll v_0$ or $E \ll Ry$. Therefore, the range of the Born applicability is independent of n for the $\Delta n = 1$ transitions for which the dipole interaction dominates.

In the case of the "impulsive interaction" the applicability range of the classical perturbation theory may be estimated with the help of formulas (8.26, 35). The closed expression for the action increment is

$$S = \hbar \frac{2ZF_m v_0}{V} \left(\frac{3q_m^2}{q_0^2}\right)^{3/2} \frac{q_0^2}{q_m} \sum_{\kappa \neq 0} \frac{1}{\sqrt{\kappa}} e^{-|\kappa|x} e^{i\kappa u}, \tag{8.41}$$

where $F_m = 0.33$, $q_m = 1.1$, $q_0^2 = q_m^2 + (k_a/v)^2$, $x = q_0\rho(Z/nn'a_0)$. It should be noted that this series diverges as $r, u \to 0$. We write the divergence in the clear form:

$$\sum_{\kappa=1} \frac{1}{\sqrt{\kappa}} e^{-(x+iu)\kappa} = \Phi(e^{-x+iu}, \frac{1}{2}, 1),$$

$$\Phi \underset{\substack{r \to 0 \\ u \to 0}}{\longrightarrow} \Gamma(1/2)(1 - e^{-x+iu})^{-1/2} \longrightarrow \frac{\sqrt{\pi}}{\sqrt{x - iu}}, \tag{8.42}$$

$$S = 2\pi\hbar Z \frac{V_0}{V} \mathbf{Re}\left\{\Phi(e^{-x+iu}, 1/2, 1)\right\} \underset{\substack{x \to 0 \\ u \to 0}}{\longrightarrow} 2\pi\hbar\sqrt{\pi}\frac{V_0}{V} \mathbf{Re}\left\{\frac{\sqrt{x + iu}}{\sqrt{x^2 + u^2}}\right\}.$$

The divergence in formula (8.42) means that for any velocity there is an impact parameter region in which the perturbation theory is not valid. We estimate its contribution to the cross section. Condition $|S| \gg 2\pi n\hbar$ may be broken if $x < \pi/n^2 Z^2 (v_0/V)^2$, i.e., the contribution of this region is $\pi^3(v_0/V)^4(Z/Z)^2$. Since the cross section is proportional to n^4, the classical perturbation theory application is justified for large n. The Born range applicability is independent of n. Nevertheless, the Born approximation yields better results than those expected. It is connected with the specific properties of the Born approximation which is not only first order of the quantum perturbation theory but includes the binary impulse approximation.

Thus, formulas of the semiclassical perturbation theory give the solution to the problem of the transitions between highly excited levels induced by charged particles with velocities $V > v_0|Z|/Zn$. A number of kinetic problems need calculating the average square of the energy transfer $\langle(\Delta E)^2\rangle$. As formula (4.82) shows, this value may be also obtained in the framework of the Born approximation.

8.2.4 Coulomb-Born Approximation

The $n - n'$ transitions between highly excited states of ions induced by electron collisions need a special approach. The Born approximation considered in detail above is a good approximation only for the transitions induced by electrons with high energies. The semiclassical approach does not take into account the Coulomb long-range effects which are important for small and middle energies of the incident electron. The most adequate method for the investigation of impact excitation of multicharged ions by electrons is the Coulomb-Born approximation, which makes it possible to consider the effects of Coulomb interaction of the incident electron with a multicharged ion. This approximation is widely used for calculation of transitions from the ground state or between weekly excited states (see, for example, in [3.8]). But these methods based on partial wave expansion encounter great calculation difficulties for highly excited states which increase drastically with the growth of the principal quantum number. Here we give general a idea and two methods applicable to transitions between Rydberg states.

a) **General Formulas.** We write the total Hamiltonian of the system ion+incident electron in the form:

$$H = H_a + H_e + \mathcal{V},$$

$$H_e = K - \frac{ze^2}{R}, \quad \mathcal{V} = \frac{e^2}{|R - r|} - \frac{e^2}{R},$$

(8.43)

where H_a is the Hamiltonian of the Rydberg ion, H_e is the Hamiltonian of the incident electron, including the operator of kinetic energy K and potential of the Coulomb interaction of an external electron with an ion, $z = Z - 1$ is the charge of the Rydberg ion, R and r are the radius-vectors of the incident and Rydberg electrons, respectively. $\mathcal{V}$ is the perturbation potential which is not included in the "unperturbed" Hamiltonian $H_0 = H_a + H_e$. In opposite of the Born approximation, in the Coulomb-Born approximation the long range potential $-ze^2/R$ is included in the "unperturbed" Hamiltonian H_0, but not perturbation $\mathcal{V}$. Similar to the Born approximation (see Sect. 4.3) the cross section of $n \to n'$ transition can be written in the form

$$\sigma_{n'n} = \frac{m^2}{4\pi^2\hbar^4} \frac{q'}{q} \frac{1}{n^2}$$

$$\sum_{l,m,l'm'} \left| \int F_{q'}(R) \psi_{n'l'm'}(r) \mathcal{V} F_q(R) \psi_{nlm}(r) \, dR \cdot dr \right|^2 d\Omega_{q'}. \quad (8.44)$$

Here $F_{q'}$ and F_q are the eigenwavefunctions of the Hamiltonian H_e with energies $(\hbar q')^2/2m$ and $(\hbar q)^2/2m$, respectively and, asymptotically at large R they correspond to plane waves. In the Born approximation it is possible to carry out the integration over coordinates of the incident electron and the $n' - n$ cross-section can be expressed in terms of the generalized oscillator strength $\mathcal{F}_{n'n}(Q)$. This way is impossible for (8.44). Expansions of the functions $F_{q'}, F_q$ over partial waves are used for calculations of the cross-section according to (8.44). For more detailed formulas and calculations of the excitation cross-sections of not highly excited levels see, for example, [3.8].

b) **Dipole approximation.** The dipole approximation within the Born method corresponds to the first nonzero term of the expansion of the exponent by powers of $\boldsymbol{Q} \cdot \boldsymbol{\tau}$. In this case its possible to integrate over the scattering angle. This approach gives the well-known Born-Bethe formula. The dipole approximation within the Coulomb-Born method corresponds to the dipole term of the interaction potential $\mathcal{V}$. Replacing this potential by the dipole potential with a pole and using the integral representation for the Coulomb wave function we can carried out the integration over the radius vector $\boldsymbol{R}$ in close form. Then, after integrating over the scattering angles one obtains (see [8.18]):

$$\sigma_{n'n} = \frac{\pi}{q^2} \left(\frac{Ry}{\Delta E} \right)^2 f_{n'n} \Phi \left(\eta, \eta' \right),$$

$$\Phi \left(\eta, \eta' \right) = 2\pi^2 \frac{\left(\eta' \right)^2 - \eta^2}{\left(\eta' \right)^2} \frac{\exp \left(2\pi\eta \right)}{\left[\exp \left(2\pi\eta \right) - 1 \right] \left[\exp \left(2\pi\eta' \right) - 1 \right]} \frac{d}{dx} F \left(i\eta, i\eta'; 1; x \right),$$

$$\eta = -z \left(\frac{Ry}{E} \right)^{1/2}, \quad \eta' = -z \left(\frac{Ry}{E'} \right)^{1/2}, \quad x = -\frac{4\eta\eta'}{\left(\eta - \eta' \right)^2}, \quad \Delta E = E - E',$$

$$(8.45)$$

where E, E' are energies of the external electron before and after collision, ΔE is the transition energy (we suggest $E > E'$), and $F \left(i\eta, i\eta', 1, x \right)$ is the hypergeometric function. The Coulomb-Born and Born approximations with nonregularized dipole potential overestimate by a factor of two the usual Born approximation cross-sections at high energies. This effect is due to the pole of the dipole potential in the vicinity of zero. To obtain the correct results its necessary to make the procedure of regularization of the cross-section. The regularization method, using the cutoff of large momentum transfer can be obtained by comparing the asymptotic formulas of the Born approximation and Born-Bethe formulas with regularization. It gives the cutoff parameter for the momentum transfer $Q_{\max} = |\Delta E / 2E_n|$ istead of $q + q'$. The use of this cutoff parameter for the dipole Coulomb-Born approximation gives the correct result for high energies of the incident electron. This parameter is independent of the energy of the incident electron. But for low energies of the incident electron $Q_{\max}$ becomes less than $q - q'$. Therefore, an interpolation formula for $Q_{\max}$ is to be used. For high energies this formula should give $Q_{\max} = \Delta E / 2E_n$ and for low energies $q + q'$. The final formula for the regularized dipole Coulomb-Born approximation is (8.45) with parameter $Q_{\max}$:

$$Q_{\max} = Q_0 + \frac{q + q' + Q_0}{1 + 2nn'E'/Z^2 Ry}, \quad Q_0 = \frac{Z \Delta n}{a_0 \, n^2},$$

$$x = 1 - \frac{Q_{\max}^2}{\left(q - q' \right)^2}.$$

$$(8.46)$$

c) **Effective Trajectories Method.** For low energies of the incident electron $(E < E_n)$ the Born approximation gives drastically incorrect results for cross-sections of electron transitions in multicharged ions. It is connected with the fact that the Born approximation doesn't take into account two effects due to the Coulomb field of the multicharged ion. One effect is the increasing of the cross-section due to the effect of focusing of the particle flux by the Coulomb field of the ion, the other is the acceleration of incident electrons by the same field. The Effective Trajectory Method (ETM) takes into account both of these effects with the help of replacing the Coulomb hyperbolic trajectory by the rectilinear trajectory with effective parameters. The trajectory of the incident electron in the Coulomb field is a hyperbola with parameter $\mathfrak{p}$ and eccentricity ϵ (see Sect. 2.1):

$$\mathfrak{p} = \frac{L^2}{mze^2} \quad , \epsilon = \sqrt{1 + 2\frac{EL^2}{mz^2e^4}}, \tag{8.47}$$

where E is the energy, and L is the angular momentum of the incident electron. The minimum distance r_{min} is equal to

$$r_{min} = \mathfrak{p}/(\epsilon+1) = \mathfrak{a}(\epsilon-1), \ \mathfrak{a} = \frac{ze^2}{E}, \tag{8.48}$$

where $\mathfrak{a}$ is the half-axis of the hyperbola. In ETM the hyperbolic trajectory is replaced by the rectilinear trajectory with effective impact parameter $\rho_{eff} = r_{min}$ and effective velocity V_{eff}:

$$\rho_{eff} = \sqrt{\frac{z^4e^4}{m^2V_\infty^4} + \rho^2} - \frac{z^2e^2}{mV_\infty^2}, \quad V_{eff} = \sqrt{V_\infty^2 - \frac{\Delta E}{m} + \frac{2ze^2}{m\rho_{eff}}}. \tag{8.49}$$

Since the particle moves through a part of the trajectory before the transition from n to n', and another part of the trajectory after the transition, and the initial energy is equal to $mV_\infty^2/2$, and final energy to $mV_\infty^2/2 - \Delta E$, (8.49) contains the part that averages the motion along the trajectory. Thus in the framework of the Effective Trajectory Method the cross-section of transition is determined by formulas (8.19, 24, 25) with effective impact parameter ρ_{eff} and effective velocity V_{eff}. For high velocities ($mV/2 \gg ze^2/\rho$):

$$\rho_{eff} \to \rho\left(1 - \frac{1}{2}\frac{z^2e^4}{mV_\infty^2}\right), \quad V_{eff} \to V_\infty\left(1 + \frac{2z}{m\rho V^2} - \frac{\Delta E}{mV^2}\right). \tag{8.50}$$

Therefore, ρ_{eff} and V_{eff} tends to values ρ, V, i.e. for high velocities the Effective Trajectories Method gives a result that automatically coincides with Born approximation given by formulas (8.19, 24, 25). At low velocities of the incident particle in the usual Born formulas $\beta = \omega\left(\rho^2 + \rho_0^2\right)^{1/2}/V$ tends to infinity, therefore the transition probability (and the corresponding cross section) tends to zero. In opposite of Born approximation, at low velocities of the incident particle in the Effective Trajectory Method β tends to a finite value, hence $A_d(r)$ is a finite value too. Amplitude $A_i(r)$ is also a finite value, because the power of the exponent is a finite value for low velocities of the incident particle. Thus, ETM yields a finite value of the cross section at threshold as Coulomb Born approximation.

d) Numerical Results and Discussion. First we consider the results given by various evaluation methods of the Coulomb-Born cross-section at large enough Z ($Z > 10$).

$\Delta n = 1$ *transitions.* The main contribution to the $\Delta n = 1$ transition cross-section is given by the dipole part of the interaction potential of the Rydberg ion with an incident electron. Therefore, use of the dipole approximation is justified in this case. The cross-sections, obtained with ETM, and other methods for transitions 3–4 and 5–6 are given in Fig 8.4 in comparison with direct calculations by the code ATOM [3.8] in the Born and Coulomb-Born approximations. We see that the behavior of the Coulomb-Born and Born cross-sections is drastically different at low energies. The dipole approximation gives much more to the results than direct calculations at low and middle energies and twice more at high energies of collisions. Behavior at large energies is connected with pole of the dipole potential in the vicinity of zero and is removed by a regularization procedure (see, for example, [8.19]), however at low energies the cross-section is too large. The extrapolating method suggested in [8.20] gives results that are too large at middle and low energies. The best result is obtained by the ETM method.

$\Delta n > 1$ *transitions.* The cross-sections obtained with the ETM method in comparison with direct calculations by the code ATOM [3.8] for transitions 3-5, 3-6 are given in the Fig. 8.5. The relations between the various evaluation methods of the cross-section are similar to $\Delta n = 1$ transitions. In the case of $3 - 6$ transitions the difference between ETM and direct evaluations is more than for case of $\Delta n = 1$. It can be connected with violation of the $\Delta n \ll n$ condition.

$n \gg 1$ *transitions.* The developed ETM method allows us to calculate the Coulomb-Born cross-sections when direct calculations are impossible. There is an approximated scaling law for the excitation cross-sections. The cross-section is proportional to Z^{-4} at the corresponding energies proportional to Z^2. Figure 8.6 shows the Coulomb-Born cross-sections for $20 - 21$ and $20 - 24$ transitions and $Z = 2$, 4, and 10 and Born cross-sections for comparison.

We see the drastic difference between the Born and the Coulomb-Born cross sections for the energies of the incident electron less than energy of the Rydberg electron $E < E_n$.

Fig. 8.4. The cross sections of the $n \to n'$ transitions in hydrogen-like ions ($Z = 10$) induced by electron impact; panel a: the $3 - 4$ and panel b: the $5 - 6$ transitions, respectively. "ETM" is the Effective Trajectories Method, "Coul.Born" and "Born" are calculations by code "ATOM" [3.25] in Coulomb-Born and Born approximations, respectively, "Dipole" is the Coulomb-Born approximation with dipole potential, "Reg. Dipole" is the Coulomb-Born approximation with regularized dipole potential, "Extr" is extrapolating formula [8.20]

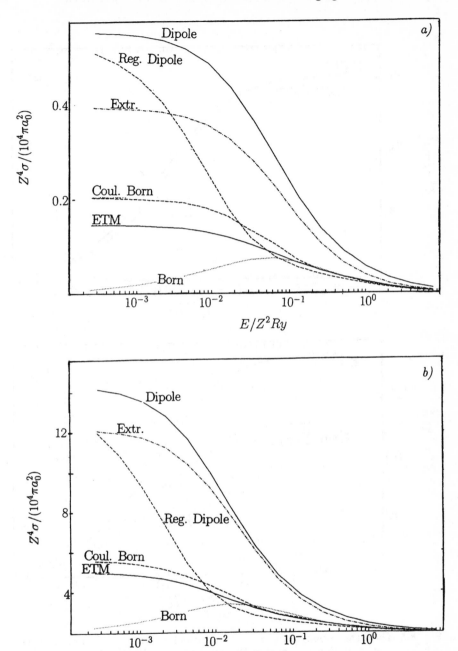

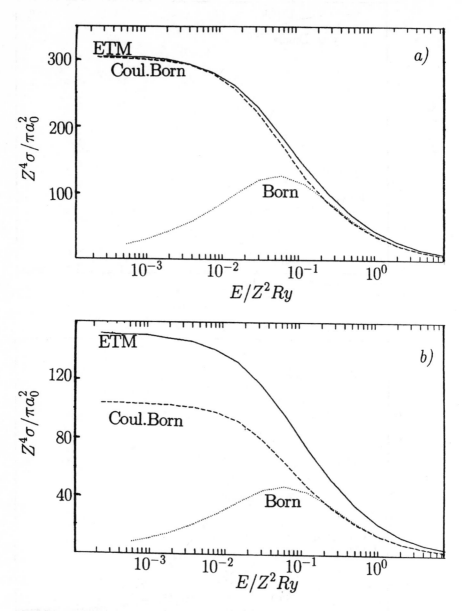

Fig. 8.5. The cross sections of the $n \to n'$ transitions in hydrogen-like ions ($Z = 10$) induced by electron impact; panel a: the $3 - 5$ and panel b: the $3 - 6$ transitions, respectively. "ETM" is Effective Trajectories Method, "Coul.Born", "Born" are calculations by code "ATOM" [3.25] in Coulomb-Born and Born approximations, respectively

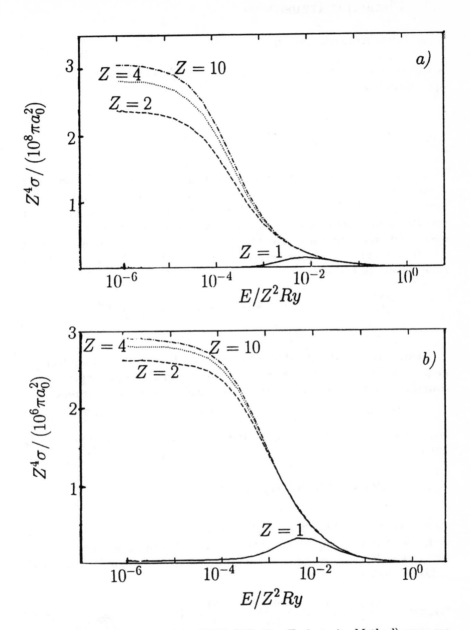

Fig. 8.6. Z-dependence of the ETM (Effective Trajectories Method) cross sections of $20 - 21$ (panel a) and $20 - 24$ (panel b) transitions in Hydrogen-like ions, $Z = 2, 4, 10$ induced by electron impact. "Born" are the Born cross sections of the corresonding transitions

8.3 l-Changing Transitions

8.3.1 Born Approximation

In this section we consider $nl - nl'$ transitions. The cross section in the Born approximation is

$$\sigma_{nl',nl} = \frac{8\pi}{(Zqa_0)^2} \frac{1}{g_l} \int\limits_{|q-q'|}^{q+q'} \mathcal{F}_{nl',nl}(Q) \frac{dQ}{Q^3}, \tag{8.51}$$

where $\hbar q$, $\hbar q'$ are the initial and final momenta of the external electron, g_l is the statistical weight of the initial state, $\mathcal{F}_{nl',nl}(Q)$ is the form factor [see formula (3.31) from the Sect. 3.3]. Formula (8.51) includes the integration over the momentum transfer Q and summation over multipoles $\varkappa$. The series over $\varkappa$ converges very rapidly, and the first term $\varkappa_{min} = \Delta l$ dominates. Below we consider only the main term with $\varkappa = \Delta l$ and one electron outside the closed shells. In this case:

$$\mathcal{F}_{nl',nl}^{(\varkappa)} = (2l+1)(2l'+1)(2\varkappa+1) \begin{pmatrix} l & \varkappa & l' \\ 0 & 0 & 0 \end{pmatrix}^2 R_\varkappa^2,$$

$$R_\varkappa = \langle nl'|j_\varkappa(Qr)|nl\rangle. \tag{8.52}$$

At first we consider the dipole transitions. At small Q the form factor is proportional to Q^2, and the cross section in formula (8.51) depends logarithmically on the lower limit. We fit the radial integral by the expression

$$\tilde{R}(Qa_0n^2) = 0.5\Lambda_1 \exp(-\alpha_1 Qa_0n^2/2), \quad \Lambda_1 = (1-l/n)^{1/2} \tag{8.53}$$

and make this dependence explicit. Parameter α_1 is determined by the condition:

$$\int\limits_0^\infty \left[R\left(Qa_0n^2\right) - \tilde{R}\left(Qa_0n^2\right) \right] \frac{dQ}{Q^3} = 0. \tag{8.54}$$

Parameter $\alpha_1 = 0.586$, if the radial integral is given by (3.49) from the Sect. 3.3. It is very important, that the value of α obtained through exact functions α depends weakly on n and l, if $l \ll n$. The fitting formula (8.54) yields for the cross section

$$\sigma_{nl\pm1,nl} = 6(\pi a_0^2 n^4/Z^2) \cdot \Lambda_1^2 \frac{\max(l, l\pm1)}{(2l+1)} \frac{Ry}{E} E_1\left[(0.5\alpha_1 n^2)\frac{\Delta E/Ry}{\sqrt{E/Ry}}\right], \tag{8.55}$$

where $E_1(x)$ is the integral exponent and we put the upper limit of the integral over Q equal to infinity. From (8.55) we obtain for the large energies and $l \ll n$:

Table 8.1. The dependence of the $K_{\ae}$ on n and l

	$\ae = 2 \ (K_2^0 = 0.45)$			$\ae = 3 \ (K_2^0 = 0.42)$		
n	$l = \ae$	$l = [n/2]$	$l = n - 1$	$l = \ae$	$l = [n/2]$	$l = n - 1$
6	0.46	0.47	0.54	0.44	0.44	0.53
9	0.46	0.47	0.58	0.43	0.44	0.61
12	0.45	0.48	0.6	0.43	0.44	0.69

The K^0 value corresponds to expression (3.49) from the Sect. 3.3. We see that in most of the cases constant $K_{\ae} \approx 0.5$.

$$\sigma_{nl\pm1,nl} = 6(\pi a_0^2 n^4 / Z^2) \frac{\max(l, l \pm 1)}{(2l + 1)} \frac{Ry}{E} \ln \left[(c/n^2) \frac{\sqrt{E/Ry}}{\Delta E/Ry} \right], \quad c = 1.93.$$

$$(8.56)$$

Similar formulas may also be obtained from the Born calculations [8.9] and [8.21], a slight difference being the constants inside the logarithmic functions $c = 1.36$ in [8.21] and $c = 1.7$ in [8.9].

In the case of transitions with multipole $\Delta l \geq 2$ the lower limit of the integral (8.51) may be put to zero at large energies, and the cross section is proportional to $1/E$. The coefficient is determined by the integral:

$$I_{\varkappa} = \iint dr dr' \rho(r) \rho(r') \int_0^\infty \frac{dQ}{Q^3} j_{\varkappa}(Qr) \cdot j_{\varkappa}(Qr'),$$

$$(8.57)$$

$$\rho(r) = r^2 \mathcal{R}_{nl}(r) \mathcal{R}_{nl'}(r).$$

It is seen that the region $r \approx r'$ dominates, since outside this region the integral is oscillating. The internal integral may be expressed in closed form and according to [8.30] we have

$$I_{\varkappa} = \iint dr dr' \rho(r) \rho(r') rr' f_{\varkappa}(r, r'),$$

$$f_{\varkappa}(r, r') = \frac{\pi}{32\Gamma(5/2)} \frac{\Gamma(\varkappa + 1)}{\Gamma(\varkappa + 3/2)} \frac{(rr')^{\varkappa - 1}}{(r + r')^{2\varkappa - 2}} \qquad (8.58)$$

$$F \left[\varkappa - 1, \varkappa + 2, 2\varkappa + 2; 4rr'/(r + r')^2 \right].$$

The asymptotic analysis shows, that the main term over $\varkappa$ is proportional to $\Gamma(\varkappa - 1)/\Gamma(\varkappa + 3/2)] \cdot \varkappa^{-3/2} \sim \varkappa^{-4}$. Therefore, we find $I_{\varkappa}$ in the form:

$$I_{\varkappa} \approx \Lambda_{\varkappa}^2 K_{\varkappa}(n, l) \left(\frac{n}{\varkappa}\right)^4, \quad \Lambda_{\varkappa}^2 = \prod_{i=0}^{\varkappa - 1} \left[1 - \left(\frac{l - i}{n}\right)^2 \right], \qquad (8.59)$$

where the $K_{\varkappa}(n, l)$ is the weak function of $\varkappa, n, l$. We present numerical calculations of $K_{\varkappa}$ for the values of $\varkappa = 2, 3$ and some values of n, l in the Table 8.1.

The Born approximation may be justified at large velocities of the incident particles $(V \gg v_0)$. Nevertheless, the surprising simplicity of the Born approximation has made attractive the idea to extrapolate this approach to the region of mean velocities (energies). However, direct extrapolation leads to a bad result in a number of cases. One of the reason is that, the corresponding Born transition probability may be greater than unity. It is possible to modify the Born approach to provide a physically reasonable inequality: the transition probability must be less than unity. One of these improvements is the K-matrix method, which is rather fruitful for the processes between the low excited levels. The simple modifications of the Born approach based on changing the Born probability W^B to "normalized" W^N in the impact parameter representation and the corresponding extension to the momentum transfer representation are given by formulas (4.48, 58) from the Sect. 4.4. In opposite of their simplicity, these modifications improve significantly the results for the transitions between the near levels.

8.3.2 Close Coupling Method

In this section we consider the close coupling method for the transition of an atom from then nl to then nl' state $(l' > 1)$. We use the impact parameter representation. The cross sections are determined by the region of large distances between the incident particle and the atom: $\rho \gg n^2 a_0$. We consider only the dipole interaction, which is dominant at such distances. Since the energy splitting of the levels decreases rapidly with increasing l, we ignore the energy difference between level $l + 1$ and states $l' > l + 1$. The close coupling system is (ω is the frequency of the transition between the l level and other levels, $l' > l_0$)

$$i\hbar \dot{a}_{l,m}(t) = \sum_{m'} e^{-i\omega t} V_{l+1,m;l,m'} a_{l,m}(t), \quad |m| \leq 1,$$

$$i\hbar \dot{a}_{l+1,m}(t) = \sum_{m'} e^{i\omega t} V_{l+1,m;l,m'} a_{l,m}(t)$$

$$+ \sum_{m'} V_{l+1,m;l+2,m'} a_{l+2,m}(t), \quad |m| \leq l+1,$$

$$i\hbar \dot{a}_{l',m}(t) = \sum_{m'} V_{l',m;l'-1,m'} a_{l'-1,m}(t)$$

$$+ \sum_{m'} V_{l',m;l'+1,m'} a_{l'+1,m}(t), \quad l' > l, \ |m'| \leq l'$$

$$V_{l',m';l,m} = \delta_{l',l\pm1} \{ \langle l', m'|z|l, m \rangle \rho \delta_{m',m} + \langle l', m'|x|l, m \rangle (vt)\delta_{m',m\pm1} \}/R^3(t),$$

$$\langle l', m|z|l, m \rangle = \left[\frac{l_m^2 - m^2}{(2l_m + 1)(2l_m - 1)} \right]^{1/2} \frac{3}{2} n^2 \Lambda_1(a_0/Z),$$

Table 8.2. Cross sections for 28*d*–28*l* transitions and the total *l*-mixing cross section σ_t in the Na atom induced by collisions with charged particles ($Z = 1$). (The values $\sigma/10^4 \pi a_0^2 n^4$ are presented, $x = Vn/v_0$)

l	$x = 0.5$ compl	av	$x = 1.$ compl	av	$x = 2.$ compl	av
f	0.094	0.08	0.223	0.202	0.152	0.141
g	0.124	0.115	0.069	0.075	0.017	0.018
h	0.1	0.152	0.025	0.043	0.005	0.009
tot	0.318	0.347	0.317	0.32	0.174	0.168

$$\langle l', m' | x | l, m \rangle = (\mu\nu/2) \left[\frac{(l_m + \nu + \mu\nu m)(l_m + \mu\nu m)}{(2l_m + 1)(2l_m - 1)} \right]^{1/2} \frac{3}{2} n^2 \Lambda_1(a_0/Z),$$

$$\mu \operatorname{sign}(m' - m), \quad \nu = \operatorname{sign}(l' - l), \quad l_m = \max(l, l'). \tag{8.60}$$

This system should be solved under the initial condition

$$a_{l', m'}(-\infty) = \delta_{l'l} \cdot \delta_{m'm}$$

and then averaged over m. Even at small values of n, there are many equations in (8.60). The dependence on magnetic quantum numbers is not of major importance in this problem. We will accordingly study the system of equations "averaged over m":

$$i\hbar\dot{a}_l(t) = e^{-i\omega t} V_{l,l+1} a_{l+1}(t),$$
$$i\hbar\dot{a}_{l+1}(t) = e^{i\omega t} V_{l+1,l} a_l(t) + V_{l+1,l+2} a_{l+2}(t), \tag{8.61}$$
$$i\hbar\dot{a}_{l'}(t) = V_{l',l'-1} a_{l'-1}(t) + V_{l',l'+1} a_{l'+1}(t),$$

with the initial condition $a_{l'}(-\infty) = \delta_{l',l}$ and the "averaged" potential

$$V_{l,l-1} = \langle d \rangle \cdot (\rho + vt)/(\rho^2 + v^2 t^2)^{3/2}, \tag{8.62}$$

where the averaged dipole moment $\langle d \rangle$ is defined by

$$\langle d \rangle = \left[\frac{1}{2l + 1} \sum_m | \langle l \pm 1, m | ez | l, m \rangle |^2 \right] = \Lambda_1 n^2 \cdot e a_0 \left[\frac{3}{4} \frac{l_>}{2l + 1} \right]. \tag{8.63}$$

This expression for the potential gives a qualitatively correct description of the interaction at large distances, and a first-order perturbation theory treatment of a system with this potential gives an exact expression for the transition probability. To illustrate the situation, Table 8.2 shows the cross sections for transitions from 28*d* level of Na which have been calculated from the complete system (8.60) and from the averaged system (8.61) for four values of *l*.

The difference between the results is seen to be insignificant for the total cross section and for the transition $d \to f$, while for $d \to g$ and $d \to h$, system (8.61) still describes the basic qualitative features. Below we use the average system (8.61) which is simpler.

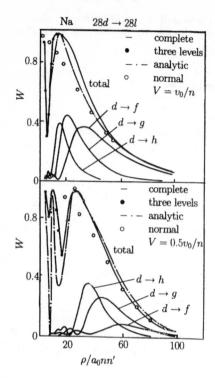

Fig. 8.7. Probabilities for the l-mixing transitions from the 28d state of Na, induced by collision with singly charged ion. Close coupling calculations represent the results for complete system involving 26 levels, and for three level system

Figure 8.7 shows the probabilities for 28d–28l transitions versus impact parameter ρ for velocities $V = v_n$ and $V = (1/2)v_n$ of the external particle according to the solution of system (8.61) for all $l \geq 2$ (26 levels). We see that in the region of impact parameters dominating the cross section, the normalized Born approximation for the $d \rightarrow f$ actually describes not transition $d \rightarrow f$ but the total probability for transitions from the d level. At large values of ρ, as expected, the probability for transition $d \rightarrow f$ which is found from system (8.61) agrees with the Born probability. We also see that in the region $\rho < 30n^2a_0$, where the dipole interaction is still dominant, the probability for the transition with $\Delta l > 1$ exceeds the probability for the direct transition $d \rightarrow f$. It should be noted that relative importance of transitions with $\Delta l > 1$ in the overall probability increases with decreasing V.

We can also show that the total transition probability (for all Δl) down to velocities $x \geq 0.3$ can be described simply by a three-level system, and we give an approximate analytic solution for this system. Let us consider a system consisting of the three levels $l, l+1$ and $l+2$. Diagonalizing the interaction with respect to the levels $l+1$ and $l+2$ by means of the transformation

$$|0\rangle = |l\rangle, \quad |-\rangle = (1/\sqrt{2})(-|l+1\rangle + |l+2\rangle),$$
$$|+\rangle = (1/\sqrt{2})(|l+1\rangle + |l+2\rangle),$$

$$(8.64)$$

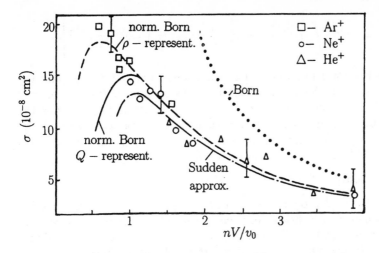

Fig. 8.8. Total cross section for transitions $28d \rightarrow 28l$, (summed over $l > 2$) in Na induced by collision with singly charged ion. The points are experimental data from [8.29]. Born approximation and normalized Born approximation in ρ and Q representations are from [8.10]. Sudden approximation are calculations [8.5, 9]

and ignoring the difference between $\mathcal{V}_{l,l+1}$ and $\mathcal{V}_{l+1,l+2}$, we can rewrite (8.64) as follows, introducing dimensionless distance $\tau = Vt/\rho$ and parameter $\lambda = \langle d \rangle /\rho V$:

$$i\dot{a}_0(\tau) = (1/\sqrt{2})\mathcal{V}(\tau)\{-a_-(\tau) \cdot \exp[-i\Phi_-(\tau)] + a_+(\tau) \cdot \exp[-i\Phi_+(\tau)]\},$$
$$i\dot{a}_+(\tau) = (1/\sqrt{2})\mathcal{V}(\tau)a_0(\tau) \cdot \exp[i\Phi_+(\tau)],$$
$$i\dot{a}_-(\tau) = -(1/\sqrt{2})\mathcal{V}(\tau)a_0(\tau) \cdot \exp[i\Phi_-(\tau)],$$
$$\Phi_\pm(\tau) = \int_0^\infty \alpha_\pm(\tau')d\tau', \ \alpha_\pm(\tau') = \beta \pm \mathcal{V}(\tau),$$
$$\mathcal{V}(\tau) = \lambda(1 + \tau)/(1 + \tau^2)^{3/2}. \tag{8.65}$$

This system describes the interaction of the $|0\rangle$ level with the levels $|+\rangle$ and $|-\rangle$, which do not interact with each other.

The total probability for transitions from the $28d$ level found through a numerical solution of system (8.64) is shown in Fig. 8.8 In the ρ region which dominates the cross section, this probability is essentially equal to the corresponding probability found from (8.61).

An approximate analytic solution of system (8.64) using an asymptotic expansion in parameter λ was derived in [8.22]. It can be described by

$$W_t = W_1 + W_2, \ W_{1,2} = w_\pm(1 - w_\pm)/P,$$
$$P = w_+(1 - w_-) + w_-(1 - w_+) + (1 - w_-)(1 - w_+), \tag{8.66}$$

$$w_\pm = \sin^2 \left| \int\limits_{-\infty}^{\infty} d\tau \frac{\mathcal{V}(\tau)}{\sqrt{2}} \exp\left\{ i \int\limits_{0}^{\tau} d\tau' [\alpha_\pm^2(\tau') + 2\mathcal{V}^2(\tau')]^{1/2} \right\} \right|,$$

where W_t is the total probability for transitions from the d level, and the probabilities $w_\pm$ are the two-level probabilities for the interaction of only two states. It is rather difficult to evaluate the integrals in (8.66) for the potential (8.62). An approximate analytic expression for $w_\pm$ giving results for W_t which are in satisfactory agreement with the numerical solutions of the system (8.65) can be written in the form

$$w_\pm = \sin^2[(1/2)W_B^\pm(\rho, v)].$$

Here $W_B^\pm$ is the Born probability, where β is replaced by $\beta_\pm = [(\beta \pm \lambda)^2 + 2(\gamma\lambda)^2]^{1/2}$, where γ is a fitting parameter ($\gamma \approx 0.64$). If $\gamma\lambda > \beta$, we can set $\beta_- = \sqrt{2}\,\beta$. As can be seen from Fig. 8.7, the total transition probability found in this manner reproduce quite well the probability found from (8.61), (8.65). As a result, the corresponding cross section is essentially the same as that found from complete system (8.61).

8.3.3 Method of Effective Magnetic Field

Within the dipole approximation for the interaction potential, the Rydberg atom is affected by the uniform electric field $\mathbf{F}(t)$. This approach was suggested in [8.23]. We introduce the rotating frame which fixes the direction of the electric field, but the Coriolis force emerges. According to the Larmor theorem it is equivalent to the effective magnetic field $H(t)$. It is important that both fields (and precession frequency Ω) have the same time dependence

$$F(t) = \mathcal{Z}e^2/R^2(t), \quad \Omega = \rho V/R^2(t), \quad H = \frac{2mc}{e}\Omega. \tag{8.67}$$

The vector $\mathbf{\dot{F}}$ is directed along the internuclear radius vector, and $\mathbf{H}$ is parallel to Ω and perpendicular to the collision plane.

The problem can be solved in closed form in the framework of both classical and quantum mechanics for the case of the hydrogen atom. The additional symmetry of the Coulomb field allows us to introduce a new motion integral in addition to angular momentum. In classical mechanics it is the Runge-Lentz vector (see, for example, [2.2]):

$$\mathbf{A}(t) = \mathcal{Z}e^2\frac{\mathbf{r}}{r} - [\mathbf{p}, \mathbf{l}], \tag{8.68}$$

which corresponds to the quantum mechanical operator:

$$\mathbf{\hat{A}}(t) = \mathcal{Z}e^2\frac{\mathbf{r}}{r} - \frac{1}{2}\left([\mathbf{\hat{p}}, \mathbf{\hat{l}}] - [\mathbf{\hat{l}}, \mathbf{\hat{p}}]\right). \tag{8.69}$$

According to [8.24, 25] evolution of the atom in cross electric and magnetic fields can be described as the uniform precession of the vectors:

$$\mathbf{J}_1 = \frac{1}{2}(1 + \mathbf{A}), \quad \mathbf{J}_2 = \frac{1}{2}(1 - \mathbf{A}),$$

in terms of effective time, which is the angle Φ of rotation of the internuclear vector $\mathbf{R}(t)$:

$$\Phi = \int\limits_{-\infty}^{t} \Omega(t') \, dt'.$$

The axes of precession are directed along the vectors:

$$\boldsymbol{\omega}_1 = \frac{3}{2}n\mathbf{F} + \boldsymbol{\Omega}; \quad \boldsymbol{\omega}_2 = \frac{3}{2}n\mathbf{F} - \boldsymbol{\Omega}.$$

In the framework of classical mechanics a simple solution was obtained for transitions from s-states in [8.26]. The differential cross-section of transition from s-state to state with angular momentum l ($l = \ell n$) and projection of angular momentum m ($m = \mu l$) is given by:

$$\frac{d\sigma}{d\ell d\mu} = 2\pi \int\limits_{0}^{\infty} W^{\mathrm{cl}}(\rho, \ell, \mu)\, \rho d\rho, \quad \frac{d\sigma}{d\ell} = 2\pi \int\limits_{0}^{\infty} W^{\mathrm{cl}}(\rho, \ell)\, \rho d\rho,$$

$$W^{\mathrm{cl}}(\rho, \ell, \mu) = W^{\mathrm{cl}}(\rho, \ell) \cdot \frac{1}{\pi}\frac{1}{\sqrt{\mu_m^2 - \mu^2}}, \quad W^{\mathrm{cl}}(\rho, \ell) = \frac{\ell}{\ell_m\sqrt{\ell_m^2 - \ell^2}},$$

$$(8.70)$$

where values ℓ_m and μ_m are maximal possible angular momentum transfer and projection of angular momentum transfer, respectively, at fixed velocity and impact parameter of the projectile V, ρ. It is suggested that classical probabilities W^{cl} are equal to zero if $\ell > \ell_m$ or $\mu > \mu_m$. The values ℓ_m, μ_m are functions of one dimensionless parameter

$$\eta = \frac{3Zn}{Z} \cdot \frac{a_0 v_0}{\rho V}$$

and are determined by relations:

$$\ell_m = 2\sin\gamma \, |\sin(\varpi/2)| \, \sqrt{\cos^2\gamma \sin^2(\varpi/2) + \cos^2(\varpi/2)}, \quad \mu_m = \cos\zeta,$$

$$\tan\gamma = \eta, \quad \varpi = \Delta\Phi/\cos\gamma, \quad \tan\zeta = \cot(\varpi/2)/\cos\gamma. \qquad (8.71)$$

8.3.4 Fitting Formulas

The total l-mixing cross section δ of transitions from nl level to all levels with $l' > l$, as was shown above, in the case of $\delta \equiv \delta_l \gg \delta_{l'}$ are satisfactory described by the normalized Born approximation with dipole potential. The results for neutral Rydberg atom may be fitted by the formulas [8.27]:

$$\sigma_t = \pi a_0^2 n^4 \cdot F \cdot D \cdot \ln[1 + (x/\delta)^2/(1+D)], \quad D = (2d_0n/x)^2,$$
$$d_0 = \{3/4 \cdot [1 - (l+1)^2/n^2](l+1)/(2l+1)\}^{1/2}, \quad x = Vn/v_0, \tag{8.72}$$
$$F = 1 + 0.19 \cdot \ln^2[1 + 3.55(x_m/x)^4], \quad x_m = (2d_0n\delta)^{1/2}$$

where δ is the quantum defect of the nl state (we assume that the quantum defects of the l' states may be neglected), x is the reduced velocity of the incident particle, x_m corresponds to the position of the cross section maximum. In this formula and below, up to the end of this Chapter we will suggest Rydberg atom is neutral ($Z = 1$).

The cross section of the given angular momentum transfer ($\sigma_{nl', nl}$) is described by the model with infinite number of levels presented in the previous section. The numerical results may be fitted by formulas ($\Delta l = |l - l'|$):

$$\sigma_{nl', nl} = \begin{cases} \sigma_t/\{1 + 0.19[(x_m/x)^8 + 2(x_m/x)^4](x/d_0n)^2(\sigma_t/\pi a_0^2 n^4)\}, & \Delta l = 1, \\[2ex] \pi a_0^2 n^4 (d_0n/x)^2(24/\Delta l)^3 \dfrac{1 + a(x/n)^2}{1 + (x_{\Delta l}/x)^8 + (x_{\Delta l}/x) + (x/n)^2}, & \Delta l > 1, \end{cases}$$
$$\tag{8.73}$$

where

$$a = (0.14/d_0^2 \Delta l)(2\Delta l + 1)(2l + 1) \begin{pmatrix} l & l' & \Delta l \\ 0 & 0 & 0 \end{pmatrix},$$
$$x_{\Delta l} = (1.2/\Delta l^{5/8})(2d_0n\delta)^{1/2},$$

whereas $x_{\Delta l}$ corresponds to the position of the cross section maximum. Formulas (8.73) include the Born results as the limiting case of large velocities.

The total rate coefficient $\langle V\sigma_t \rangle$ for transition from nl level to all levels with $l' > l$ is obtained by averaging (8.73) over the Maxwellian distribution. The related fitting formulas are

$$\langle V\sigma_t \rangle = \pi a_0^2 n^4 v_0 \frac{8d_0^2}{(\pi\theta)^{1/2}g(\theta)\ln[1 + 0.54(n\theta\delta)^2/(1+\theta)]},$$
$$g(\theta) = \frac{1 + 0.09(\delta/d_0^3 n\theta)^2}{1 + 0.054(\delta/d_0^3 n\theta)^{3/2}}, \quad \theta = (m/\mu)(kT/Ry), \tag{8.74}$$

where T is the temperature, and μ is the reduced mass of the perturbing particles and Rydberg atoms.

Similarly, the rate coefficients for the given angular momentum transfer ($\langle V\sigma_{nl', nl} \rangle$) may be fitted by formulas:

$$\langle v\sigma_{nl', nl} \rangle = \begin{cases} \dfrac{\langle V\sigma_t \rangle}{1 + \pi y^2/2 + (0.008y^8 + 0.009y^7)(\sqrt{\pi\theta}/2d_0^2)(\langle V\sigma_t \rangle/\pi a_0^2 n^4 v_0)}, & \Delta l = 1 \\[3ex] \pi a_0^2 n^4 v_0 \cdot 2d_0^2(\pi\theta)^{-1/2}\left(\dfrac{24}{\Delta l}\right)^3 \dfrac{1 + a\theta^2}{1 + \theta^2 + \sqrt{\pi}y_1 + 0.072y_1^7 + 0.055y_1^8}, & \Delta l > 1 \end{cases}$$
$$y = (2d_0\delta/n\theta)^{1/2}, \quad y_1 = 1.17y/(\Delta l)^{5/8};$$

$$\tag{8.75}$$

8.3.5 Comparison with Experiment

An extensive development of the collision theory of the Rydberg atoms with charged particles began in the early seventies. It was stimulated both by internal theoretical requirements and astrophysical applications. However, the first direct measurements of the inelastic Rydberg atom transition cross sections induced by charged particles were published by *MacAdam* et al. [8.8] in 1980. Later on, MacAdam and coworkers (see paper [8.11] and references therein) developed experimental apparatus, improved recording techniques, and methods of control. In particular, since the Rydberg atoms are extremely sensitive to electric fields, original methods of the reduction and measurement of the stray fields were developed. Thus far, all direct collisional experiments concerning inelastic transitions between Rydberg states induced by charged particles have been carried out by MacAdam's group.

The scheme of the experiments [8.8, 11] has been as follows. A sodium atomic beam was photoexcited to nd states $(21 \leq n \leq 28)$ by two pulsed dye lasers. The heavy ion (Na^+, Ar^+, He^+) beam passed between two parallel plates that formed a condenser around a small volume of the $Na(nd)$ Rydberg atoms. The ion beam covered the entire Rydberg target volume and was received in a Faraday cup located beyond the target. A variable slope linear positive high voltage ramp was applied to the upper plate. Time resolved selective field ionization was used to analyze distribution of the final states. The apparatus was described in detail in [8.8, 11].

Measurements of the inelastic width due to l-changing induced by slow electrons were carried out by *Sorokin* et al. [8.28].

a) Total Cross Section: Dependence on Energy. Figure 8.8 shows the comparison of various theoretical total cross sections

$$\sum_{l>2} \sigma(28l \leftarrow 28d)$$

for the Na atom with experimental data from [8.29]. This cross section is actually determined by the dipole transition. We might note that the Born calculations from (8.55) of Sect. 8.2 yield results which are essentially the same as the dipole Born cross sections from [8.30]. We see that the Born cross section overestimates the result and that, as expected for transitions between highly excited states, the Born approximation is good only if the dimensionless velocity of the perturbing particle satisfies $x = V/v_n \gg 1$.

Figure 8.8 shows the Born cross sections normalized in the ρ and Q- representations according to expressions (4.48, 58) of Sect. 4.5.1 and the sudden approximation from [8.5, 6]. We see that these cross sections give a satisfactory description of the experimental data at velocities down to $V \sim v_n$.

Figure 8.9 shows the cross sections for the transitions $d \to f$, $d \to g$, and $d \to h$ for $n = 28$ versus velocity, in comparison with the total cross section from experiments. Over the entire energy range which has been studied experimentally, the total cross section calculated from system (8.61) agrees with

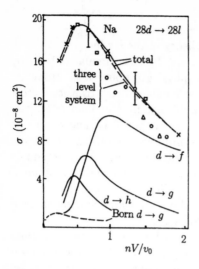

Fig. 8.9. Cross sections for transitions from the 28d state of Na induced by collision with singly charged ion. Total cross section for transitions $28d \rightarrow 28l$, with $l > 2$: the points are experimental data [8.29]. Full curve is the close coupling calculations involving 26 levels

the experimental data. At low velocities, however, the total cross section is dominated by transitions with $\Delta l > 1$. In particular, at $V = 0.5v_n$ the cross section for transition $d \rightarrow g$ is greater than that for transition $d \rightarrow f$. For comparison, Fig. 8.9 shows the Born cross section for transition $d \rightarrow g$; we see that it can be ignored.

There is also an experiment on fast electron collision with the Rydberg atoms [8.15] too. The cross sections of $30s - 30p$ and $30s - 29p$ transitions in Na induced by electrons with energy from 105 to 660 eV are given in [8.15]. In this energy region the usual Born approximation is valid. Really, the Born results are in agreement with the experimental ones within experimental errors.

b) Total Cross Section: Dependence on Principal Quantum Number. At charged particle velocities above the average velocity of an atomic electron $V > v_n$, the total cross section for transitions from the nl level is, as we have shown above, described by the normalized Born cross section. Figure 8.10 compares theoretical and experimental results on the n dependence of the cross sections in the interval 21–28 from [8.29]. The n dependence of the cross sections was approximated in [8.29] by a function n^α, where $\alpha = 5.41 \pm 0.13$, 5.04 ± 0.16 and 5.15 ± 0.17 for charged particle energies of 400, 1000, and 2000 eV, respectively. We see that, within the experimental errors we can assume $\sigma \sim n^4 \ln(Cn)$ ($C \sim V^2/\delta$, where δ is the quantum defect of the d-states) in accordance with the normalized Born approximation.

The broadening of the Rydberg levels by the thermal electrons was investigated in experiments [8.28]. For low temperatures of the plasma and moderately large principal quantum numbers, the width of the Rydberg state is determined by the total l-changing cross section averaged over the thermal (usually, Maxwellian) electron distribution. These values (rate coefficients of

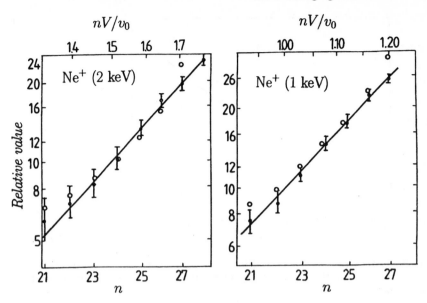

Fig. 8.10. Total cross sections for transitions from nd levels of Na as a function of n for fixed beam energy of Ne$^+$. Filled circles are experimental data [8.29]. Full curves are the experimental dependence obtained by the mean square method [8.29]. Open circles are normalized Born approximation

the inelastic broadening) for nf states of cesium ($13 < n < 20$) in a thermal low-temperature plasma were measured in [8.28]. The Cesium vapor was heated in a heat tube up to $1500\,\mathrm{K}$ and then the width of the absorption $5D_{5/2} - nf$ lines were measured by the Faraday spectroscopy method. The magnetic field was equal to 100 Gauss. The experimental setup is described in detail in paper [8.31]. The width of the nf levels dominates in the total line width. The electron concentration was calculated from the thermal equilibrium condition. Since the electron concentration is very sensitive to the value of temperature ($10\,\mathrm{K}$ temperature change corresponds to a 20% concentration change), the electron concentration was additionally controlled by the line asymmetry measurements. The total error is determined by the uncertainty of the electron concentration and has been estimated as 20%. Figure 8.11 shows the comparison of the experimental rate coefficients with the theoretical results.

c) Angular Momentum Dependence of the Final-State Distribution.
The *l*-changing cross section dependence on the angular momentum dependence of the final states is the critical feature of the theory. The perturbation theory or various perturbation theory modifications (for example, normalization) yield the cross section which decreases monotonically with increasing angular momentum transfer (Δl) at all velocities. The specific feature of the close coupling theory is the "redistribution" over angular momentum of the

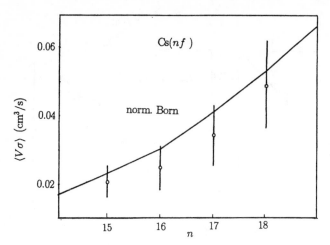

Fig. 8.11. Electron impact broadening rate coefficients for nf states of Cs. Experimental data are those from [8.27]; Full curve corresponds to normalized Born approximation

final states at small velocities of the charged particles ($v < v_n$). The close coupling theory predicts, it has been mentioned above that the cross sections with large angular momentum transfer and velocities $V < v_n$ increase drastically and grow up to values more than the allowed transition cross sections (with $\Delta l = 1$). It should be noted, that no similar effect has been observed for the n-changing cross sections. We have seen above that there is a region of the impact parameter where the probability of the $\Delta n > 1$ transitions may exceed that of the $\Delta n = 1$ transitions. However, integration over the impact parameter gives cross sections of the $\Delta n > 1$ transition being less than the corresponding cross sections with $\Delta n = 1$. Likely, the reason of the dramatic difference between l- changing and n- changing cross sections is the extremely small resonance defect for the l-changing in opposite to the n-changing processes. That is why the angular momentum dependence of the l- changing cross section at small velocities is of great interest for checking the theory.

For the first time, measurements of the dependence of the final l-states distribution under single-collision conditions were reported in [8.32]. Effects of close coupling were found, i.e., growing significance of the $\Delta l > 1$ transition in the final state distribution for slow collisions on Na($28d$) was confirmed. A comparison of the experimental results with theoretical calculations gave qualitative agreement. However, there was significant quantitative disagreement, in particular, for $V/v_n < 1$ the observed final populations of $28g$ did not exceed $28f$ as theory had predicted.

Various attempts to refine the close coupling theory had been made. Using the nonaveraged system results in a drastic increase of computer time, but the cross sections thus obtained did not change significantly. The nondipole

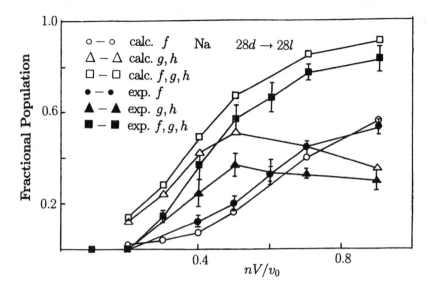

Fig. 8.12. Comparison of experimental fractional populations [8.11] with calculations from the complete system (26 levels) for *l*-changing induced by $Na^+ + Na(28d)$ collisions

interactions were included into the equation system, but at velocities $V < v_n$ the results were the same. It is accounted for by the large impact parameters $(\rho > a_0 n^2)$ dominating in the cross section, and the dipole approximation is valid at such large ρ. Only an additional weak electric field may smooth the distribution over final angular momenta and increase the cross sections with small angular momentum transfer (Δl), due to mixing of the various *l*-states.

A detailed study of dipole forbidden $(\Delta l > 1)$*l*-changing slow collisions of Na^+ with $Na(26d)$ and $Na(28d)$ Rydberg atoms has been reported in [8.11]. The f, g and h state populations for the targets $Na(26d)$ and $Na(28d)$ were measured. The techniques developed in this study, such as constantly heating the plates, using soot-covered plates and microwave resonance to monitor the stray electric fields were proved successful. An experimental procedure to detect and correct the stray electric field was established.

A comparison of the theory with experiment for the process

$$Na^+ + Na(28d) \rightarrow Na^+ + Na(28l, \; l > 3)$$

is shown in Fig. 8.12. The measured fractional f-state populations for different reduced velocities (V/v_n) are in good agreement with calculations. The observed fractional populations of the sum of g and h states are smaller than the calculated results by about 15%, but they both reach maximum values at the same velocity $V = 0.5v_n$. The observed fractional populations of the sum of f, g, and h states are smaller than the calculation results in average by less than 10%.

9. Spectral-Line Broadening and Shift

This chapter is devoted to the impact broadening and shift of spectral lines corresponding to radiative transitions involving Rydberg states. We give a brief account of the semiclassical and quantum approaches to impact broadening and shift and consider various mechanisms of identified processes based on the results of the collision theory presented in the preceding chapters. Main attention is paid to studies of the broadening width and shift behavior for Rydberg atomic series perturbed by thermal collisions with the rare gas and alkali-metal atoms, for which there are extensive experimental and theoretical data. Another goal is to describe the spectral-line broadening for transitions between highly excited levels induced by inelastic electron collisions in a plasma.

9.1 Classical and Quantum Treatments of Impact Broadening

There are a number of physical effects leading to the broadening and shift of spectral lines: radiative damping, Doppler effect, interaction of the atom with external fields and perturbing particles. Here our attention will be focused only on collisional broadening and shift processes. It is explained by two reasons. First, collisional mechanism is predominant in the most of real observations of spectral lines from Rydberg levels. Second, the considerable part of available experimental data about Rydberg atoms and their interaction with environment was obtained from the analysis of impact broadening and shift of spectral lines. We briefly remember the main results of impact approximation in the theory of line broadening and shift. A detailed consideration of general impact broadening theory and its validity conditions is given in monographs [3.8] and reviews [9.1, 2].

9.1.1 Impact-Parameter Method

We start the theoretical description of impact broadening processes by considering a widely used time-dependent approach in the impact-parameter representation which is similar to the method presented in Sect. 4.3. It allows us to express the width and shift of spectral lines in terms of the collision theory quantities such as the scattering phase shifts and S-matrix, which are functions of the impact parameter ρ and relative velocity of colliding particles V. Within the framework of this approach the relative motion of the atom and perturbing particle is usually taken following the straight-line trajectory, so that the interaction $\mathcal{V}$ between the atom and perturbing particle depends on time t through the position vector $\mathbf{R}(t) = \rho + \mathbf{V}t$. The impact approximation is applicable if the major contribution to the broadening of spectral line is a result of only binary interactions with the nearest perturbing particle and the multiparticle interactions do not play any significant role in contrast to the quasi-static approximation (see [3.8, 9.1] for more details).

a) **Model of Classical Oscillator with Random Phase.** The main ideas of the impact approximation can be illustrated within the framework of a simple model of the classical oscillator. In this model the frequency ω of the atomic oscillator depends on time as a result of the interaction with a perturbing particle. Let $f(t)$ describe a classical oscillator with unperturbed frequency ω_0 and random phase of oscillations $\eta(t)$:

$$f(t) = \exp\left[i\omega_0 t + i\eta(t)\right], \quad \eta(t) = \int_{-\infty}^{t} \delta(t')\, dt'. \tag{9.1}$$

Here $\delta(t')$ denotes the frequency shift induced by collision with the perturbing particle. The spectrum of this process [the intensity distribution function $I(\omega)$] can be expressed in terms of the correlation function $\Phi(\tau)$:

$$
\begin{aligned}
I(\omega) &= \lim_{T\to\infty} \frac{1}{2\pi T} \left| \int_{-T/2}^{T/2} f(t) \exp\left(-i\omega t\right) dt \right|^2 \\
&= \frac{1}{\pi} \mathbf{Re} \left\{ \int_{0}^{\infty} \Phi(\tau) \exp\left(-i\left(\omega - \omega_0\right)\tau\right) d\tau \right\},
\end{aligned}
\tag{9.2}
$$

which is given by

$$\Phi(\tau) = \lim_{T\to\infty} \frac{1}{T} \int_{-T/2}^{T/2} f^*(t) f(t+\tau)\, dt = \langle f^*(t) f(t+\tau) \rangle. \tag{9.3}$$

It is important to stress that according to the ergodic hypothesis time averaging can be replaced by averaging over the statistical ensemble, which is denoted by angle brackets.

Let us consider a perturbation of the oscillator by instantaneous collisions following the *Anderson* [9.3] model. If the collision duration $\Delta\tau$ is much smaller than the time between the collisions, this perturbation is equivalent to a random phase jump. Each phase jump $\Delta\eta$ induced by a collision during time interval $\Delta\tau$ yields the increment of the correlation function

$$\Delta\Phi(\tau) = \langle[\exp(i\Delta\eta) - 1]\exp[i\eta(\tau)]\rangle. \tag{9.4}$$

The next important step consists of separate averaging of two exponential factors in (9.4) that is possible because they are statistically independent from each other [i.e., the phase jump $\Delta\eta$ is not dependent on the value of $\eta(\tau)$]. Hence, it can be rewritten as

$$\Delta\Phi(\tau) = -\Phi(\tau)\langle 1 - \exp[i\Delta\eta(\tau)]\rangle. \tag{9.5}$$

For given values of ρ and V and the density of perturbing particles N, the number of collisions during the time interval $\Delta\tau$ is given by $[NVW(V)\,dV2\pi\rho d\rho]\,\Delta\tau$, where $W(V)$ is the relative velocity distribution function (e.g., the Maxwellian distribution for a given temperature T of the perturbing particles). As a result, the required equation for the correlation function is

$$\frac{d\Phi}{d\tau} = -N\langle V\sigma\rangle\Phi. \tag{9.6}$$

It has the following solution:

$$\Phi(\tau) = \exp[-N\langle V\sigma\rangle\tau], \quad \langle V\sigma\rangle = \int_0^\infty V\sigma(V)W(V)\,dV. \tag{9.7}$$

Here the angular brackets mean averaging over the velocity distribution, and the effective cross section σ is determined by the integral

$$\sigma(V) = 2\pi\int_0^\infty \{1 - \exp[i\eta(\rho, V)]\}\rho d\rho. \tag{9.8}$$

As is evident from (9.8), the cross section σ contains the real and imaginary parts

$$\sigma = \sigma^{br} - i\sigma^{sh},$$

$$\sigma^{br}(V) = 2\pi\int_0^\infty \{1 - \cos[\eta(\rho, V)]\}\rho d\rho, \tag{9.9}$$

$$\sigma^{sh}(V) = 2\pi\int_0^\infty \sin[\eta(\rho, V)]\rho d\rho,$$

which are the broadening σ^{br} and shift σ^{sh} effective cross sections, respectively.

Substitution of result (9.7) into (9.2) leads to the Lorentzian shape for the intensity distribution function

$$I\left(\omega\right) = \frac{1}{\pi}\frac{\left(\gamma/2\right)}{\left(\omega - \omega_0 - \Delta\right)^2 + \left(\gamma/2\right)^2} , \tag{9.10}$$

$$\gamma = 2N\left\langle V\sigma^{\mathrm{br}}\right\rangle, \quad \Delta = N\left\langle V\sigma^{\mathrm{sh}}\right\rangle, \tag{9.11}$$

The width of this distribution (at half-height) is called the impact width γ. The peak is shifted from the unperturbed frequency ω_0 on the Δ value. It should be noted that the Lorentzian distribution is universal for the impact mechanism of broadening. The specific features of the interaction determine only the values of the width γ and shift Δ, but do not change the form of the distribution.

We have suggested above that the random phase $\eta\left(t\right)$ is real that corresponds to the elastic scattering. Extension to the inelastic scattering may be given using the complex phase, imaginary part of which describes the oscillator damping. Further we shall express the impact broadening and shift cross sections in terms of the quantities of the collision theory. At first we present the semiclassical formulas for the broadening and shift cross sections within the framework of the general impact parameter approach, and then give the final results obtained using a consistent quantum treatment of spectral-line broadening.

b) Expressions for Broadening and Shift through S-Matrix. According to the general impact-parameter approach the relative motion of colliding particles is considered as a classical with given magnitude of ρ and V, while the atom is described in terms of the quantum-mechanical values. Usually, the density matrix technique is used for description of quantum atom perturbation by collisions with an external particle instead of the random phase jump model. It should be emphasized that in the general case the broadening of the spectral line can not be reduced to the broadening of an isolated level. Therefore, the perturbation of the initial and final levels must be considered simultaneously. It can be described with the help of the so called "double density matrix" $\widetilde{\rho}_{\alpha_0,\alpha_0'}^{\alpha,\alpha'}$ which is equal to $\delta_{\alpha,\alpha_0}\delta_{\alpha',\alpha_0'}$ for the initial α_0, α_0' states and is increased as a result of collision on the value of $\Delta\widetilde{\rho} = \mathsf{S}^\dagger\widetilde{\rho}\mathsf{S} - \widetilde{\rho}$. For an isolated $\alpha \to \alpha'$ line a general consideration (see [3.8]) of the problem is somewhat similar to that given above for the classical oscillator and yields the Lorentzian distribution (9.7) with the effective cross section

$$\sigma^{\mathrm{br}} - i\sigma^{\mathrm{sh}} = 2\pi \int\limits_{0}^{\infty} [1 - \mathsf{S}_{\alpha\alpha}^* \mathsf{S}_{\alpha'\alpha'}]\,\rho d\rho. \tag{9.12}$$

The S-matrix is complex, whereas in the presence of both the elastic and inelastic scattering processes its diagonal element can be presented as $\mathsf{S}_{\alpha\alpha}\left(\rho\right) = \exp\left[-\Gamma_\alpha\left(\rho\right) + i\eta_\alpha\left(\rho\right)\right]$. Here the η_α and Γ_α values are real and $\Gamma_\alpha \geq 0$. Note also that $\Gamma_\alpha \to 0$ and $\mathsf{S}_{\alpha\alpha} \to \exp\left(i\eta_\alpha\right)$ if there is only elastic scattering. Therefore, the width and shift effective cross sections can be rewritten in the form

$$\sigma^{\mathrm{br}} = 2\pi \int_0^\infty \{1 - \exp\left[-\left(\Gamma_\alpha + \Gamma_{\alpha'}\right)\right] \cos\left(\eta_\alpha - \eta_{\alpha'}\right)\} \rho d\rho, \qquad (9.13)$$

$$\sigma^{\mathrm{sh}} = 2\pi \int_0^\infty \exp\left[-\left(\Gamma_\alpha + \Gamma_{\alpha'}\right)\right] \sin\left(\eta_\alpha - \eta_{\alpha'}\right) \rho d\rho, \qquad (9.14)$$

which clarifies the physical meaning of these formulas. As is evident from (9.9), the broadening and shift depend on the difference of the additional phases η_α and $\eta_{\alpha}{}'$ of an atom as a result of the collision and on the sum of the Γ_α and $\Gamma_{\alpha'}$ values describing the devastation of the atomic α, α' states. Note that Γ_α is determined by all inelastic transitions from the α level, while the contribution to the phase η_α is given for both the elastic and inelastic scattering.

The basic semiclassical expression for the broadening cross section of the spectral line $\alpha \to \alpha'$ is reduced to a rather simple form

$$\sigma_\alpha^{\mathrm{br}}(V) = 2\pi \int_0^\infty \mathrm{Re}\,\{1 - S_{\alpha\alpha}(\rho, V)\}\,\rho d\rho, \qquad (9.15)$$

$$\sigma_\alpha^{\mathrm{sh}}(V) = 2\pi \int_0^\infty \mathrm{Im}\,\{S_{\alpha\alpha}(\rho, V)\}\,\rho d\rho \qquad (9.16)$$

if the perturbation of the lower level α' can be neglected. Formula (9.15) for the impact broadening cross section is equal to one-half of the corresponding result (4.46) for the total scattering cross section (see discussion below for more details).

In the general case of transitions between degenerate levels, the spectral distribution is reduced to a superposition of the Lorentzian distributions with various constants γ and Δ. This case was considered in detail in [3.8, 9.1].

9.1.2 Quantum Formulas

We assumed above that the motion of the perturbing particle relative to the atom can be described by classical mechanics. A consistent quantum approach to the impact broadening can be built with the help of the density matrix for the system including the Rydberg atom and perturbing particle. As follows from the general theory [3.8, 9.1], the quantum treatment of impact broadening processes also leads to the Lorentzian shape (9.7) of spectral intensity distribution. However, the quantum-mechanical expression for the effective cross section takes the form

$$\sigma^{\mathrm{br}} - i\sigma^{\mathrm{sh}} = \frac{2\pi i}{q}\left[f_{\alpha'\alpha'}^*(\mathbf{q},\mathbf{q}) - f_{\alpha\alpha}(\mathbf{q},\mathbf{q})\right] - \int f_{\alpha'\alpha'}^*(\mathbf{q},\mathbf{q}')\,f_{\alpha\alpha}(\mathbf{q}',\mathbf{q})\,d\Omega_{\mathbf{q}'\mathbf{q}}, \qquad (9.17)$$

where $f_{\alpha\alpha}(\mathbf{q}',\mathbf{q})$ and $f_{\alpha'\alpha'}(\mathbf{q}',\mathbf{q})$ are the elastic scattering amplitudes of the perturbing particle by the Rydberg atom in the α and α' states, respectively. According to the optical theorem (4.24) for the elastic scattering amplitude in the forward direction $\theta_{\mathbf{q}'\mathbf{q}} = 0$ (when $\mathbf{q}' = \mathbf{q}$), the imaginary parts of the expressions in the square brackets on the right-hand side of (9.17) can be expressed in terms of the total cross sections $\sigma_\alpha^{tot} = \sum\limits_f \sigma_{f\alpha}$ and $\sigma_{\alpha'}^{tot} = \sum\limits_f \sigma_{f\alpha'}$ for the Rydberg atom-perturbing particle collision

$$\mathbf{Im}\,\{f_{ii}(\mathbf{q},\mathbf{q})\} = \frac{q}{4\pi} \sum_f \sigma_{fi} = \frac{q}{4\pi}\left(\sigma_\alpha^{el} + \sigma_\alpha^{in}\right). \tag{9.18}$$

Using this relation, the final formula for the impact broadening cross section of the spectral line $\alpha \to \alpha'$ can be written in the form:

$$\sigma^{br} = \frac{1}{2}\left[\sigma_\alpha^{in} + \sigma_{\alpha'}^{in} + \int |f_{\alpha'\alpha'}^*(q,\theta_{\mathbf{qq}'}) - f_{\alpha\alpha}(q,\theta_{\mathbf{q}'\mathbf{q}})|^2\, d\Omega_{\mathbf{q}'\mathbf{q}}\right]. \tag{9.19}$$

Here σ_α^{in}, $\sigma_{\alpha'}^{in}$ are the total inelastic cross sections of projectile by the Rydberg atom in the α, α' states, respectively.

As is apparent from (9.19), if the perturbation of one level is negligible (i.e., $|f_{\alpha'}| \ll |f_\alpha|$) the line broadening is determined by total scattering cross section σ_α^{tot}

$$\sigma_\alpha^{br} = \frac{1}{2}\sigma_\alpha^{tot} = \frac{1}{2}\left(\sigma_\alpha^{el} + \sigma_\alpha^{in}\right), \tag{9.20}$$

which includes the elastic $\sigma_\alpha^{el} \equiv \sigma_{\alpha\alpha}$ and inelastic $\sigma_\alpha^{in} = \sum\limits_{f\neq\alpha} \sigma_{f\alpha}$ cross sections for a given state α. Thus, the line broadening for the $\alpha \to \alpha'$ transition is reduced to the broadening of one isolated level α. This situation is realized for spectral lines between highly excited and lower excited (or ground) states. The major part of the experimental studies of the width and shift of the Rydberg series in gases corresponds to such transitions.

In the opposite case, when the scattering amplitudes for the initial and final states are close to each other, the elastic scattering becomes negligible as compared to the inelastic one. Therefore, in accordance with (9.19), the broadening cross section is equal to one-half of the sum of the total inelastic cross sections for the initial and final states of the Rydberg atom

$$\sigma_{\alpha\alpha'}^{br} = \frac{1}{2}\left(\sigma_\alpha^{in} + \sigma_{\alpha'}^{in}\right). \tag{9.21}$$

This situation takes place for spectral lines corresponding to transitions between closely spaced or neighboring Rydberg levels. These spectral lines were observed from the astronomical objects such as planetary nebulas, interstellar medium, etc. The broadening of these lines is usually determined by collisions with charged particles.

9.2 Theory of Width and Shift of Rydberg Levels in Gas

Let us consider in detail the line-broadening process for transitions between an isolated highly excited $|i\rangle$ and lower excited $|0\rangle$ (or ground) states of an atom A. In accordance with basic the quantum formulas (9.11, 17) the impact width γ_i and shift Δ_i of the Rydberg level $|i\rangle$ are given by

$$\gamma_i = 2N \left\langle V\sigma_i^{\mathrm{br}}\right\rangle, \qquad \sigma_i^{\mathrm{br}} = \frac{2\pi}{q}\, \mathrm{Im}\,\{\mathfrak{f}_{ii}(\mathbf{q},\mathbf{q}\},$$

$$\Delta_i = N \left\langle V\sigma_i^{\mathrm{sh}}\right\rangle, \qquad \sigma_i^{\mathrm{sh}} = -\frac{2\pi}{q}\, \mathrm{Re}\,\{\mathfrak{f}_{ii}(\mathbf{q},\mathbf{q}\}, \tag{9.22}$$

i.e., they are determined by the imaginary and real parts of the amplitude for elastic scattering of a perturbing neutral particle B by the atom A in the forward direction ($\theta_{\mathbf{q'q}} = 0$), respectively. Here N is the gas density, and $q = \mu V$ is the wave number for the perturber-atom relative motion.

It is possible to rewrite these formulas in terms of the $T-$matrix elements involving all possible transitions $i \to f$ from an isolated i-level. Using expression (4.17) for the amplitude $\mathfrak{f}_{fi}(\mathbf{q'},\mathbf{q})$, the Lippman-Schwinger equation (4.21), and relation (2.94), the broadening cross sections can written as

$$\sigma_i^{\mathrm{br}} = \frac{1}{2}\left(\frac{\mu}{2\pi\hbar^2}\right)^2 \sum_f \frac{q'}{q} \int d\Omega_{\mathbf{q'q}}\, |\langle \mathbf{q'}, f|\, T(E)\, |i, \mathbf{q}\rangle|^2, \tag{9.23}$$

where the total energy E of system A+B should be taken on the energy shell (4.6). A similar result for the shift cross section is given by

$$\sigma_i^{\mathrm{sh}} = \frac{\mu}{q\hbar^2}\left[\langle \mathbf{q}, i|\, V\, |i, \mathbf{q}\rangle + \sum_f \int d\Omega_{\mathbf{q''q}}\mathcal{P}\int \frac{(q'')^2 dq''}{(2\pi)^3}\, \frac{|\langle \mathbf{q''}, f|\, T(E)\, |i, \mathbf{q}\rangle|^2}{\hbar^2\left[(q')^2 - (q'')^2\right]/2\mu}\right]. \tag{9.24}$$

Here the final wave number of the colliding particle relative motion q' is determined from the law of energy conservation $(q')^2 = q^2 + 2\mu(E_i - E_f)/\hbar$, and the symbol $\mathcal{P}\int$ means the principal value of the integral over q'' in accordance with (2.91). In contrast to (9.13) the matrix elements in (9.14) should be taken both on and off the energy shell.

Basic formulas (9.13, 14) have a fundamental importance in the broadening theory of one isolated level $|i\rangle$ because they reduce the calculation of its impact width and shift to solution of the problem for projectile B scattering by the Rydberg atom A(i). As we have already seen, the simple relationship (9.20) appears between the broadening cross section σ_i^{br} and total scattering cross section σ_i^{tot}. Note also that if the contribution of all inelastic transitions is predominant (i.e., $\sigma_i^{\mathrm{in}} \gg \sigma_i^{\mathrm{el}}$), then (9.20) demonstrates a definite correspondence between the impact broadening and collisional quenching processes, considered in detail in Sect. 6.5.

Thus, the physical approaches and theoretical techniques presented in the preceeding Chaps. 5–7 combined with the general formulas (9.20) or (9.22) of spectral-line broadening may be applied for the width and shift calculations of Rydberg levels in gases. We shall present in this section only a short summary of simple analytical results in this field which are needed for understanding the broadening and shift behavior and their dependence on the major physical parameters such as the principal quantum number n, and relative velocity of collision V, the electron scattering length L, and polarizability of the perturber, etc. It provides a basis for further discussion in Sects. 9.3, 4 of more complex theoretical calculations and available experimental results on the broadening and shift rates. Note that the most interesting experimental applications is the case of thermal energies and the range of $n \ll v_0/V$ corresponding to slow collisions.

If the principal quantum number n is not too small the total broadening and shift cross sections can be presented as a superposition of contributions determined by scattering of the perturbing atom on the quasi-free electron and on the ion core:

$$\sigma_{nl}^{\mathrm{br}} = \sigma_{nl}^{\mathrm{br}} \left(\mathrm{e-B}\right) + \sigma_{nl}^{\mathrm{br}} \left(\mathrm{A^+-B}\right), \quad \sigma_{nl}^{\mathrm{sh}} = \sigma_{nl}^{\mathrm{sh}} \left(\mathrm{e-B}\right) + \sigma_{nl}^{\mathrm{sh}} \left(\mathrm{A^+-B}\right). \quad (9.25)$$

Thus, it is convenient to discuss separately the contributions of two possible scattering mechanisms to the broadening and shift of Rydberg levels in a gas.

9.2.1 Mechanism of Core-Perturber Scattering

At large enough internuclear separations R the core-perturber interaction can be successfully described by the polarization interaction $V_{\mathrm{A+B}} = -\alpha e^2/2R^4$. The contributions of this "polarization" mechanism are given by simple formulas [5.59, 9.4]

$$\sigma_{nl}^{\mathrm{br}} \left(\mathrm{A^+-B}\right) = \frac{1}{2}\sigma_{\mathrm{A+B}}^{\mathrm{el}}\left(V\right) = \frac{\pi^{5/3}}{2^{7/3}}\Gamma\left(\frac{1}{3}\right)\left(\frac{\alpha e^2}{\hbar V}\right)^{2/3} \approx 3.58 \left(\frac{\alpha v_0}{V}\right)^{2/3}, \quad (9.26)$$

$$\sigma_{nl}^{\mathrm{sh}} \left(\mathrm{A^+-B}\right) = -\sqrt{3}\sigma_{nl}^{\mathrm{br}}\left(\mathrm{A^+-B}\right) \approx -6.20 \left(\frac{\alpha v_0}{V}\right)^{2/3}, \quad (9.27)$$

which have been derived in the impact approximation. Formula (9.26) is in full correspondence with the result (7.21) for the elastic scattering contribution (7.24) to the total cross section of the Rydberg atom-neutral collision induced by the mechanism of the core-perturber scattering (polarization effect in the impact limit). Thus, at thermal energies the impact broadening of the Rydberg level is mainly determined by the elastic scattering. In accordance with simple formula (7.18), the contribution of all possible inelastic $nl \to n', l \pm 1$ transitions due to the B-A$^+$ scattering is very small in practically the whole range of n, which is interesting for experimental applications.

The final formulas for the impact width and shift directly follows from (9.26, 27) after averaging over the Maxwellian velocity distribution

$$\gamma_{A+B} = 7.16 \left(\alpha e^2/\hbar\right)^{2/3} \left\langle V^{1/3} \right\rangle N, \qquad \Delta_{A+B} = -\frac{\sqrt{3}}{2}\gamma_{A+B} , \quad (9.28)$$

where $\left\langle V^{1/3} \right\rangle = \left\langle V \right\rangle^{1/3} \Gamma\left(5/3\right) \approx 0.9 \left\langle V \right\rangle^{1/3}$ and $\left\langle V \right\rangle = (8kT/\pi\mu)^{1/2}$ is the mean thermal velocity of perturbing atoms or molecules. As is apparent from (9.28), the broadening and shift rates for the core-perturber scattering mechanism (polarization effect) are independent of the principal n and orbital l quantum numbers. The shift of spectral line is negative and its absolute value is about the impact width.

9.2.2 Mechanism of Electron-Perturber Scattering

For the alternative mechanism of spectral-line broadening induced by the electron-perturber scattering, the specific values of the width and shift rates turn out to be quite different in various ranges of the principal quantum number n, and quantum defect δ_l of the Rydberg nl-level.

a) Elastic and Inelastic Scattering Contributions to Impact Width.
If the noninteger part of the quantum defect δ_l of an isolated Rydberg nl-level is sufficiently large and the principal quantum number n is not too large $n < (v_0/V)^{1/2}$ then the contribution of the inelastic transitions is small and usually can be neglected. As a result, the impact width is determined by the elastic scattering alone and may be described by the formula [5.49]

$$\gamma_{nl}^{el}(e-B) = N\left(v_0 a_0^2\right) 4\sqrt{\pi c C_{ll}} \left(\frac{|L|}{a_0}\right) \mathcal{F}(\eta) , \quad \eta = \frac{C_{ll}}{4cn_*^8} \left(\frac{v_0 L}{V_T a_0}\right)^2 . \quad (9.29)$$

This expression directly follows from (6.14). Here the $\mathcal{F}(\eta)$ function is given by (6.15); $V_T = (2kT/\mu)^{1/2}$; C_{ll} is the constant of the order of unity, see (6.8), whose value is somewhat dependent on the orbital angular momentum l. The impact width determined by (9.29) grows with n roughly as n_*^4 in the close coupling range and reaches its maximum at $n_* \sim \left(|L| v_0/a_0 V_T\right)^{1/4}$. The specific value of the normalized constant c should probably be chosen from the comparison with experiment or the close coupling calculations at low enough n (usually its magnitude is taken to be equal to $5/8$ that corresponds to normalization on $\sigma_{geom} = \left(5\pi a_0^2/2\right) n_*^4$ at $n_* \ll n_{max}$). The asymptotic behavior of the width $\gamma_{nl}^{el}(e-B)$ at high principal quantum numbers is given by

$$\gamma_{nl}^{el}(e-B) = 8C_{ll}\sqrt{\pi} \left(\frac{v_0^2 L^2}{V_T n_*^4}\right) N, \quad n_* \gg n_{max} \sim \left(|L| v_0/a_0 V_T\right). \quad (9.30)$$

It is seen that the broadening rate reveals a rapid drop $\gamma_{nl}^{el} \propto n_*^{-4}$ with n-growing, while its temperature dependence is $T^{-1/2}$. Note that this result is independent of the specific choice of the normalized constant c as usually in the weak-coupling limit.

For an arbitrary value of quantum defect δ_l, the contribution of all inelastic $nl \to n'$ transitions with change in the orbital and principal quantum numbers into the impact width may be evaluated in a wide range of n on the basis of semiclassical formulas of Sect. 6.3, see (6.21–6.23, 35),

$$\gamma_{nl}^{\text{in}}(\text{e} - \text{B}) = 2N \left\langle V\sigma_{nl}^{\text{br}}(V) \right\rangle_T , \quad \sigma_{nl}^{\text{br}} = \sigma_{nl}^{\text{in}} = \frac{1}{2} \sum_{n'} \sigma_{n',nl} \zeta_{l0,n'} , \qquad (9.31)$$

Thus, calculation of the broadening rate is reduced to summation of separate contributions of the $nl \to n'$ transitions over all n'-levels and to integration over the velocity distribution function. In the range of the weak coupling formula (6.21) for $\sigma_{n',nl}$ reduces to the semiclassical impulse-approximation result (5.59), which is applicable for an arbitrary form of the amplitude $f_{\text{eB}}(k, \theta)$ for the electron-perturber scattering. It takes a rather simple form (5.38) provided the scattering length approximation is valid.

In the asymptotic region of high enough principal quantum numbers $n \gg (v_0/V)^{1/2}$ (i.e., $n \gg 40$–50 at thermal collisions) the broadening cross section $\sigma_{nl}^{\text{br}}(\text{e} - \text{B})$ is determined by the contribution of a great number of inelastic $nl \to n + \Delta n$ transitions with $(\Delta n = 0, \pm 1, \pm 2, ...)$. Then, summation over Δn in (9.31) can be replaced by integration over the transition energy $\Delta E = (n' - n_*)/(n')^3$. This allows us to perform the integration in analytical form (see [1.3] for more details) with the use of general semiclassical formula of the impulse approximation (5.59). As a result, we obtain the following asymptotic expressions for the broadening cross section and impact width in terms of the elastic scattering amplitude f_{eB} and the momentum distribution function $W_{nl}(k) = k^2 |g_{nl}(k)|^2$ of the Rydberg electron

$$\sigma_{nl}^{\text{br}}(\text{e} - \text{B}) = \frac{2\pi\hbar}{mV} \int_0^\infty \text{Im}\left\{ f_{\text{eB}}(k, \theta = 0) \right\} W_{nl}(k)\, dk, \quad \sqrt{\frac{v_0}{V}} \ll n \ll \frac{v_0}{V},$$

$$\gamma_{nl}^{\text{in}}(\text{e} - \text{B}) = N \int_0^\infty \left(\frac{\hbar k}{m} \right) \sigma_{\text{eB}}^{\text{el}}(k) W_{nl}(k)\, dk \equiv N \left\langle v\sigma_{\text{eB}}^{\text{el}}(v) \right\rangle_{nl} , \quad v = \hbar k/m.$$

$$(9.32)$$

These results were derived first [5.59] directly from (9.22) using the impulse-approximation amplitude in the forward direction (5.70). Note that in the most interesting case of small $l \ll n$ the distribution function $W_{nl}(k)$ in (9.32) is described by the simple semiclassical expression (2.137).

In the scattering length approximation ($\sigma_{\text{eB}}^{\text{el}} = 4\pi L^2$) formulas (9.32) take a rather simple form

$$\sigma_{nl}^{\text{br}}(V) = \frac{4L^2}{n} \left(\frac{e^2}{\hbar V} \right), \quad \gamma_{nl}^{\text{in}}(\text{e} - \text{B}) = \frac{8v_0 L^2}{n} N , \quad l \ll n \qquad (9.33)$$

that corresponds to the *Omont's* [5.40] expression for the total cross section of all $nl \to n'$ transitions (5.75). Comparison of (9.30, 33) clearly demonstrates that at high $n \gg (v_0/V)^{1/2}$, the impact width of the spectral line due to the electron-perturber scattering mechanism is determined by the total contribution of all inelastic $nl \rightarrow n'$ transitions, while the elastic scattering can be neglected.

b) Shift Behavior at Large and Intermediate n. The opposite situation occurs for the shift of high-Rydberg levels. According to (9.22, 5.57) the shift cross section in the impulse approximation can be presented as [5.59]

$$\sigma_{nl}^{sh}(V) = -\frac{2\pi\hbar}{mV} \int\limits_{0}^{\infty} \mathbf{Re} \{ f_{eB}(k, \theta_{\mathbf{k'k}} = 0) W_{nl}(k)dk. \tag{9.34}$$

It is mainly determined by the elastic scattering of the Rydberg atom by a perturbing particle. Note that in spite of the formal criterion $n \gg (v_0/V)^{1/2}$ for applicability of expressions (9.32, 34) the contribution of the electron-perturber scattering mechanism to the impact shift is reasonably reproduced by (9.34) not only at very high $n \gg 40$–50 but also at the intermediate range of n ($n \gtrsim 15$–20 for the rare gas, and $n \gtrsim 25$–30 for the alkali-metal atoms).

In the scattering length approximation ($\mathbf{Re}\{f_{eB}\} = -L$) expression (9.34) leads to the well-known *Fermi* [5.1] law:

$$\sigma_{nl}^{sh}(V) = \frac{2\pi\hbar L}{mV}, \qquad \Delta_{nl} = \frac{2\pi\hbar L}{m} N. \tag{9.35}$$

According to (9.35), the electron-perturber scattering yields the n- and l-independent value for the shift of high-Rydberg level in a gas. The sign of Δ_{nl} is determined by the sign of the electron-scattering length L. Note also that the quasi-static approximation also leads to the Fermi law for the shift of Rydberg series in the asymptotic region of high n [9.5].

The correction terms to the Fermi law due to the long range polarization and quadrupole interactions between the highly excited atomic electron e and perturbing particle (atom or nonpolar molecule) B has been calculated by *Ivanov* [9.6] with the use of the modified effective range theory [5.10]

$$\Delta_{nl} = \frac{2\pi\hbar}{m} \left[L - \left(\frac{4L\alpha_0}{3a_0^3} \right) \frac{\ln(n)}{n^2} + \frac{\chi}{n^2} \right] N, \tag{9.36}$$

where $\alpha_0 = (\alpha_\| + 2\alpha_\perp)/3$ is the symmetric part of the polarizability of perturbing particle B, and χ is the constant determined by the terms of low-energy expansion of the scattering amplitude of the order of k^2 (see Sect. 5.2.2 for more details).

Let us now discuss the combined influence of the electron-perturber and core-perturber scattering on the limiting values of the shift and width at very high n. In accordance with the results presented above, they tend asymptotically to the n- and l- independent constants equal to

$$\Delta_{nl} \rightarrow \Delta_{eB} + \Delta_{A+B} = const, \qquad \gamma_{nl} \rightarrow \gamma_{A+B} = const. \tag{9.37}$$

Thus, at very high n, when the contribution of all possible $nl \rightarrow n'$ transitions induced by the quasi-free electron-perturber scattering to the impact width becomes small, it is determined only by scattering of the perturbing atom on the ion core of the Rydberg atom, i.e., by the polarization effect [see (9.28)]. On the other hand, the limiting value of the shift of the high-Rydberg level is determined by a superposition of contributions (9.35, 28) due to the electron-perturber and core-perturber scattering mechanisms.

A simple approximate result for the elastic scattering contribution to the shift cross section due to the combined effect of the e-B and A^+-B scattering can be written as

$$
\sigma_{ns}^{sh} \approx
\begin{cases}
\dfrac{2\pi\hbar L}{mV}\left[1-\left(\dfrac{n_1}{n_*}\right)^8\right], & 1.1n_1 < n_* < 0.73n_2, \\[3mm]
\dfrac{2\pi\hbar L}{mV} - \dfrac{\pi^2\alpha v_0}{2a_0 V}\left[\left(\dfrac{2Va_0^3}{v_0\alpha}\right)^{1/3} - \dfrac{1}{n_*^2}\right], & n_* > 0.73n_2,
\end{cases}
\tag{9.38}
$$

where n_1 and n_2 are given by

$$
n_1 = \left(\frac{v_0\,|L|}{4a_0 V}\right)^{1/4}, \qquad n_2 = \left[\frac{|L|}{(\alpha a_0^3)^{1/6}}\left(\frac{v_0}{V}\right)^{5/6}\right].
$$

This expression was derived [5.43] in the scattering length approximation within the framework of the semiclassical impact broadening theory [3.8] using the JWKB-approximation for the phase shifts. One can see that at high principal quantum numbers $n_* \gg n_2$ expression (9.38) yields the aforementioned result for the sum of the Fermi shift and the shift due to the polarization effect. However, expression (9.38) may also be used for estimation of the elastic scattering contribution to the shift cross section of ns-level at the intermediate magnitudes of the principal quantum number if $n > 1.1n_1$. Dependence of the shift of the Rydberg state on its angular momentum was discussed by *Hermann* et al. [9.7].

Below we present some results of theoretical calculations for the cross sections and the rate constants of the broadening and shift of Rydberg atomic levels perturbed by the ground-state rare gas (Sect. 9.2) and alkali-metal (Sect. 9.3) atoms. Our main purpose is to analyze the characteristic physical features of one or another process. Considerable attention will be paid to the comparison between theory and experiment. However, we do not claim to consider all the available experimental and theoretical data.

9.3 Comparison of Theory with Experiment

9.3.1 Broadening and Shift in Rare Gases

Among the great number of early and recent experimental data on impact broadening of the Rydberg atomic series perturbed by the rare gas atoms, we have chosen here only several results for ns and nd states of Na [9.8, 9]; ns states of Rb [5.15]; and $5sns\,^1S_0$, $5snd\,^1D_2$ states of Sr I [5.14, 9.10]. These detailed measurements are selected in order to make comparisons with recent theoretical calculations of spectral line widths and shifts. Analysis of these data is also of considerable interest because of qualitatively different influence of elastic and inelastic scattering of Rydberg atoms by neutral perturbers on the broadening cross sections.

The observed cross sections [9.5, 9] for impact broadening of the n^2S and n^2D levels of Na perturbed by the rare gas atoms He, Ne, Ar, Kr, and Xe are plotted in Fig. 9.1 (panel a) and (panel b) versus n. Due to the large difference in the quantum defect magnitudes $\delta_S^{Na} = 1.35$ and $\delta_D^{Na} = 0.015$ there is a significant distinction between these two cases. In particular, the broadening cross sections of the well isolated n^2S levels are much smaller than the corresponding values for the n^2D levels. Broadening of Na(ns) levels is mainly due to the elastic scattering of the perturbing rare gas atom both on the quasi-free electron and on the parent core of the Rydberg atom (see, for example, [5.43]). In accordance with available experimental [9.11, 12] and calculated [5.44] results, the total contribution of all inelastic transitions is very small and, hence, does not affect the broadening width. Thus, as is apparent from Fig. 9.1a, calculations [5.43] of the broadening cross sections $\sigma_{ns}^{br} \approx \sigma_{ns}^{el}/2$ taking into account elastic scattering alone provide a successful

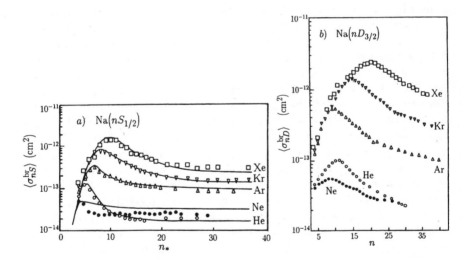

Fig. 9.1. Cross sections $\left\langle \sigma_{nS}^{br} \right\rangle$ and $\left\langle \sigma_{nD}^{br} \right\rangle$ for impact broadening of the Rydberg Na(nS) (panel a) and Na(nD) (panel b) levels by the ground-state rare gas atoms plotted against the effective principal quantum number $n_* = n - \delta_l$ ($\delta_s^{Na} = 1.35$, and $\delta_d^{Na} = 0.015$). Experimental data (full circles – He, empty circles – Ne, empty triangles: Δ- Ar and ∇- Kr, and empty squares – Xe) are those from [9.9] at $T = 400$ K. Full curves in panel (a) present semiclassical results [5.43] for elastic scattering contribution $\left\langle \sigma_{nS}^{br} \right\rangle = (1/2) \left\langle \sigma_{nS}^{el} \right\rangle$ to the broadening cross sections of the well isolated nS-levels

description of available experimental data [9.8, 9] at high, intermediate and low enough n.

On the other hand, the broadening of the Rydberg Na(n^2D) levels by the rare gas perturbers in the experimentally studied [9.8, 9] range of n is mainly the result of the inelastic perturber-quasi-free electron scattering. Thus, we may approximately assume $\sigma_{nd}^{br} \approx \sigma_{nd}^{inel}/2$ if the magnitudes of the principal quantum number are not very large. It is evident that the dominant contribution to the cross section $\sigma_{nd}^{inel} = \sum_{n'l'}' \sigma_{n'l',nd}$ is determined by the quasi-elastic l-mixing $nd \to nl'$ transitions to a great number of hydrogen-like degenerate nl' sublevels with $l' > 2$. The elastic scattering contribution becomes important only at low enough principal quantum numbers $n \lesssim n_{max}$ and in the asymptotic region of high $n > 30$–50, where it is independent of n in accordance with the results [5.59, 9.4] for the perturber-core scattering mechanism (polarization effect). Thus, it is natural that the impact broadening data [9.8, 9] for the nd states of sodium are in correspondence with the measurements [6.4–6] and calculations [5.65, 71, 56, 52] of the quenching cross sections $\left\langle \sigma_{nd}^{Q} \right\rangle_{T}$ presented in Fig. 9.1 (panel b).

The detailed calculations of the impact broadening widths for the Rydberg Rb(ns) levels perturbed by He, Ne, Ar, Kr, and Xe has been performed by *Sun* and *West* [5.45] on the basis of the *Kaulakys* [5.43] theory for elastic scattering and *Lebedev* and *Marchenko* [5.44] theory for inelastic scattering. The results of the calculations [5.45] and available experimental data [9.13] are shown in Fig. 9.2 for Rb(ns) +Xe. It can be seen that the elastic broadening rates (dashed curves) are in reasonable agreement with the measured ones at low enough n and in the vicinity of the maximum $n \lesssim n_{max}$. However, in the range of $n > n_{max}$ the inelastic contribution due to scattering of the quasi-free electron on the perturbing atom Xe becomes predominant and should be taken into account to explain the available experimental results. The polarization effect [$\sigma_{ns}^{pol} = \sigma_{el}^{A^+B}/2$, see (7.24)] induced by scattering of the Xe atom on the ionic core Rb$^+$ gives a major contribution to the broadening cross section only in the asymptotic range of $n > 60$. As follows from the results [5.45], presented in Fig. 9.2, semiclassical theories of elastic [5.43] and inelastic [5.44] scattering based of Rydberg-atom–neutral-atom collisions together with the basic equation (9.20) provide a good description of the total broadening widths on the whole investigated range of n.

The shift rates Δ_{ns} and Δ_{nd} for the Rydberg n^2S and n^2D series of Rb by Ne and Kr atoms has been measured by *Thompson* et al. [5.15] in a wide range of n (see Fig. 9.3). The observed dependence on the principal quantum number n is similar to that obtained in other experimental works with the rare gases (see review [1.3]). For example, for the Kr atom as a perturber the absolute value of the shift rate increases rapidly up to $n \sim 15$ and, then, tends to the asymptotic limit defined by the sum of the Fermi mechanism (9.35) and the polarization effect (9.28). This behavior agrees reasonably with

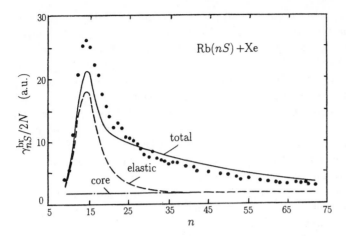

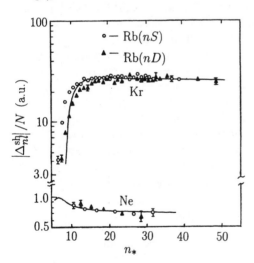

Fig. 9.2. Comparison of theoretical and experimental data for the impact broadening of the Rydberg Rb(nS) levels by Xe ($T = 530$ K). Full curve present the results [5.45] for the total broadening cross section $\left\langle \sigma_{nS}^{\mathrm{br}} \right\rangle = \left\langle \sigma_{\mathrm{el}}^{\mathrm{br}}(nS) \right\rangle + \left\langle \sigma_{\mathrm{in}}^{\mathrm{br}}(nS) \right\rangle$ calculated using semiclassical formulas for the elastic [5.43] and inelastic [5.44] scattering contributions. Dashed curve is the contribution of elastic scattering alone. Dashed-dotted curve indicates the contribution of the perturber-core elastic scattering (polarization effect). Full circles are the experimental data [5.15]

Fig. 9.3. Experimental data [5.15] and corresponding results of calculations (full curves) for the shift rates $\Delta_{nl}^{\mathrm{sh}}/N = \left\langle V \sigma_{nl}^{\mathrm{sh}} \right\rangle_T$ of the Rydberg Rb(nl) levels perturbed by Ne ($T = 525$ K) and Kr ($T = 530$ K). Empty cirles and full triangles are the nS and nD levels, respectively

expression (9.38). However, at low principal quantum numbers $n < 15$ there is little difference between the shift rates for the n^2S and n^2D levels.

The detailed experimental studies of the impact broadening and shift also have been performed by *Weber* and *Niemax* [9.10] and *Heber* et al. [5.14] for the Rydberg $5sns\,^1S_0$, $5snd\,^1D_2$ and $5snd\,^3D_2$ levels of strontium perturbed by the rare gas atoms. High order of accuracy was achieved in these measurements for a wide range of the principal quantum numbers. The experimental data [5.14] for the pressure shift and broadening of the $Sr(5snd^1D_2)$ levels perturbed by different rare gas atoms He, Ne, Ar, Kr, and Xe are presented in Figs. 9.4a,b at $11 \leq n \leq 100$. One can see (see Fig. 9.4a) that the pressure shift is positive for He and Ne perturbers and becomes negative for the heavy atoms Ar, Kr, and Xe, for which the scattering lengths are negative. Thus, the sign of the pressure shift Δ_{nl} is determined by the corresponding sign of the scattering length L for the electron-perturber elastic scattering (see Table 5.1). On the other hand, it is evident from Fig. 9.4b that the asymptotic magnitudes of the impact broadening width at high enough principal quantum numbers n are determined by the elastic scattering contribution of the perturbing atom by the parent core of the Rydberg atom (polarization effect). Thus the measurements [5.14] at very high n ($35 \leq n \leq 100$) successfully confirm the first theoretical predictions by *Fermi* [5.1], *Reinsberg* [9.4] and *Alekseev* and *Sobelman* [5.59] for the asymptotic behavior (constant values independent of n and l) of the impact broadening and shift of the Rydberg series in the rare gases.

At low enough $8 \leq n \leq 25$ *Weber* and *Niemax* [9.10] and *Heber* et al. [5.14] have found the irregular dependence of the impact broadening widths and shifts of the $5snd^1D_2$, $5snd\,^3D_2$, and $5sns^1S_0$ series on the principal quantum number n (see, for example, Fig. 9.5). This phenomenon is the result of the specific electronic structure of the Rydberg $5snd^1D_2$, $5snd^3D_2$, and $5sns^1S_0$ levels of strontium (having two valence electrons in the inner shell) and the inelastic transitions induced by collisions with the rare gas atoms. As was established in [5.14, 9.10] the peaks in the broadening widths appear for those values of n, for which the moduli (noninteger parts) of the quantum defects of different level series become close to each other. *Sun* et al. [5.46] give a quantitative explanation of the irregular spectral line broadening of $Sr(5sns^1S_0)$ by xenon. They extended the semiclassical theories of inelastic [5.44] and elastic [5.43] scattering of the Rydberg atom by a neutral atom on the systems having more than one valence electron with the aid of the multichannel-quantum defect theory [1.4]. As follows from Fig. 9.5, the results of calculation [5.46] are in good agreement with the experimental data [5.14].

9.3.2 Broadening and Shift in Alkali-Metal Vapors

Due to the large values of the triplet scattering lengths and, particularly, to the presence of the ^3P-resonance in electron scattering by the ground–state alkali atom (see Table 5.1), the impact broadening and shift rates of high Rydberg levels in alkali vapors turn out to be considerably greater than in the case of rare gas atoms. In other words, we have the situation, which is similar to the quenching processes of high-Rydberg levels by alkali-metal atoms (see Sect. 6.4.2). Another observed phenomenon [5.27, 28], new as compared to the rare gas perturbers, consists in the appearance of the oscillatory structure in the broadening and shift of highly excited states ($15 < n < 25$)

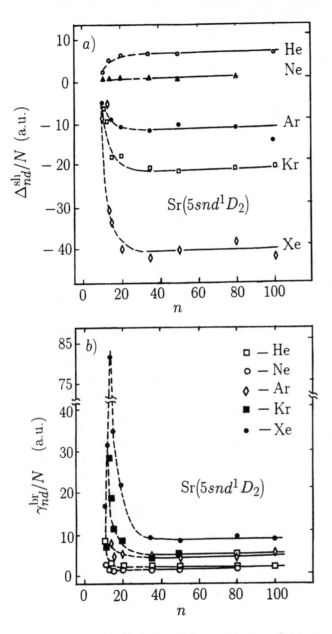

Fig. 9.4. Experimental data [5.14] for the shift Δ_{nd}^{sh}/N (panel a) and broadening γ_{nd}^{br}/N (panel b) rates of the Rydberg $5snd^1D_2$ levels of Sr I induced by thermal collisions ($T = 950$ K) with the rare gas atoms. Full curves represent the asymptotic magnitudes obtained in [5.14] by averaging the measured shift and broadening at $n = 35$, 50, 80, and 100. Dashed curves show their behavior in the range of $11 < n < 35$

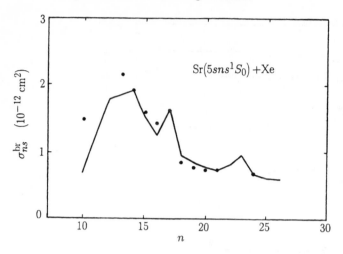

Fig. 9.5. Broadening cross sections σ_{ns}^{br} of the Rydberg $5sns\,^1S_0$ levels of Sr I at thermal collisions with Xe. Full curve represents theoretical results [5.46]. Full circles are the experimental data [5.14]

induced by thermal collisions with $K(4S)$, $Rb(5S)$, and $Cs(6S)$ (see Fig. 9.6). At present two physical approaches have been developed for the theoretical description of the impact broadening process. The first one is based on the general impulse-approximation treatment [4.16, 5.21] of the Rydberg atom–neutral collisions involving the effects of resonance and potential scattering (see Sect. 5.4.6). It gives the opportunity to provide a successful quantitative description of the experimental data [5.27, 28] on impact broadening of the Rydberg atomic series in alkali-metal vapors in the range of applicability of the quasi-free electron model, i.e., at high enough n ($n > 15$–20 for Li and Na, $n > 25$–30 for K and Rb, and $n > 30$–35 for Cs). The second quasi-molecular approach [5.78, 26, 79] is based on the theory of nonadiabatic transitions between the ionic and covalent terms of the colliding particles. It is applicable at intermediate n ($\sim$ 15–30 for K, Rb, and Cs) and allows us to explain the oscillatory behavior of the broadening widths.

a) **Impulse-Approximation Results.** Here we consider in detail the self-broadening process of the $K(nS)$ Rydberg levels at high principal quantum numbers. The cross sections $\langle\sigma_{nS}^{br}\rangle_{ex} = \gamma_{ex}/2N\,\langle V\rangle_T$ of this process measured by *Heinke* et al. [5.27] and *Thompson* et al. [5.28] are plotted in Fig. 9.7 versus n. The estimates [5.21] of the contribution of elastic scattering of the perturbing $K(4S)$ atom by the quasi-free electron of the Rydberg $K(nS)$ atom show that the corresponding cross sections $\langle\sigma_{el}^{br}\rangle_T = (1/2)\,\langle\sigma_{nS}^{el}\,(B-e)\rangle_T$ prove to be much lower than the experimental values in the range of high n ($25 < n$). The contribution of the elastic scattering of the $K(4S)$ atom by the Rydberg $K(nS)$ atom in the perturber-core scattering mechanism (polarization effect) is equal to $\sigma_{el}^{br}\,(K - K^+) \approx 10^{-12}$ cm^2 at $T = 510$ K. This

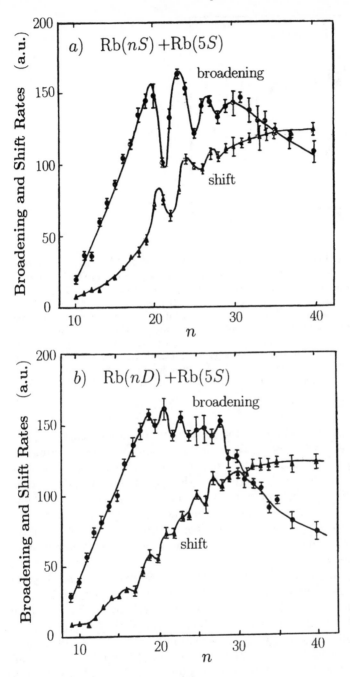

Fig. 9.6. Experimental data [5.28] for the impact broadening and shift rates of the Rydberg Rb(nS) (panel a) and Rb(nD) (panel b) states perturbed by the ground-state Rb($5S$) atoms ($T = 510$ K) plotted against n in a wide range of the principal quantum number

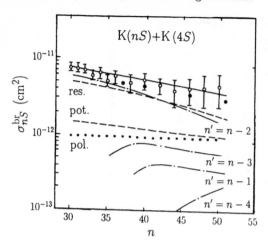

Fig. 9.7. Cross sections for self-broadening of $K(nS) + K(4S)$ with the collision energy $\mathcal{E} = 0.044$ eV ($V_T = 3 \cdot 10^{-4}$ a.u.) plotted against n. Full curve is the impulse-approximation result [5.21] for the total broadening cross section σ_{nS}^{br} at high $n \geq 30$. Dashed curves present the contributions of all inelastic $nS \to n'$ transitions induced by the ^3P-resonance σ_r^{br}K$-$e) and potential σ_p^{br}(K$-$e) scattering of quasi-free electron by the perturbing atom, respectively. Dotted curve is the contribution of elastic perturber-core scattering σ_{el}^{br}(K$-$K$^+$) (polarization effect). Dashed-dotted curves indicate the contributions of the separate inelastic $nS \to n'$ transitions with $n' = n - 1$, $n - 2$, $n - 3$, and $n - 4$ to the broadening cross section. Full and empty circles are the experimental data [5.27] ($T = 513$ K) and [5.28] ($T = 510$ K), respectively

cross section turns out to account for 15–25% of the experimental $\left\langle \sigma_{nS}^{br} \right\rangle_{ex}$ values within the range $n \sim 30$–50 under investigation. The contribution $\left\langle \sigma_{in}^{br} (K - K^+) \right\rangle_T$ of the inelastic $nS \to n'$ transitions caused by the scattering of the perturber $K(4S)$ by the parent core K^+ of the highly excited atom $K(nS)$ to the total broadening cross section can also be neglected, as follows from the estimates of this mechanism in the shake-up model (see Sect. 7.2).

Thus, at high enough n the self-broadening cross sections of the Rydberg levels in alkali vapors are mainly accounted for by the contribution of the inelastic $nl \to n'$ transitions [4.16, 5.2], which are primarily due to the resonance scattering of the quasi-free electron by the perturbing alkali atom in the ground state. The contributions of the most important mechanisms to the impact broadening of the Rydberg $K(nS)$ levels by $K(4S)$ atoms are shown in Fig. 9.7. It is seen that the experimental data [5.27, 28] are described successfully with the ^3P-resonance parameters of $\epsilon_0 = 8.5 \cdot 10^{-4}$ a.u. and $\gamma = 20$ a.u., which correspond to the energy $E_r = 0.02$ eV and full width at half-height $\Gamma_r = 0.021$ eV (see Table 5.2). The role of the ^3P-resonance electron-potassium atom scattering makes up about 70–50% of the total broadening cross section in the region $30 \leq n \leq 50$. The role of the

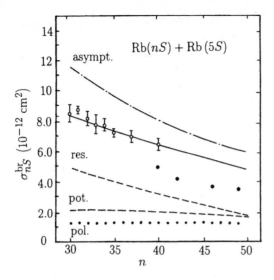

Fig. 9.8. Cross sections for self-broadening of Rb(nS) + Rb($5S$) with the collision energy $\mathcal{E} = 0.044$ eV ($V_T = 2\cdot10^{-4}$ a.u.) plotted against n. Full curve is the impulse-approximation result [5.21] for the total broadening cross section $\sigma_{nS}^{\mathrm{br}}$ at high $n \geq 30$. Dashed curves present the contributions of all inelastic $nS \rightarrow n'$ transitions induced by the ^3P-resonance $\sigma_{\mathrm{r}}^{\mathrm{br}}(\mathrm{K}-\mathrm{e})$ and potential $\sigma_{\mathrm{p}}^{\mathrm{br}}(\mathrm{K}-\mathrm{e})$ scattering of quasi-free electron by the perturbing atom, respectively. Dotted curve is the contribution of elastic perturber-core scattering $\sigma_{\mathrm{el}}^{\mathrm{br}}(\mathrm{K}-\mathrm{K}^+)$ (polarization effect). Full and empty circles are the experimental data [5.27] ($T = 513$ K) and [5.28] ($T = 510$ K), respectively

potential electron-atom scattering is increased for high values of $n > 50$–60. Notice also that the main contribution to the broadening cross section (see Fig. 9.7) is determined by the inelastic $nS \rightarrow n - 2$ and $nS \rightarrow n - 3$ transitions with the minimum energy defects $\Delta E_{n',nS} = 2Ry\,|\delta_S + \Delta n|\,/n^3$ (where $\delta_S = 2.18$).

The results of analogous calculations [5.21] for the self-broadening process of Rb(nS) +Rb($5S$) are shown in Fig. 9.8. In this case, the corresponding experimental data [5.28] are well described (full curve) with the ^3P-resonance parameters of the Rb$^-$ ion being $\epsilon_0 = 1.2\cdot10^{-3}$ a.u. and $\gamma = 18$ a.u. ($E_{\mathrm{r}} = 0.028$ eV and $\Gamma_{\mathrm{r}} = 0.031$ eV). The contributions of the different scattering mechanisms to the total broadening cross section are denoted by the broken curves in Figs. 9.6, 7. As is evident from these figures, the use of the ^3P-resonance parameters for K$^-$ and Rb$^-$ ions presented above gives the opportunity to explain reliably the experimental data for the impact broadening processes of K(nS) +Rb and Rb(nS) +K in all range of the impulse approximation validity.

Figure 9.8 also illustrates the comparison between the impulse-approximation results [4.16, 5.2] for the broadening cross sections

$\langle\sigma_{nS}^{br}\rangle_T$ calculated with the use of the general formula (5.59) and those made on the basis of the asymptotic formula [5.59] taking into account both the potential and ^3P-resonance scattering. One can see that the use of the asymptotic expressions for the perturber-electron scattering contribution (with the resonance parameters obtained in [5.21]) and for the perturber-core scattering contribution leads to satisfactory values for the total broadening cross section in the region $30 \le n \le 50$ under investigation. As has been also shown in [5.19], the asymptotic theory [5.59] supplemented by the contribution of the ^3P-resonance scattering, provides a reasonable description of the available experimental data [5.27, 28] for the self-broadening of $K(nS) + K$, $Rb(nS) + Rb$, and $Cs(nS) + Cs$ at $n > 30$. However, this fact is accounted for by the specific magnitudes of the quantum defects $\delta_S^K = 2.18$, $\delta_S^{Rb} = 3.15$, and $\delta_S^{Cs} = 4.057$ with sufficiently low values of noninteger parts. Hence, it is, to a certain degree, fortuitous because the asymptotic theory [5.59] becomes formally valid at $n \gg (v_0/V)^{1/2}$ (i.e., $n \gg 50$ at thermal collisions between the heavy alkali atoms, when $V = 4 \cdot 10^{-4}$ a.u.). In particular, the observed [5.27, 28] decrease of the self-broadening widths for the nD-levels of alkali atoms (as compared to the nS-levels, see Fig. 9.6) can not be explained using the asymptotic theory.

As follows from the available calculations, the impulse-approximation results are sensitive to the choice of the ^3P-resonance parameters E_r and Γ_r. Moreover, the small variations of the energy E_r and width Γ_r lead to a considerable difference in the magnitudes of the quenching [4.16, 5.75] and broadening [4.16, 5.21, 19, 26] cross sections. Therefore, the semiempirical method [4.16, 5.21] of detection of the resonance parameters from the monotone components of the available experimental data on broadening [5.27, 28] or quenching [5.29] of Rydberg states by alkali-metal atoms provides reasonable values for E_r and Γ_r. In particular, both the energies E_r and widths Γ_r obtained in [4.16, 5.21] for K^- and Rb^- ions turn out to be close to those calculated using the modified effective range theory developed by *Fabrikant* [5.18, 19].

b) Oscillatory Behavior of Broadening Cross Section. At intermediate principal quantum numbers ($15 < n < 25-30$ for collisions with K, Rb, and Cs) the impulse approximation and, hence, the theory [4.16, 5.21] does not hold (see Sect. 5.4.5). An appropriate theoretical explanation of the oscillatory behavior of the broadening rates in this range of n (see Fig. 9.6) was given in a series of papers by Borodin and coworkers [5.78, 26, 79]. These authors investigated the influence of the quasi-discrete ^3P-level of the negative alkali-metal ion B^- in the impact broadening process of the Rydberg alkali atom A^* perturbed by the ground-state K, Rb, and Cs atoms within the framework of a quasi-molecular approach. In contrast to the impulse approximation ($n > 25-30$), this approach takes into account the interference phenomena accounted for by the multiple scattering of the highly excited atomic electron by a perturbing atom B. Due to the multicollisional interaction between these particles the adiabatic energy curves are formed. Thus, the inelastic process of the Rydberg atom-neutral collision can be considered on the basis of the theory [9.14] of nonadiabatic transitions.

The nonadiabatic transitions between the ionic $(A^+ - B^-)$ and covalent $(A^* - B)$ terms of the colliding particles is an interesting example of application of the quasi-molecular approach to the Rydberg atom-neutral collision problem. Such transitions occur if the perturbing atom or molecule can produce the negative B^- ion in the weakly bound state $(E_b < 0)$ or in the quasi-stationary state $(E_r > 0)$. One of the possible mechanism of inelastic transitions between the Rydberg states induced by collisions with neutral atom was discussed by *Presnyakov* [9.15]. He considered the influence of nonadiabatic transitions between the weakly bound ionic term of the $(A^+ - B^-)$-system and the covalent hydrogen-like terms (split by the Fermi pseudopotential) upon the behavior of the broadening width. Further development of the quasi-molecular approach was made by *Golubkov* and *Ivanov* [9.16], who have investigated the harpoon mechanism of energy transfer processes (excitation and ionization) of Rydberg atom A^* in thermal collision with homonuclear molecule $B = X_2$ possessing a quasi-discrete level of the negative X_2^- ion. A specific feature of the interaction between the ionic $A^+X_2^-$ configuration and the covalent nlm-states was taken into account using a procedure based on the expansion of the atomic wave function in the pseudo-crossing region of molecular terms. The transition probability was calculated within the framework of the *Demkov* and *Osherov* [9.17] model.

Below we focus only on the main idea of the quasi-molecular approach [5.78] to the calculation of the triplet part $\sigma_r^{br} \equiv (3/4)\,\sigma_+^{br}$ of the total broadening cross section

$$\sigma_\alpha^{br} = \frac{3}{4}\sigma_+^{br}(\alpha) + \frac{1}{4}\sigma_-^{br}(\alpha) \tag{9.39}$$

for the simplest process of collisions between the ground-state alkali atom $B(^2S)$ and the Rydberg alkali atom A^* in the nS-state. The quasi-discrete ^3P-resonance state of the negative B^- ion and the ^1S-state of the positive A^+ ion lead to the formation of the two $^3\Sigma$ and $^3\Pi$ two-center ionic terms of the quasi-molecule A+B. These ionic quasi-discrete terms (with the asymptotic energy $E_r = \hbar^2 k_r^2/2m$ as $R \to \infty$) become stable, when the internuclear distance R of the colliding particles is smaller than $R_r = e^2/E_r$. An additional decreasing of the internuclear distance provides pseudo-crossing (at $R = R_{cr}(n_*) < R_r$) of the ionic $^3\Sigma$ and $^3\Pi$ terms with both the degenerate manifold of the Coulomb (hydrogen-like) n, l $(l \gtrsim 2)$ terms and with nonhydrogen-like nS, nP, and nD covalent terms of quasi-molecule $A(nl)+B$ (see Fig. 9.9). For example, a parallel system of the covalent $^1\Sigma$ and $^3\Sigma$ terms is generated by the n^2S-state of the Rydberg atom A^* and the ^2S-state of the perturbing atom B. Note that only the $^3\Sigma$ configurations are predominant in the process of nonadiabatic transitions under consideration because the small effects of the spin-orbit and Coriolis interactions can be neglected.

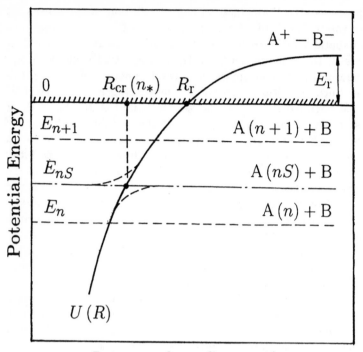

Internuclear Separation

Fig. 9.9. Schematic diagram of the potential energy curves for Rydberg atom-neutral $A^* + B$ collision in the presence of the resonance on the quasi-discrete level $(E_r > 0)$ of the negative B^- ion. Full curve is the ionic term $U(R) = E_r - e^2/R$ of the quasi-molecule correlated with the $A^+ - B^-$ state of separated ions with asymptotic energy $U(\infty) = E_r > 0$. Dashed-dotted and dashed curves represent the covalent terms of quasi-molecule $A(nl) + B$ correlated with nonhydrogen-like nS state $(E_{nS} = -Ry/n_*^2, n_* = n - \delta_S)$ and with degenerate manifold of hydrogen-like nlm and $n+1, l, m$ states with $l \gtrsim 23$ $(E_n = -Ry/n^2)$, respectively

The position of the crossing point $R_{cr}(n_*)$ via the effective principal quantum number $n_* = n - \delta_l$ is given by

$$R_{cr}(n_*) = \frac{2n_*^2 a_0}{1 + n_*^2 (E_r/Ry)}. \tag{9.40}$$

Meanwhile, the position of the first zero of the Rydberg electron wave function $R_{ns}(r)$ can be approximately estimated as

$$r_1^{(0)} \approx (2n_*)^2 a_0 \left[1 - (2/n_*^2)^{1/3}\right]. \tag{9.41}$$

Thus, in the region of $n_* \gtrsim n_*^{(0)}$ (where $n_*^{(0)} = \left[2(Ry/E_r)^3\right]^{1/8}$) the pseudo-crossing points $R_{cr}(n_*)$ of the ionic and covalent $^1\Sigma$ terms lie inside the classically allowed range of the Rydberg electron motion on the left from the right turning point, i.e., $R_{cr}(n_*) < r_1^{(0)}$. In this case the splitting of these terms (or the mixing parameter) at $R = R_{cr}(n_*)$ is proportional to the square of the radial derivative of the Rydberg electron wave function.

In the semiclassical approximation the impact broadening cross section of the atomic α state can be expressed in terms of the diagonal element of the S-matrix [see (9.15)]. The two-state approximation leads to the following expression for the S-matrix element [9.14]

$$S_{\alpha\alpha} = \left(1 - W^{(a)}\right) \exp\left(2i\eta^{(a)}\right) + W^{(a)} \exp\left(2i\eta^{(d)}\right). \tag{9.42}$$

Here $W^{(a)}$ is the probability of the inelastic transition between the adiabatic states under consideration, while $\eta^{(a)}$ and $\eta^{(d)}$ are the phases of the quasi-molecular BA system evolution along the adiabatic and diabatic paths. Thus, the triplet part of the broadening cross section can be estimated by formula

$$\sigma_+^{br} = 4\pi \left[\int_0^\infty \left(1 - W^{(a)}\right) \sin^2\left(\eta^{(a)}\right) \rho d\rho + \int_0^\infty W^{(a)} \sin^2\left(\eta^{(d)}\right) \rho d\rho\right]. \tag{9.43}$$

At small values of ρ, the phases $\eta^{(a)}$ and $\eta^{(d)}$ become large enough, so that $\sin^2\left[\eta^{(a,d)}(\rho)\right] \approx 1/2$. According to [5.78] the adiabatic $^1\Sigma$ and diabatic $^3\Sigma$ terms coincide and, hence, the phase $\eta^{(d)}$ is small for the ρ values greater than the Weisskopf radius $\rho_W = (\pi\alpha v_0/4V)^{1/3}$ corresponding to the polarizational interaction of the alkali atom B with the ionic core A^+.

The adiabatic $^3\Sigma$ phase is large in the range of $\rho < R_{cr}(n_*)$ and reveals strong damping with the impact parameter ρ greater than $R_{cr}(n_*)$. Thus, the total contribution of the resonance (triplet) and potential scattering as well as of elastic $B \to A^+$ scattering to the broadening cross section (9.39) is given by [5.78]

$$\sigma_{nS}^{br} = \sigma_{pot}^{br}(nS) + \pi\rho_W^2 + \frac{3\pi}{2} \int_{\rho_W}^{R_{cr}(n_*)} \left(1 - W^{(a)}\right) \rho d\rho. \tag{9.44}$$

The probability $W^{(a)}$ of the inelastic nonadiabatic transition between the ionic and covalent terms is approximately unity, when the mixing parameter of these terms is small enough. Then, the total broadening cross section reaches its minimum, whose magnitude is approximately equal to

$$\sigma_{min}^{br} \approx \sigma_{pot}^{br}(nS) + \pi\rho_W^2.$$

In the opposite case, when the mixing parameter is large the transition probability $W^{(a)}$ is small, so that

$$\sigma_{max}^{br} \approx \sigma_{pot}^{br}(nS) + \frac{\pi}{4}\left[\rho_W^2 + 3R_{cr}^2(n_*)\right].$$

Thus, the oscillatory structure of the broadening width of Rydberg nS states versus the principal quantum number is a result of the oscillatory dependence of the mixing parameter on n_*.

A simple estimate for the resonance part of the broadening cross section may be obtained by the Landau-Zener formula. As follows from [9.17], this formula is applicable for calculating the transition probability when the sloping ionic term crosses the system of the parallel covalent terms. Hence (9.44) provides appropriate results for collisions of the Rydberg atoms with the alkali atoms in spite of the multichannel character of the problem. Nevertheless, in the standard Landau-Zener approach, the mixing parameter (or the interaction of the pseudo-crossing terms) is a sufficiently slow function of the internuclear distance in the vicinity of the transition point $R_{cr}(n_*)$. For the Rydberg atom-alkali atom collision the effective matrix element of this interaction is the oscillatory function of the internuclear distance, i.e., this assumption does not hold. However, the probability $W^{(d)} = 2\pi\xi$ of the inelastic transition between the quasi-adiabatic ionic term and the covalent nS-state (i.e, between the diabatic energy curves) is small enough. Therefore, the calculation of the Massey parameter ξ was carried out in [5.78] by the distorted wave method while the transition probability was evaluated by the Landau-Zener formula using this parameter.

The resultant probability $W^{(a)} = \exp\left(-W^{(d)}\right)$ of inelastic transition between the adiabatic terms takes the form

$$W^{(a)} = \exp\left[-2\pi\xi(n_*)\right], \quad \xi(n_*) = \frac{3\Gamma_r v_0 \cos^2\left(\eta_1^{(0)}\right)}{4\pi n_*^3 E_r V(R_{cr})}\sin^2[\Phi(n_*)], \quad (9.45)$$

$$\Phi(n_*) = 2n_*\left\{\frac{\pi}{2} - [s_{cr}(1-s_{cr})]^{1/2} - \arctan\left[\left(\frac{s_{cr}}{1-s_{cr}}\right)^{1/2}\right] - \frac{\pi}{4}\right\}, \quad (9.46)$$

where $s_{cr} = R_{cr}(n_*)/2n_*^2 a_0$; $V(R_{cr})$ is the radial velocity at $R = R_{cr}$; $\eta_1^{(0)}$ is the partial phase of the potential scattering; and $\Phi(n_*)$ is the phase of semiclassical wave function $\mathcal{R}_{ns}(r)$ of the Rydberg electron at $r = R_{cr}(n_*)$. Thus, the transition probability between the ionic and covalent terms reveals the oscillatory dependence on n, which provides the oscillatory behavior of the broadening cross section [5.78]

$$\sigma_{ns}^{br} \approx \frac{3\pi}{4}R_{cr}^2(n_*)\left\{1 - 2\left(\frac{\Lambda}{V}\right)^2\Gamma\left(-2, \frac{(\Lambda/V)}{(1-\rho_W^2/R_{cr}^2)^{1/2}}\right)\right\}$$

$$+\frac{\pi}{4}\rho_W^2 + \sigma_p^{br}(nS), \quad \Lambda = 2\pi\xi V(R_{cr}), \quad (9.47)$$

where $V \equiv V_\infty$ is the relative velocity of colliding atoms as $R \to \infty$, and $\Gamma(x,y)$ is the incomplete gamma function. It can be seen from (9.47) that the minimum magnitudes of the broadening cross section appear, when the $\xi(n_*)$ parameter becomes equal to zero, i.e., $\xi\left(n_*^{min}\right) = 0$. The maximum magnitudes of σ_{nS}^{br} correspond to the values of n_*, for which $\sin^2[\Phi(n_*)] \sim 1$.

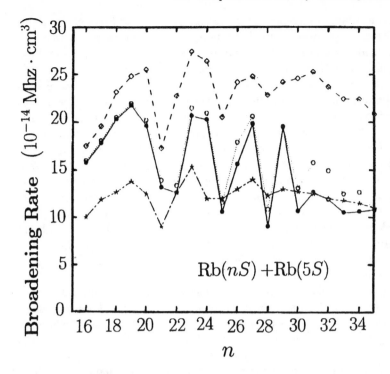

Fig. 9.10. Comparison of theoretical calculations of the broadening rates for the Rb(nS)+Rb($5S$) thermal collisions with experiment at the intermediate range of n ($16 \leq n \leq 34$). Full circles are the adiabatic-theory results [5.79] based on calculations of transition probabilities in the complex-plane of internuclear distances. Empty circles are the diabatic perturbation theory results [5.26]. Stars and diamonds are the experimental data [5.27, 28], respectively

The oscillatory structure of $\sigma_{ns}^{\mathrm{br}}$ appears only in the range of $n_* > n_*^{(0)} = \left[2\left(Ry/E_{\mathrm{r}}\right)^3\right]^{1/8}$ [see (9.40, 41)]. A simple estimate shows that the $n_*^{(0)}$ value is approximately equal to 10–11 for K and Rb (when $E_{\mathrm{r}} \sim 10^{-3}$ a.u.) in accordance with the available experimental data. The physical meaning of the disappearance of the oscillations at high enough $n_* \gtrsim 25$–30 was already discussed in Sect. 5.4.5 within the framework of the impulse approximation. However, the mechanism of their disappearance within the framework of the quasi-molecular approach is less evident.

The results presented above give the opportunity to explain the oscillatory behavior (see Fig. 9.10) of the broadening widths observed [5.27, 28] at the intermediate range of n ($15 < n < 25$–30) for collisions of Rydberg atoms with K, Rb, and Cs. A quantitative description [5.26, 78] of such phenomenon is based on the Wigner parametrization of the ^3P-resonance scattering phase shift. The authors of [5.26, 78] used the adiabatic approach in describing the

$^3\Sigma$ quasi-molecular state (generated by the quasi-discrete ^3P-resonance level) while all other partial-wave contributions to the broadening width were approximately calculated using the asymptotic expression [5.59] of the impulse approximation. The ^3P-resonance and nonresonant scattering phase shifts η_ℓ^+ and η_ℓ^- for the electron-alkali atom scattering were taken from the Fabrikant's results [5.19] obtained by extrapolation of the close coupling calculations into the range of low energies. As was established in [5.26], the combination of the impulse and adiabatic approaches give the opportunity to obtain satisfactory agreement for the monotone part of the broadening widths and qualitative agreement for their oscillatory parts at intermediate n. However, the calculations [5.26, 78] yield overestimated magnitudes for the amplitude of these oscillations compared with experiment [5.27, 28]. The adiabatic theory results [5.79] based on calculations of transition probabilities in the complex-plane of internuclear distances of the Rydberg Rb(nS)-state broadening by Rb are presented in Fig. 9.10. It can be seen that these calculations provides quite reasonable quantitative description of the oscillatory behavior of the impact width at the intermediate range of n.

The problem of inelastic transitions and impact-broadening in collisions of Rydberg atoms with the alkali atoms needs further theoretical analysis, since the adiabatic quasi-molecular approach does not hold for large n, while the impulse approximation considered in Sect. 5.4.5 is broken down at low and intermediate n. Thus, it is necessary to develop a new common approach for describing such collisions in a wide range of the principal quantum number.

9.4 Broadening of $n - n'$ Lines in a Plasma

The spectral $n - n'$ lines between highly excited levels are observed from the astronomical objects (planetary nebulas, interstellar medium) last several decades. It is a powerful diagnostic method of a plasma. In this section we give expressions for the broadening cross sections, their comparison with available laboratory experimental data, and present some fitting formula convenient for applications. The consideration is based on the semiclassical theory of transitions between Rydberg states induced by charged particles, presented in the preceding chapter. We will suggest also that the plasma temperature is much larger energy difference between neighboring levels, i.e. $kT \gg \Delta E_{n,n\pm1}$.

Since the scattering electron amplitudes by the n and n' states are close to each other (i.e., we assume $\Delta n \ll n$), hence, in accordance with (9.17) the main contribution into the broadening of the $n - n'$ lines is given by inelastic electron collisions, while the elastic scattering is negligible (see also [9.18, 19]). Therefore, in the case of large enough principal quantum numbers we find from formulas (9.21) for the width on the half-height of the $n \to n+\kappa$ line

$$\gamma = N_e[\langle V\sigma_n\rangle + \langle V\sigma_{n+\kappa}\rangle], \qquad \langle V\bar\sigma_n\rangle = \sum_{\kappa \neq 0} \langle V\sigma_{n+\kappa,n}\rangle. \qquad (9.48)$$

Here N_e is the electron density, the brackets mean the averaging over the Maxwellian electron distribution. Really, the terms with small κ dominate in the sum. Using formulas (8.26, 27) of Chap. 8, we can write the expression for the total broadening cross section of the n level as

$$\sigma_n = \left(\pi a_0^2/Z^4\right) n^4 s_n, \quad s_n = \sum_{\kappa \neq 0} s_{n+\kappa,n}/\kappa^3$$

$$= 2Z^2 \int_0^\infty \frac{\rho d\rho}{n^4 a_0^2} \sum_{\kappa \neq 0} W_{n+\kappa,n} \left(\rho\right). \tag{9.49}$$

The normalization conditions (4.79, 49) give

$$\sum_{\kappa \neq 0} W_{n+\kappa,n} = 1 - \frac{1}{2\pi} \int_0^{2\pi} du \exp\left[-iS\left(u\right)/\hbar\right]. \tag{9.50}$$

Formulas (9.49, 50) are convenient for calculations of the the cross section σ_n and rate coefficient $\langle V\sigma_n \rangle$. We write $\langle V\sigma_n \rangle$ in the form:

$$\langle V\sigma_n \rangle = \left(n^4/Z^3\right) K_n, \tag{9.51}$$

$$K_n = \left(\pi a_0^2 v_0\right) \left(\frac{m}{\mu}\right)^2 \theta^{-3/2} \int_0^\infty e^{-E/\kappa T} s_n \frac{EdE}{\left(Z^2 Ry\right)^2}, \quad \theta = \frac{m}{\mu} \frac{kT}{Z^2 Ry},$$

where μ is the reduced mass of the Rydberg atom and perturbing particle. The K_n value depends weakly on temperature and principal quantum number. In the Born approximation (see Sect. 8.2) the s_n value ($\varepsilon_e = mE/\mu Z^2 Ry$, $\varepsilon = |E_n|/Z^2 Ry$) is

$$s_n^B = \left(\frac{Zv_0}{V}\right)^2 \left[\frac{1}{2} \ln\left(1 + \varepsilon_e/\varepsilon\right) \sum_{\kappa \neq 0} \left(1 - \frac{0.25}{\kappa}\right)/\kappa^4 \right.$$

$$\left. +\frac{4}{3} \frac{\varepsilon_e/\varepsilon}{1 + \varepsilon_e/\varepsilon} \sum_{\kappa \neq 0} \left(1 - \frac{0.6}{\kappa}\right)/\kappa^3 \right] \tag{9.52}$$

$$= \left(\frac{Zv_0}{V}\right)^2 \left[0.82 \ln\left(1 + \varepsilon_e/\varepsilon\right) + 1.47 \frac{\varepsilon_e/\varepsilon}{1 + \varepsilon_e/\varepsilon}\right].$$

Similarly, for the K_n value, we obtain

$$K_n^B = \left(\pi a_0^2 v_0\right) \frac{1}{\sqrt{\theta}} \left\{0.82\varphi\left(\varepsilon/\theta\right) + 1.47 \cdot \left[1 - \left(\varepsilon/\theta\right)\varphi\left(\varepsilon/\theta\right)\right]\right\}. \tag{9.53}$$

$$\varphi(x) = -e^x \, \mathbf{Ei}\left(-x\right) = \int_0^\infty \frac{e^{-t}}{x+t} dt.$$

Function $\varphi(x)$ may be fitted with an error less than 5% by the expression

$$\varphi(x) \approx \ln\left(1 + \frac{1 + 2.49x}{1.78x\,(1 + 1.4x)}\right). \tag{9.54}$$

The K_n value and the σ_n cross section obtained in the framework of the semiclassical approach may be written (see Sect. 8.3) in the form:

$$\sigma_n = \sigma_n^{\mathrm{B}} f_E(Z,n), \quad K_n = K_n^{\mathrm{B}} f_\theta(Z,n). \tag{9.55}$$

Factors $f_E(Z,n)$ and $f_\theta(Z,n)$ may be fitted with an accuracy of 10–20% by the expressions:

$$f_E(Z,n) = \frac{\ln[1 + (Vn/Z^2 v_0)/(1 + (1.25 v_0/V)^{3/2})]}{\ln[1 + (Vn/Z^2 v_0)]}, \quad \begin{array}{l} v = (\varepsilon_e)^{1/2} \\ v_1 = 1.25 v_0 \end{array},$$

$$f_\theta(Z,n) = \frac{\ln[1 + (n\sqrt{\theta}/Z)/(1 + c/Z\sqrt{\theta})]}{\ln[1 + (n\sqrt{\theta}/Z)]}, \quad c = 2.5. \tag{9.56}$$

At present, there are no direct measurements of the cross sections averaged over angular quantum numbers. The width measurement offers the possibility to check the theory indirectly. The decay time measurements of the Rydberg states in the helium afterglow plasma were presented in [8.13]. The $n^3 P$-states ($8 \le n \le 17$) were excited from the $\mathrm{He}(2^3 S)$ state by a dye laser, and then the time dependences of the $2^3 P - n^3 D$ lines are investigated. The thermodynamic equilibrium over angular quantum numbers was established very quickly (this fact was under particular control), therefore the decay time gives the total (averaged over l) inelastic rate coefficient. The temperature and electron density were measured independently by microwave methods. Figure 9.11 shows the comparison of theoretical and experimental data for the

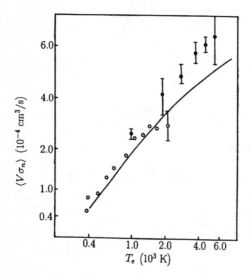

Fig. 9.11. The decay rate coefficient of the He triplet levels with $n = 10$ versus the electron temperature in afterglow of discharge. Experimental data from [8.13], Full line is the semiclassical approach

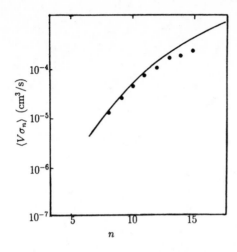

Fig. 9.12. The decay rate coefficient of the He triplet levels versus the principal quantum number in afterglow of discharge. Experimental data from [8.13]. Full line is the semiclassical approach ($T_e = 500$ K)

inelastic electron rate coefficient of $n = 10$ state versus the electron temperature. The agreement proved to be good at low temperatures. At temperature 8000 K the difference between theoretical and experimental results account for about 20–40%. However, it should be noted that experimental errors are growing in this region. Comparison of the theoretical and experimental dependencies on principal quantum number n is given in Fig. 9.12

A similar experiment was carried out for a Hydrogen plasma with $N_e = 1.75 \cdot 10^{13}$ cm^{-3} and $T_e = 1750$ K in [9.20]. The decay time of the states with principal quantum numbers $n = 3$, 4, 6, and 7 were measured. The experimental rate coefficients for levels with $n = 6$, and 7 are equal to $(1 - 2.5) \cdot 10^{-5}$ cm^3/s and $(1.5 - 4.3) \cdot 10^{-5}$ cm^3/s, respectively. The corresponding theoretical values are $0.8 \cdot 10^{-5}$ cm^3/s and $1.9 \cdot 10^{-5}$ cm^3/s, respectively.

List of Symbols

Fundamental constants

$\hbar = h/2\pi$ Planck constant ($\hbar = 1.0546 \cdot 10^{-27}$ erg $\cdot$ s)

c velocity of light ($c = 2.9979 \cdot 10^{10}$ cm/s)

e charge of electron ($e = 4.8032 \cdot 10^{-10}$ CGSE)

m mass of electron ($m = 9.1095 \cdot 10^{-28}$ g)

s electron spin ($s = 1/2$)

$Ry = me^4/2\hbar^2$ Rydberg constant ($Ry = 2.18 \cdot 10^{-11}$ erg $= 13.606$ eV)

$\alpha = e^2/\hbar c$ fine-structure constant ($\alpha = 1/137.036$)

Atomic Units

$a_0 = \hbar^2/me^2$ Bohr radius ($a_0 = 0.529 \cdot 10^{-8}$ cm)

$v_0 = e^2/\hbar$ atomic unit of velocity ($v_0 = 2.188 \cdot 10^8$ cm $\cdot$ s^{-1})

$\tau_0 = \hbar^3/me^4$ atomic unit of time ($\tau_0 = a_0/v_0 = 2.419 \cdot 10^{-17}$ s)

$2Ry$ atomic unit of energy

a_0^2 atomic unit of area

a_0^{-3} atomic unit of density

$v_0 a_0^2$ atomic unit of the rate coefficient (rate constant)

General Collisional Characteristics

ρ impact parameter

$\mathbf{Q}$ momentum transfer

f_{fi} scattering amplitude for the $|i\rangle \rightarrow |f\rangle$ transition

a_{fi} amplitude of the $|i\rangle \rightarrow |f\rangle$ transition

W_{fi} transition probability ($W_{fi} = |a_{fi}(t = +\infty)|^2$)

σ_{fi} cross section of the $|i\rangle \rightarrow |f\rangle$ transition

K_{fi} rate constant (rate coefficient $K_{fi} = \langle V\sigma_{fi}\rangle$)

Rydberg Atom

Z	spectroscopic symbol ($Z = 1$ for neutral atom)
n	principal quantum number
l	orbital quantum number
m	magnetic quantum number
J	total angular momentum quantum number
g	statistical weight ($g_n = n^2$; $g_{nl} = 2l + 1$; $g_s = 2$; $g_J = 2J + 1$)
E_{nl}	energy of the Rydberg electron ($E_{nl} = -Z^2 Ry/n_*^2$)
n_*	effective principal quantum number
δ	quantum defect ($n_* = n - \delta$)
$\varepsilon = \mid E_{nl} \mid /Z^2 Ry$	energy in Coulomb units ($\varepsilon = 1/n_*^2$)
v_n	characteristic velocity ($v_n = Zv_0/n_*$)
ϵ	eccentricity of the Rydberg electron orbit ($\epsilon = [1 - (l + 1/2)^2/n^2]^{1/2}$
	radius vector of the Rydberg electron
$\mathbf{r}$	(x, y, z and r, θ, φ are the cartesian and spherical coordinates)
$\mathbf{p}$	momentum of the electron
$\mathbf{k}$	wave vector of the electron
H	Hamiltonian
$U(r)$	interaction potential of the Rydberg electron with ionic core
$\psi_{nl}(\mathbf{r})$	wave function in the coordinate space ($\psi_{nlm} = \mathcal{R}_{nl} \cdot Y_{lm}$)
$\mathcal{R}_{nl}(r)$	radial part of the coordinate wave function
$Y_{lm}(\theta, \varphi)$	angular part of the coordinate wave function
$G_{nl}(\mathbf{p})$	wave function in the momentum space ($G_{nl} = g_{nl} \cdot Y_{lm}$)
$g_{nl}(p)$	radial part of the momentum wave function
$G(\mathbf{r}', \mathbf{r}; \mathbf{E})$	Green function
S	action function
	action variables ($\mathbf{I} = (I_1, I_2, I_3)$;
$\mathbf{I}$	relation to the quantum numbers: $I_1 = n\hbar, I_2 = l\hbar, I_3 = m\hbar,$
$\mathbf{u}$	conjugate action variables ($\mathbf{u} = (u_1, u_2, u_3)$)
γ	width of Rydberg level (full width at half maximum)
Δ	shift of Rydberg level ΔE_{fi} energy of the $\mid i \rangle \to \mid f \rangle$ transition
ω_{fi}	frequency of the $\mid i \rangle \to \mid f \rangle$ transition
$F_{fi}(Q)$	form factor of the $\mid i \rangle \to \mid f \rangle$ transition
$\mathcal{F}_{fi}(Q)$	form factor symmetrical over the initial and final states ($\mathcal{F}_{fi} = g_i F_{fi}$)
f_{fi}	oscillator strength of the $\mid i \rangle \to \mid f \rangle$ transition
$R_{nl}^{n'l'}$	radial integral ($R_{nl}^{n'l'} = \langle \mathcal{R}_{n'l'} \mid r \mid \mathcal{R}_{nl} \rangle$)

Projectile Characteristics

$\mathcal{Z}e$ charge of the incident particle

M mass of the incident particle

set of the internal quantum numbers of neutral projectile

β (v and j are the vibrational and rotational quantum numbers for molecular projectile)

E_β internal energy of neutral projectile

$\mathcal{D}$ dipole moment of neutral projectile

$\mathcal{Q}$ quadrupole moment of neutral projectile

α polarizability of neutral projectile

$\mathcal{V}$ total interaction between the Rydberg atom and projectile

$\mathbf{R}$ radius vector of projectile with respect to the center of mass of the Rydberg atom

μ reduced mass of the Rydberg atom and projectile

$\mathcal{E}$ kinetic energy of relative motion

$\mathbf{V}$ relative velocity with respect to the Rydberg atom

$\mathbf{q}$ wave vector of the incident neutral particle relative to the Rydberg atom

$\mathbf{k}$ wave vector of the Rydberg electron with respect to the neutral projectile

$\mathbf{v}$ velocity of the incident electron

E energy of the incident electron

Special Functions

$\Gamma(x)$	Gamma-function
$\Gamma(a, x)$	incomplete Gamma-function
$B(x, a)$	Beta-function $[B(x, a) = \Gamma(x)\Gamma(a)/\Gamma(x + a)]$
$B_x(a, b)$	incomplete Beta-function
$E_n(x)$	Integral exponential function of the nth order
$\mathbf{Ei}(x)$	Integral exponential function $(-\mathbf{Ei}(x) = [E_1(-x + \mathrm{i}0) + E_1(-x - \mathrm{i}0)]/2, x > 0)$
$\mathrm{erf}(x)$	probability integral
$\mathrm{erfc}(x)$	additional probability integral $[\mathrm{erfc}(x) = 1 - \mathrm{erf}(x)]$
$C(x), S(x)$	Fresnel integrals
$P_l^m(x)$	Legendre polynomial
$L_n^m(x)$	Laguerre polynomial
$C_n^m(x)$	Gegenbauer polynomial
$J_n(x), I_n(x)$	Bessel functions
$K_n(x)$	MacDonald function
$\mathbf{J}_\nu(x)$	Anger function
$j_n(x)$	spherical Bessel function
$M_{\mu\nu}(x), W_{\mu\nu}(x)$	Whittaker functions

$D_\nu(x)$ Weber parabolic cylinder function $[D_\nu(x) = U(-\nu - 1/2, x)]$
$F(a, c; x)$ confluent hypergeometrical function
$F(a, b; c; x)$ hypergeometrical function
$\zeta(x)$ Riemann zeta function
$K(k)$ complete elliptic integral
$F(\varphi, k)$ incomplete elliptic integral

Reference List

Chapter 1

[1.1] R.F. Stebbings, F.B. Dunning (eds.): *Rydberg States of Atoms and Molecules* (Cambridge University Press, Cambridge 1983)

[1.2] T.F. Gallagher: *Rydberg Atoms* (Cambridge University Press Cambridge, 1994)

[1.3] I.L. Beigman, V.S. Lebedev: Phys. Rep. **250**, 95–328 (1995)

[1.4] M.J. Seaton: Rep. Prog. Phys. **46**, 167–257 (1983)

[1.5] M. Aymar, C.H. Greene, E. Luc-Koenig: Rev. Mod. Phys. **68**, 1015–1123 (1996)

[1.6] S. Haroche, J. Raimond: Adv. Atom. Mol. Phys. **20**, 347–411 (1985)

[1.7] J.A.C. Gallas, G. Leuchs, H. Walther, H. Figger: Adv. Atom. Mol. Phys. **20**, 413–466 (1985)

[1.8] G. Casati, B.V. Chirikov, D.L. Shepelyansky, I. Guarneri: Phys. Rep. **154**, 77–123, (1987)

[1.9] P.M. Koch: *The Ubiquity of Chaos* ed. by S. Krasner (American Association for the Advancement of Science, Washington, DC 1990) pp.75–96; *Chaos/Xaoc: Soviet-American Perspectives on Nonlinear Science* ed. by D.K. Campbell American Institute of Physics, (New York 1990) pp.441–475; Lect. Notes in Phys. **411**, 167 (Springer, Heidelberg 1992).

[1.10] V.S. Lisitsa: Usp. Fiz. Nauk **153**, 379–421 (1987)

[1.11] I.Sh. Averbukh, N.F.Perel'man: Usp. Fiz. Nauk. **161**, 41–81 (1991)

[1.12] M.A. Gordon, R.L. Sorochenko (eds.): *Radio Recombination Lines: 25 years of Investigations* (Kluwer Academic Publishers, Dordrecht, Boston, London 1990)

Chapter 2

[2.1] L.D. Landau, E.M. Lifshitz: *Mechanics* (Pergamon, Oxford 1969)

[2.2] L.D. Landau, E.M. Lifshitz: *Quantum Mechanics* (Pergamon, Oxford 1977)

[2.3] H.A. Bethe, E.E. Salpeter: *Quantum Mechanics of One- and Two- Electron atoms* (Springer-Verlag, New York 1957)

[2.4] H. Bateman, A. Erdelyi: *Higher Transcendental Functions* (McGraw-Hill, New York, Toronto, London 1953); I.S. Gradshteyn, I.W. Ryzhik: *Tables of Integrals, Series and Products* (Academic, New York 1965); M. Abramovitz, I.A. Stegun: *Handbook of Mathematical Functions* (Dover, New York 1965)

[2.5] E. de Prunele: J. Phys. **B13**, 3921(1980)

[2.6] A.B. Migdal: *Qualitative Methods in Quantum Theory* (Nauka, Moscow 1975)

[2.7] R.M. More, K.H. Warren: Ann. Phys. **207**, 282(1991)

[2.8] D.A. Varshalovich, A.N. Moskalev, V.K. Khersonsky: *Quantum Theory of the Angular Momentum* (World Scientific, Singapore 1988)

[2.9]   D.R. Bates, A. Damgaard: Philos. Trans. R. Soc. **242**, 101(1949)
[2.10] M. Matsuzawa: J. Phys. **B8**, 2114(1975), corrigendum **9**, 2559(1976)
[2.11] S.W. Qian, X.Y. Huang: Phys. Lett. **A115**, 319(1986)
[2.12] L. Hostler: J. Math. Phys. **5**, 591(1964); L. Hostler, R. Pratt: Phys.Rev. **10**, 469(1963)
[2.13] V. Fock: Zeitschr. f. Phys. **98**, 145(1935)
[2.14] E.P. Wigner: Phys. Rev. **40**, 749(1932)
[2.15] M.V. Berry: Phil. Trans. Roy. Soc. **287**, 237 (1977)

Chapter 3

[3.1]   W. Gordon: Ann. Phys. **2**, 1031(1929)
[3.2]   I.I. Sobel'man: *Atomic Spectra and Radiative Transitions* (Springer-Verlag, Berlin, Heidelberg 1992)
[3.3]   V.A. Davidkin, B.A. Zon: Opt. Spectrosk. **51**, 13(1981)
[3.4]   J. Picart, A.R. Edmonds, N.Tran Minh: J. Phys. **B11**, L651(1978); A.R. Edmonds, J. Picart, N. Tran Minh, R. Pullen: J. Phys. **B12**, 2781(1979)
[3.5]   G.K. Oertel, L.R. Shomo: Astrophys. J. Suppl. **16**, 175(1968)
[3.6]   W.J. Karsas, R.Latter: Astrophys. J. Suppl. **6** 167,(1961)
[3.7]   M. Pajek, R.Schuch: Phys. Rev. **A45**, 7894(1992)
[3.8]   I.I. Sobel'man, L.A. Vainshtein, E.A. Yukov: *Excitation of Atoms and Broadening of Spectral Lines* (Springer, Berlin 1981)
[3.9]   L.D. Landau, E.M.Lifshitz: *Classical Field Theory* (Pergamon, New York 1975)
[3.10] S.P. Goreslavsky, N.B. Delone, V.P. Krainov: Zh. Eksp. Teor. Fiz. **82**, 1789(1982)
[3.11] J.G. Baker, D.M. Menzel: Astrophys. J. **88**, 52(1938)
[3.12] H.Z. Elwert: Naturforsch. **10a**, 361(1955); L.I. Podlubnyi: Opt. Spectrosk. **22**, 60(1967); ibid. **27**, 404(1969)
[3.13] G.C. McGoyd, S.N. Milford, J.J. Wah: Phys. Rev. **119**, 149(1960); G.C. McGoyd and S.N. Milford: Phys. Rev. **130**, 906(1963); L.N. Fisher, S.N. Milford, F.R. Pomilla: Phys. Rev. **130**, 153(1960)
[3.14] A.O. Barut, R. Wilson: Phys. Rev. **A13**, 918(1976); Phys. Rev. **A40**, 1340(1989)
[3.15] I. Percival, D. Richard: Adv. Atom. Mol. Phys. **11**, 1(1975)
[3.16] I.L. Beigman, A.M. Urnov, V.P. Shevelko: Zh. Eksp. Teor. Fiz. **58**, 1825(1970) [Sov.Phys.-JETP **31**, 978(1970)]; I.L. Beigman, A.M. Urnov: J. Quant. Spectrosc. Radiat. Transf. **14**, 1009(1974)
[3.17] M. Matsuzawa: Phys. Rev. **A9**, 241(1974); ibid. **20**, 860(1979)
[3.18] E. Sobeslavsky: Ann. Phys. **28**, 321(1973)
[3.19] I.L. Beigman, M.I.Syrkin: Leb. Phys. Rep. N5, 18(1983)
[3.20] M. Matsuzawa: J. Phys. **B12** 3743(1979)
[3.21] L.Y. Cheng, H.van Regemorter: J. Phys. **B14** 4025(1981)
[3.22] M. Inokuti: Rev. Mod. Phys. **43**, 297(1971)
[3.23] F. Gounand, L. Petitjean: Phys. Rev. **A30**, 61(1984)
[3.24] V.M. Borodin, A.K. Kazansky, V.I. Ochkur: J. Phys. **B25**, 445(1992)
[3.25] V.P. Shevelko, L.A. Vainshtein: *Atomic Physics for Hot Plasmas* (IOP Publishing Ltd, London 1993)

Chapter 4

[4.1]   M.L. Goldberger, K.M.Watson: *Collision Theory* (Wiley, New York (1964)
[4.2]   R.G. Newton: *Scattering Theory of Waves and Particles* (McGraw-Hill, New York 1966)

[4.3]   M.J. Seaton: Proc. Phys. Soc. **79**, 1105(1962)

[4.4]   B.L. Moiseiwitsch: Proc. Phys. Soc. **87**, 885(1966)

[4.5]   I.L. Beigman: Lebedev Phys. Rep. N8, 10(1973)

[4.6]   I.L. Beigman, L.A. Vainshtein, I.I. Sobel'man: Zh. Eksp. Teor. Fiz. **57**, 1703(1969) [Sov.Phys.-JETP **30**, 926(1970)]

[4.7]   D. Richards: J. Phys. B**5** L53(1972)

[4.8]   L.P. Presnyakov, A.M. Urnov: J. Phys. B**3**, 1267(1970)

[4.9]   I.L. Beigman: Zh. Eksp. Teor. Fiz. **73**, 1730(1977) [Sov. Phys. -JETP **46**, 908(1977)]

[4.10] R.P. Feynmann: Rev. Mod. Phys. **20**, 367(1948); Phys. Rev. **84**, 108(1951)

[4.11] K. Husimi: Progr.Theor.Phys. **9**, 381(1953)

[4.12] V.S. Popov, A.M. Perelomov: Zh. Eksp. Teor. Fiz. **56**, 1375(1969)

[4.13] E. Fermi: Ric. Sci. **VII**-11, 13 (1936)

[4.14] G.F. Chew: Phys. Rev. **80** 196(1950); G.F. Chew, G.C. Wick: Phys. Rev. **85**, 636(1952); G.F. Chew, M.L. Goldberger: Phys. Rev. **87**, 778(1952)

[4.15] M. Matsuzawa: in *Rydberg States of Atoms and Molecules*, eds. R.F. Stebbings and F.B. Dunning (Cambridge University Press, Cambridge 1983) Chap. 8, p.267

[4.16] V.S. Lebedev, V.S. Marchenko: Zh. Eksp. Teor. Fiz. **91**, 428(1986) [Sov.Phys. -JETP **64**, 251(1986)]

[4.17] V.S. Lebedev: J. Phys. B**24**, 1977(1991)

[4.18] M.R. Flannery: in *Rydberg States of Atoms and Molecules*, eds. R.F. Stebbings and F.B. Dunning (Cambridge University Press, Cambridge 1983) Chap. 11, p.393

[4.19] M.R. Flannery: Phys. Rev. A**22** 2408(1980)

Chapter 5

[5.1]   E. Fermi: Nuovo Cimento **11**, 157(1934)

[5.2]   A.P. Hickman, R.E. Olson, J. Pascale: in *Rydberg States of Atoms and Molecules*, eds. R.F. Stebbings and F.B. Dunning (Cambridge University Press, Cambridge 1983) Chap. 6, p.187

[5.3]   D.W. Norcross, L.A. Collins: Adv. Atom. Mol. Phys. **8**, 341(1982)

[5.4]   Y. Itikawa: Phys. Rep. **46**, 117(1978)

[5.5]   N.F. Lane: Rev. Mod. Phys. **52**, 28(1980)

[5.6]   I.Shimamura: in *Electron-Molecule Scattering*, eds. I. Shimamura and K. Takayanagi (Plenum, New York 1984)

[5.7]   M.A. Morrison: Adv. Atom. Mol. Phys. **24** 51(1988)

[5.8]   I.I. Fabrikant: Comments At. Mol. Phys. **32**, 267(1996)

[5.9]   L. Spruch, T.F. O'Malley, L. Rosenberg: Phys. Rev. Lett. **5**, 347(1960); T.F. O'Malley, L. Spruch, L. Rosenberg: Phys. Rev. **125**, 1300(1962)

[5.10] T.F. O'Malley: Phys. Rev. **130**, 1020(1963); Phys. Rev. A**134**, 1188(1964)

[5.11] K.S. Golovanivsky, A.P. Kabilan: Zh. Eksp. Teor. Fiz. **80**, 2210(1981)

[5.12] R.J. Gulley, D.T. Alle, M.J. Brennan, M.J. Brunger, S.J. Buckman: J. Phys. B**27**, 2593(1994)

[5.13] M. Weyhreter, B. Barzick, A. Mann, F. Linder: Z. Phys. D**7** 333(1988)

[5.14] K.D. Heber, P.J. West, E. Matthias: Phys. Rev. A**37**, 1438(1988); J. Phys. D**21**, 63(1988)

[5.15] D.C. Thompson, E. Kammermayer, B.P. Stoicheff, E. Weinberger: Phys. Rev. A**36**, 2134(1987)

[5.16] A.A. Radzig, B.M. Smirnov: in *Reference Data on Atoms, Molecules and Ions*, Vol. 31 of Springer Ser. in Chemical Physics, eds. V.I. Goldanskii et al. (Springer-Verlag, Berlin 1985)

[5.17] T.F. O'Malley, R.W. Crompton: J. Phys. B**13**, 3451(1980)

[5.18] I.I. Fabrikant: Opt. Spectrosk. **53**, 223(1982)
[5.19] I.I. Fabrikant: J. Phys. B**19**, 1527(1986)
[5.20] A.L. Sinfailam, R.K. Nesbet: Phys. Rev. A**7**, 1987(1973)
[5.21] V.S. Lebedev, V.S. Marchenko: J. Phys. B**20**, 6041(1987)
[5.22] U. Thumm, D.W. Norcross: Phys. Rev. A**45**, 6349(1992)
[5.23] E.M. Karule: J. Phys. B**5**, 2051(1972)
[5.24] A. Dalgarno: Proc. Phys. Soc. London **89**, 503(1966)
[5.25] A.R. Johnston, P.D. Burrow: J. Phys. B**15** L745(1982)
[5.26] V.M. Borodin, I.I. Fabrikant, A.K. Kazansky: Phys. Rev. A**44**, 5725(1991)
[5.27] H. Heinke, J. Lawrenz, K. Niemax, K.H. Weber: Z. Phys. A**312**, 329(1983)
[5.28] D.C. Thompson, E. Weinberger, G.X.Xu, B.P. Stoicheff: Phys. Rev. A**35**, 690(1987)
[5.29] M. Hugon, F. Gounand, P.R. Fournier, J. Berlande: J. Phys. B**16**, 2531(1983)
[5.30] C.W. Walter, J.R.Peterson: Phys. Rev. Lett. **68**, 2281(1992)
[5.31] H.W. van der Hart, C. Laughlin, J.E. Hansen: Phys. Rev. Lett. **71**, 1506(1993)
[5.32] D. Sundholm: J. Phys. B**28**, L399(1995)
[5.33] K.W. McLaughlin, D.W. Duquette: Phys. Rev. Lett. **72**, 1176(1994)
[5.34] H.S.W. Massey: Proc. Cambridge Phil. Soc. **28**, 99(1931)
[5.35] S. Altshuler: Phys. Rev. **107**, 114(1957)
[5.36] M.T. Frey, S.B. Hill, X. Ling, K.A. Smith, F.B. Dunning, I.I. Fabrikant: Phys. Rev. A**50**, 3124(1995)
[5.37] M.T. Frey, S.B. Hill, K.A. Smith, F.B. Dunning, I.I. Fabrikant: Phys. Rev. Lett. **75**, 810(1995)
[5.38] S.B. Hill, M.T. Frey, F.B. Dunning: Phys. Rev. A**53**, 3348(1996)
[5.39] J.I. Gersten; Phys. Rev. A**14**, 1354(1976)
[5.40] A. Omont: J. de Phys. **38**, 1343(1977)
[5.41] J. Derouard, M. Lombardi: J. Phys. B**11**, 3875(1978)
[5.42] E de Prunele, J.Pascale: J. Phys. B**12**, 2511(1979)
[5.43] B. Kaulakys: J. Phys. B**17**, 4485(1984)
[5.44] V.S. Lebedev and V.S. Marchenko: Zh. Eksp. Teor. Fiz. **88**, 754(1985) [Sov. Phys. -JETP **61**, 443(1985)]
[5.45] J.Q. Sun, P.J. West: J. Phys. B**23**, 4119(1990)
[5.46] J.Q. Sun, E. Matthias, K.D. Heber, P.J. West, J. Gudde: Phys. Rev. A**43**, 5956(1991)
[5.47] L. Sirko, K.Rosinski: J. Phys. B**24**, L75(1991)
[5.48] V.S. Lebedev: J. Phys. B**25**, L131(1992)
[5.49] V.S. Lebedev: Zh. Eksp. Teor. Fiz. **103**, 50(1993) [Sov. Phys. -JETP **76**, 27(1993)
[5.50] V.S. Lebedev: J. Phys. B**31**, 1579(1998)
[5.51] V.S. Lebedev, I.I. Fabrikant: Phys. Rev. A**54**, 2888(1996)
[5.52] V.S. Lebedev, I.I.Fabrikant: J. Phys. B**30**, 2649(1997)
[5.53] B. Kaulakys: J. Phys. B**18**, L167(1985)
[5.54] D.R. Bates, S.P. Khare: Proc. Phys. Soc. London **85**, 2331(1965)
[5.55] M.R. Flannery: Ann. Phys. **61**, 465(1970); ibid. **79**, 480(1973)
[5.56] L. Petitjean, F.Gounand: Phys. Rev. A**30**, 2946(1984)
[5.57] L.P. Pitaevsky: Zh. Eksp. Teor. Fiz. **42**, 1326(1962)
[5.58] E.M. Lifshitz, L.P. Pitaevsky: *Physical Kinetics* Nauka (Moscow, 1979)
[5.59] V.A. Alekseev, I.I. Sobel'man: Zh. Eksp. Teor. Fiz. **49**, 1274(1965) [Sov. Phys. -JETP **22**, 882(1966)]
[5.60] M. Matsuzawa: J. Chem. Phys. **55**, 2685(1971); errata **58**, 2674(1973)
[5.61] G.N. Fowler, T.W. Preist: J. Chem. Phys. **56**, 1601(1972)
[5.62] T.W. Preist: J. Chem. Soc. Farad. Trans. I **68**, 661(1972)

[5.63] A.P. Hickman: Phys. Rev. **A18**, 1339(1978)

[5.64] A.P. Hickman: Phys. Rev. **A19**, 994(1979)

[5.65] A.P. Hickman: Phys. Rev. **A23**, 87(1981)

[5.66] M. Hugon, F. Gounand, P.R. Fournier, J. Berlande, J. Phys. **B13**, 1585(1980); ibid. **12**, 2707(1979)

[5.67] M. Hugon, P.R. Fournier, E. de Prunele: J. Phys. **B14**, 4041(1981)

[5.68] M. Hugon, B. Sayer, P.R. Fournier, F. Gounand: J. Phys. **B15**, 2391(1982)

[5.69] K. Sasano, Y. Sato, M. Matsuzawa: Phys. Rev. **A27**, 2421(1983)

[5.70] Y. Sato, M. Matsuzawa: Phys. Rev. **A31**, 1366(1985)

[5.71] Y. Hahn: J. Phys. **B14**, 985(1981); ibid. **15**, 613(1982)

[5.72] B.P. Kaulakys: Zh. Eksp. Teor. Fiz. **91**, 391(1986) [Sov. Phys. -JETP, **64**, 229(1986)]

[5.73] S.T. Butler, R.A.May: Phys. Rev. **A137**, 10(1965)

[5.74] B.M. Smirnov: *Ions and Excited Atoms* (Atomizdat, Moscow 1974)

[5.75] I.I. Fabrikant: Phys. Rev. **A45**, 6404(1992)

[5.76] M. Matsuzawa: J. Phys. **B8**, L382(1975); ibid. **B10**, 1543(1977)

[5.77] B.P. Kaulakys, L.P. Presnyakov, P.D. Serapinas: Pis'ma Zh. Eksp. Teor. Fiz. **30**, 60(1979) [JETP Lett. **30**, 43(1980)]

[5.78] V.M. Borodin, A.K. Kazansky: Zh. Eksp. Teor. Fiz. **97**, 445(1990) [Sov. Phys.-JETP **70**, 252(1990)]; J. Phys. **B25**, 971(1992)

[5.79] V.M. Borodin, A.K. Kazansky, D.B. Khrebtukov, I.I. Fabrikant: Phys. Rev. **A48**, 479(1993)

[5.80] M. Hugon, F. Gounand, P.R. Fournier: J. Phys. **B13**, L109(1980)

[5.81] X. Ling, K.A. Smith, F.B. Dunning: Phys. Rev. **A47**, R1(1993)

[5.82] X. Ling, M.T. Frey, K.A. Smith, F.B. Dunning: Phys. Rev. **A48**, 1252(1994)

Chapter 6

[6.1] A. Kumar, N.F. Lane, M. Kimura: Phys. Rev. **A39**, 1020(1989)

[6.2] M. Kimura, N.F.Lane: Phys. Rev. **A42**, 1258(1990)

[6.3] H. Liu and B. Li: Phys. Rev. **A46** 1391(1992)

[6.4] T.F. Gallagher, S.A. Edelstein, R.M. Hill, Phys. Rev. **A15**, 1945(1977)

[6.5] M. Chapelet, J. Boulmer, J.C. Gauthier, J.F. Delpech: J. Phys. **B15**, 3455(1982)

[6.6] R. Kachru, T.F. Gallagher, F. Gounand, K.A. Safinya, W. Sandner: Phys. Rev. **A27**, 795(1983)

[6.7] J. Supronovicz, J.B. Atkinson, J. Krause: Phys. Rev. **A30**, 112(1984)

[6.8] M. Lukaszewski, I. Jackowska: J. Phys. **B21**, L659(1988); ibid. **24**, 2047(1991)

[6.9] J.W. Parker, H.A. Schuessler, R.H. Hill, Jr. B.G. Zollars: Phys. Rev. **A29**, 617(1984)

[6.10] F.B. Dunning, R.F.Stebbings: in *Rydberg States of Atoms and Molecules*, eds. R.F. Stebbings and F.B. Dunning (Cambridge University Press, Cambridge 1983), Chap .9, p.315

[6.11] L. Petitjean, F. Gounand and P.R. Fournier: Phys. Rev. **A30**, 71(1984); ibid. **30**, 736(1984); ibid. **30**, 143(1986)

[6.12] S. Preston, N.F. Lane: Phys. Rev. **A33**, 148(1986)

[6.13] T. Yoshizawa, M. Matsuzawa: J. Phys. Soc. Jpn., **54**, 918(1985)

[6.14] A. Kalamarides, L.N. Goeller, K.A. Smith, F.B. Dunning, M. Kimura, N.F. Lane: Phys. Rev. **A36**, 3108(1987)

[6.15] B.G. Zollars, C. Higgs, F. Lu, C.W. Walter, L.G. Gray, K.A. Smith, F.B. Dunning and R.F. Stebbings: Phys. Rev. **A32**, 3330(1985); B.G. Zollars, C.W. Walter, F. Lu, C.B. Johnson, K.A. Smith, F.B.Dunning: J. Chem. Phys. **84**, 5589(1986)

[6.16] I.M. Beterov, G.L. Vasilenko, B.M. Smirnov, N.V. Fateyev: Sov. Phys. -
        JETP **93**, 31(1987)
[6.17] X. Ling, B.G. Lindsay, K.A. Smith, F.B. Dunning: Phys. Rev. A**45**,
        252(1992)

Chapter 7

[7.1]   M.R. Flannery: J. Phys. B**13**, L657(1980)
[7.2]   A.P. Hickman: J. Phys. B**14**, L419(1981)
[7.3]   V.S. Lebedev, V.S. Marchenko, S.I. Yakovlenko: Izv. Akad. Nauk SSSR, ser.
        Fiz. **45**, 2395(1981)
[7.4]   V.S. Lebedev, V.S. Marchenko: Zh. Eksp. Teor. Fiz. **84**, 1623(1983) [Sov.
        Phys. -JETP. **57**, 946(1983)]
[7.5]   B.P. Kaulakys: Litov. Fiz. Sb. **22**, 3(1982)
[7.6]   F. Gounand, L. Petitjean: Phys. Rev. A**32** 793(1985)
[7.7]   V.S. Marchenko: Khim. Fiz. **4**, 595(1985) [Sov. J. Chem. Phys. **4** 963(1987)]
[7.8]   V.A. Smirnov: Opt. Spectrosk. **37**, 407(1974) [Opt. Spectosc. **37**, 231(1974)]
[7.9]   E. de Prunele: Phys. Rev. A**27**, 1831(1983)
[7.10] A.Z. Devdariani, A.N. Klucharev, A.V. Lazarenko, V.A. Sheverev: Pis'ma
        Zh. Tekh. Fiz. **4**, 1013(1978) [Sov. Tech. Phys. Lett. **4**, 408(1978)]
[7.11] R.K. Janev, A.A. Mihajlov: Phys. Rev. A**21**, 819(1980); J. Phys. B**14**,
        1639(1981)
[7.12] E.L. Duman, I.P. Shmatov: Zh. Eksp. Teor. Fiz. **78**, 2116(1980)
[7.13] V.P. Zhdanov, M.I. Chibisov: Zh. Eksp. Teor. Fiz. **70**, 2087(1976)
[7.14] V.S. Lebedev, V.S. Marchenko: Khim. Fiz. **3**, 210(1984) [Sov. J. Chem. Phys.
        3, 311(1986)]
[7.15] V.S. Lebedev: J. Phys. B**24**, 1993(1991)
[7.16] V.A. Smirnov, A.A. Mihajlov: Opt. Spectrosk. **30**, 984(1971) [Opt. Spec-
        trosc. **5**, 525(1971)]
[7.17] R.K. Janev, A.A. Mihajlov: Phys. Rev. A**20**, 1890(1979)
[7.18] A.N. Klucharev, M.L. Yanson: *Elementary Processes in the Plasmas of Alkali
        Metals* (Energoatomizdat, Moscow 1988)
[7.19] V.A. Ivanov, V.S. Lebedev, V.S. Marchenko: Zh. Eksp. Teor. Fiz. **94**,
        86(1988) [Sov. Phys. -JETP **67**, 2225(1988)] Pis'ma Zh. Tekh. Fiz. **14**,
        1575(1988) [Sov. Tech. Phys. Lett. **14**, 686(1988)]
[7.20] M. Matsuzawa: J. Phys. B**17**, 795(1984)
[7.21] A.M. Dykhne, G.L. Yudin: Usp. Fiz. Nauk **125**, 377(1978) [Sov. Phys. Usp.
        21, 549(1978)]
[7.22] A.N. Klucharev, A.V. Lazarenko, V. Vujnovic: J. Phys. B**13**, 1143(1980)
[7.23] M. Cheret, L. Barbier, W. Lindinger, R. Deloche: J. Phys. B**15**, 346(1982)
[7.24] J. Boulmer, R. Bonanno, J. Weiner: J. Phys. B**16** 3015(1983)
[7.25] S.B. Zagrebin, A.V. Samson: Pis'ma Zh. Tekh. Fiz. **11**, 680(1985)
[7.26] M.T. Djerad, H. Harima, M. Cheret: J. Phys. B**18**, L815(1985)
[7.27] M.T. Djerad, M. Cheret, F. Gounand: J. Phys. B**20**, 3789(1987); ibid. **20**,
        3801(1987)
[7.28] C. Gabbanini, M. Biagani, S. Gozzini, A. Lucchesini, L. Moi: J. Phys. B**24**,
        3807(1991)

Chapter 8

[8.1]   L.H. Thomas: Proc. Camb. Phil. Soc. **23**, 713(1927); ibid. **23**, 826(1927)
[8.2]   R.C. Stabler: Phys. Rev. A**133**, 1268(1964)
[8.3]   E. Gerjuoy: Phys. Rev. A**148**, 54(1966)
[8.4]   I.C. Percival, D.Richards: J. Phys. B**3** 1035(1970)

[8.5]   I. Percival: in *Atoms in Astrophysics* eds. P.G. Burke et al. (Plenum, New York, London 1983)

[8.6]   I.L. Beigman, M.G.Matusovsky: J. Phys. B**24**, 4117(1991)

[8.7]   M.I. Syrkin: Phys.Rev. A**50**, 2284(1994)

[8.8]   K.B. MacAdam, D.A. Crosby, R. Rolfes: Phys. Rev. Lett. **44**, 980(1980)

[8.9]   I.C. Percival, D.Richards: J. Phys. B**10**, 1497(1977)

[8.10] I.L. Beigman, M.I. Syrkin: Zh. Eksp. Teor. Fiz. **89**, 399(1985); Sov. Phys. -JETP **62**, 226(1985)

[8.11] X. Sun, K.B. MacAdam: Phys. Rev. A**47**, 3913(1993)

[8.12] M.I. Syrkin: Phys. Rev. A**53**, 825(1996)

[8.13] J.F. Delpech, J. Boulmer, F. Devos: Phys. Rev. Lett. **10**, 1400(1977)

[8.14] T.R. La Salle, T.J. Nee, H. Griem: Phys. Rev. Lett. **30**, 944(1973)

[8.15] R.G. Rolfes, L.G. Gray, O.P. Makarov, K.B. MacAdam: J. Phys. B**26**, 2191(1993)

[8.16] V.M. Borodin, A.K. Kazansky: J.Phys. B**26** 1863(1993)

[8.17] I.L. Beigman: Proc. Lebedev Phys. Inst. **119**, 130(1980); ibid **119**, 143(1980)

[8.18] K. Alder, O. Bohr, T. Huus, M. Mottelson, A. Winther: Rev. Mod. Phys. **28**, 432(1956)

[8.19] I.L. Beigman, S.A. Chernyagin: Rep. Leb. Inst., **3-4**, 64(1995)

[8.20] V.P. Shevelko: Phys. Scr. **46**, 531(1992)

[8.21] A.V. Vinogradov: Proc. Lebedev Phys. Inst. **51**, 44(1970)

[8.22] L.P. Presnyakov: Proc. Lebedev Phys. Inst. **119**, 52(1980)

[8.23] V.S. Lisitsa, G.V. Sholin: Zh. Eksp. Teor. Fiz. **61**, 912(1992)

[8.24] P.S. Epstein: Phys. Rev. **22**, 202(1923)

[8.25] W. Pauli: Z. Phys. **36**, 336(1926)

[8.26] A.K. Kazansky, V.N.Ostrovsky: Phys. Rev. A**52**, R1811(1995)

[8.27] I.L. Beigman, M.I. Syrkin: Lebedev Phys. Rep. N**8**, 21(1986)

[8.28] G.N. Birich, Yu.V. Bogdanov, S.I. Kanorsky, I.I. Sobel'man, V.N. Sorokin, I.I. Struck, E.A. Yukov: Opt. Spectrosk. **65**, 824(1988); G.N. Birich, Yu.V. Bogdanov, S.I. Kanorsky, I.I. Sobel'man, V.N. Sorokin, I.I. Struck, E.A. Yukov: Proc. Lebedev Phys. Inst. **195**, 217(1989)

[8.29] K.B. MacAdam, D.A. Crosby, R. Rolfes: Phys. Rev. A**24**, 1286(1981)

[8.30] A.V. Vinogradov, I.Yu. Skobelev, A.M. Urnov, V.P. Shevelko: Proc. Lebedev Phys. Inst. **119**, 120(1980)

[8.31] Yu.V. Bogdanov, S.I. Kanorsky, I.I. Sobel'man, V.N. Sorokin, I.I. Struck, E.A. Yukov: Zh. Eksp. Teor. Fiz. **87**, 776(1984)

[8.32] K.B. MacAdam, R. Rolfes, X. Sun, J. Singh, W.L. Fuqua III, D.B. Smith: Phys. Rev. A**36**, 4245(1987)

Chapter 9

[9.1]   M. Baranger: in *Atomic and Molecular Processes* ed. D.R. Bates (Academic, New York 1958) Chap. 13

[9.2]   N. Allard, J. Kielkopf: Rev. Mod. Phys. **54**, 1103(1982)

[9.3]   P. Anderson: Phys. Rev. **76**, 647(1949)

[9.4]   C.Z. Reinsberg: Z. Phys. **93**, 416(1935); ibid. **105**, 460(1937)

[9.5]   O.B. Firsov: Zh. Eksp. Teor. Fiz. **21**, 627(1951); ibid. **21**, 634(1951)

[9.6]   G.K.Ivanov: Opt. Spectrosk. **40**, 965(1976) [Opt. Spectrosc. **40**, 554(1976)]

[9.7]   G. Hermann, B. Kaulakys, G. Lasnitschka, G. Mahr, A. Scharmann: J. Phys. B**25** L407(1992)

[9.8]   A. Flusberg, R. Kachru, T. Mossberg, S.R. Hartmann: Phys. Rev. A**19**, 1607(1979)

[9.9]   R. Kachru, T.W. Mossberg, S.R. Hartmann: Phys. Rev. A**21** 1124(1980)

[9.10] K.H. Weber, K. Niemax: Z. Phys. A**309** 19(1982)

[9.11] T.F. Gallagher, W.E.Cooke: Phys. Rev. A19, 2161(1979)
[9.12] J. Boulmer, J.F. Delpech, J.C. Gauthier, K. Safinya: J. Phys. B14, 4577(1981)
[9.13] K. Weber, K.Niemax: Z. Phys. A307 13(1982); ibid. 312, 339(1983)
[9.14] E.E. Nikitin, S.Ya. Umansky: *Theory of Slow Atomic Collisions* (Springer, New York 1984)
[9.15] L.P. Presnyakov: Phys. Rev. A2, 1720(1970)
[9.16] G.V. Golubkov, G.K. Ivanov: Z. Phys. A319, 17(1984)
[9.17] Yu.N. Demkov, V.I. Osherov: Zh. Eksp. Teor. Fiz. 53, 1589(1967)
[9.18] L.A. Minaeva, I.I. Sobel'man: J. Quant. Spectrosc. Radiat. Transf. 8, 783(1968)
[9.19] H.P. Griem: Astrophys. J. 148, 547(1967)
[9.20] G. Himmel, F. Pinnekamp: J. Phys. B10, 1457(1977)

Subject Index

Springer
and the
environment

At Springer we firmly believe that an international science publisher has a special obligation to the environment, and our corporate policies consistently reflect this conviction.
We also expect our business partners – paper mills, printers, packaging manufacturers, etc. – to commit themselves to using materials and production processes that do not harm the environment. The paper in this book is made from low- or no-chlorine pulp and is acid free, in conformance with international standards for paper permanency.

 Springer

Printing: COLOR-DRUCK DORFI GmbH, Berlin
Binding: Buchbinderei Lüderitz & Bauer, Berlin